b

‖‖‖‖‖‖‖‖‖‖‖‖‖‖‖‖‖‖
D0850053

Chemical Reactions in Organic and Inorganic Constrained Systems

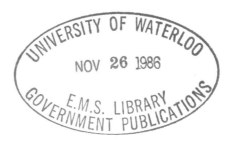

NATO ASI Series

Advanced Science Institutes Series

A series presenting the results of activities sponsored by the NATO Science Committee, which aims at the dissemination of advanced scientific and technological knowledge, with a view to strengthening links between scientific communities.

The series is published by an international board of publishers in conjunction with the NATO Scientific Affairs Division

A Life Sciences	Plenum Publishing Corporation
B Physics	London and New York
C Mathematical	D. Reidel Publishing Company
and Physical Sciences	Dordrecht, Boston, Lancaster and Tokyo
D Behavioural and Social Sciences	Martinus Nijhoff Publishers
E Engineering and	The Hague, Boston and Lancaster
Materials Sciences	
F Computer and Systems Sciences	Springer-Verlag
G Ecological Sciences	Berlin, Heidelberg, New York and Tokyo

Chemical Reactions in Organic and Inorganic Constrained Systems

edited by

R. Setton

Research on Imperfectly Crystallized Solids,
National Centre for Scientific Research,
Orleans, France

D. Reidel Publishing Company

Dordrecht / Boston / Lancaster / Tokyo

Published in cooperation with NATO Scientific Affairs Division

Proceedings of the NATO Advanced Research Workshop on
Chemical Reactions in Organic and Inorganic Constrained Systems
Les Bézards, France
June 24-28, 1985

Library of Congress Cataloging in Publication Data

NATO Advanced Research Workshop on Chemical Reactions in Organic and Inorganic Con-
    strained Systems (1985 : Château des Bézards)
    Chemical reactions in organic and inorganic constrained systems.

    (NATO ASI series. Series C, Mathematical and physical sciences ; vol. 165)
    "Proceedings of the NATO Advanced Research Workshop on Chemical Reactions in
Organic and Inorganic Constrained Systems, Bézards, France, June 24–28, 1985" – T.p. verso.
    "Published in cooperation with NATO Scientific Affairs Division."
    Includes index.
    1. Clathrate compounds – Congresses. 2. Stereo-chemistry – Congresses. 3.
Intermolecular forces – Congresses. I. Setton, R. (Ralph), 1923– II. North Atlantic Treaty
Organization. Scientific Affairs Division. III. Title. IV. Series: NATO ASI series. Series C,
Mathematical and physical sciences ; no. 165.
    QD474.N38 1985   541.2'2   85–30045
    ISBN 90–277–2175–0

Published by D. Reidel Publishing Company
P.O. Box 17, 3300 AA Dordrecht, Holland

Sold and distributed in the U.S.A. and Canada
by Kluwer Academic Publishers,
190 Old Derby Street, Hingham, MA 02043, U.S.A.

In all other countries, sold and distributed
by Kluwer Academic Publishers Group,
P.O. Box 322, 3300 AH Dordrecht, Holland

D. Reidel Publishing Company is a member of the Kluwer Academic Publishers Group

TABLE OF CONTENTS

Dr. J.A. BALLANTINE

Department of Chemistry
University College of Swansea
Singleton Park,
SWANSEA SA2 8PP
UNITED KINGDOM

Prof. H.P. BOEHM

Institut für Anorganische Chemie
Universität München
Meiserstrasse 1,
D-8000 MÜNCHEN 2
GERMANY (Fed. Rep.)

Dr. G. BRAM

Laboratoire des Réactions Sélectives
sur Supports
UA C.N.R.S. 478
Bât. 410, Université Paris-Sud
91405 - ORSAY CEDEX
FRANCE

Prof. R. BRESLOW

Department of Chemistry
Columbia University, New York
NEW-YORK 10027
U.S.A.

Dr. P. BROCHETTE

Laboratoire de Chimie Physique
Université Pierre et Marie Curie
11, rue Pierre et Marie Curie
75005 - PARIS
FRANCE

Dr. R. BÜSCHL

Institut für Organische Chemie
Universität Mainz
Postfach 3980
D-6500 MAINZ
GERMANY (Fed. Rep.)

Dr. S. CARTLIDGE

B.P. Research Centre
Sunbury on Thames,
Mddx, TW16 7LN
U.K.

Dr. A. CORNELIS

Institut de Chimie
Université de Liège au Sart-Tilman
B-4000, LIEGE 1,
BELGIQUE

Prof. A. FOUCAUD

Groupe de Physicochimie Structurale
Université de Rennes I
Avenue du Général Leclerc
35042 - RENNES CEDEX
FRANCE

Prof. J.J. FRIPIAT

C.R.S.O.C.I.-C.N.R.S.
1B, rue de la Férollerie
45071 - ORLEANS CEDEX 2
FRANCE

Prof. K. FUJITA

Faculty of Pharmaceutical Sciences
Kyushu University 62,
Maidashi, Higashi-ku, FUKUOKA 812
JAPAN

Prof. M. GUISNET

UA C.N.R.S. 350
U.E.R. Sciences
86022 - POITIERS CEDEX
FRANCE

Prof. C.D. GUTSCHE

Department of Chemistry
Washington University
ST LOUIS, MO 63130
U.S.A.

Dr. H. HEINEMANN

Lawrence Berkeley Laboratory
University of California
1, Cyclotron Road,
BERKELEY CA 94720
U.S.A.

Prof. P.A. JACOBS

Katholieke Universiteit Leuven
Laboratorium voor Oppervlaktechemie
Kardinal Mercierlaan 92
B-3030 LEUVEN (HEVERLEE)
BELGIUM

Prof. G. LAGALY

Institut für Anorganische Chemie
Universität Kiel,
Olshausenstrasse 40, D-2300 KIEL
GERMANY (Fed. Rep.)

Dr. P. LAGRANGE

Laboratoire de Chimie du Solide
Minéral
LA 158 - Université de Nancy I
B.P. 239
54506 -VANDOEUVRE-LES-NANCY
FRANCE

Prof. J.M. LEHN

Collège de France
11, Place Marcelin Berthelot
75005 - PARIS
FRANCE

Dr. J.A. MARTENS

Katholieke Universiteit Leuven
Laboratorium voor Oppervlaktechemie
Kardinal Mercierlaan 92
B-3030 LEUVEN (HEVERLEE)
BELGIUM

Dr. D. MÖBIUS

Karl-Friedrich-Bonhoeffer Institut
Max-Planck-Institut für Biophysika-
lische Chemie
Postfach 2841
D-3400 GÖTTINGEN-NIKOLAUSBERG
GERMANY (Fed. Rep.)

Dr. C. NACCACHE

Institut de Recherches sur la
Catalyse
C.N.R.S.
2, Avenue Albert Einstein
69626 - VILLEURBANNE CEDEX
FRANCE

Prof. T.J. PINNAVAIA

Department of Chemistry
Michigan State University
East Lansing, Michigan 48824
U.S.A.

Prof. G. PONCELET

Groupe de Physico-Chimie Minérale
et de Catalyse
Place Croix du Sud 1,
B-1348 LOUVAIN-LA-NEUVE
BELGIUM

Prof. F. RAMOA RIBEIRO

Grupo de Estudos de Catálise
Heterogénea
Instituto Superior Técnico
1096 LISBOA CODEX
PORTUGAL

Dr. M. RIVIERE

Laboratoire I.M.R.C.P.
UA 470, C.N.R.S.
Université Paul Sabatier
31062 - TOULOUSE CEDEX
FRANCE

Dr. E. RUIZ-HITZKY

Instituto de Fisico-Quimica Mineral
CSIC, c/Serrano 115 bis,
28006 MADRID
SPAIN

Dr. SCHÖLLHORN

Technische Universität Berlin
Institut für Anorganische Chemie
Strasse des 17. Juni 135
1000 BERLIN 12
WEST GERMANY

Dr. P.A. SERMON

Department of Chemistry
Brunel University
UXBRIDGE Mddx, UB8 3PH
U.K.

Dr. R. SETTON

C.R.S.O.C.I.-C.N.R.S.
1B, rue de la Férollerie
45071 - ORLEANS CEDEX 2
FRANCE

Prof. P. SINAY

Laboratoire de Biochimie Structurale
Université d'Orléans
45046 - ORLEANS CEDEX
FRANCE

Dr. J. SJÖBLOM

Institute for Surface Chemistry
P.O. Box 5607
S-11486 STOCKHLOM
SWEDEN

Prof. J. SOMMER

Institut de Chimie
Université Louis Pasteur
1, rue Blaise Pascal
67000 - STRASBOURG CEDEX
FRANCE

Prof. I. TABUSHI

Department of Synthetic Chemistry
Kyoto University
Sakyo-ku Yoshida
KYOTO 606
JAPON

Prof. J.K. THOMAS

Department of Chemistry
University of Notre Dame
NOTRE DAME, INDIANA 46556
U.S.A.

Prof. J.M. THOMAS

Department of Physical Chemistry
University of Cambridge
Lensfield Road
CAMBRIDGE CB2 1EP
U.K.

Dr. H. VAN DAMME

C.R.S.O.C.I.-C.N.R.S.
1B, rue de la Férollerie
45071 - ORLEANS CEDEX 2
FRANCE

Dr. D.E.W. VAUGHAN

Exxon Research and Engineering
Company
Annandale, NEW JERSEY 08801
U.S.A.

Dr. T. WARNHEIM

Institute for Surface Chemistry
P.O. Box 5607
S-11486 STOCKHOLM
SWEDEN

PREFACE

    The basic idea of the NATO International Exchange Program for funding an Advanced Research Workshop on "Chemical Reactions in Organic and Inorganic Constrained Systems" was to contribute to a better understanding of the influence of configurational constraints on reaction mechanisms, as imposed on reagents by organic or inorganic templates. The original character of the Workshop was to bring together organic and inorganic chemists with this common interest in order to promote the exchange of ideas and, eventually, interdisciplinary research.
    All the participants to the Workshop agreed that the discussions were stimulating and fruitful. The judgement of the reader of the Proceedings may perhaps be more restrictive because the director (Professor J.J. FRIPIAT) and co-director (Professor P. SINAŸ), faced with the impossible task of covering such an enormous domain, were obliged to select, somewhat arbitrarily, a limited number of topics which seemed to them to be the most important. Their choice may be discussed and there surely are important gaps, with fields which were not considered. However, both organisers believe that, within the limited span of time and number of contributors, most of the exciting areas were addressed. Dr. WÄRNHEIM was kind enough to write a commentary on the Workshop; his summary, written with the hindsight of a few weeks, supports, we believe, this opinion.
    Dr. SETTON has accepted the burden of collecting and shaping (not selectively) the manuscripts. This book would not be what it is without his efficient contribution as scientific secretary of the Workshop.
    For the last lecture, we had the privilege of having a contribution by Professor J.M. LEHN who gave a very complete review of his work on molecular recognition and supramolecular organisation. Lack of time did not allow Professor LEHN to prepare the written version of his contribution but the organisers wish to thank him for the "fireworks" which closed the conference.
    Last but not least, both organisers would like to thank the participants, who all thoroughly played the game; their efforts are a worthy contribution to this domain of the physical sciences.

                       J.J. FRIPIAT   P. SINAŸ

                       Director   Co-Director

REACTIONS IN CONSTRAINED SYSTEMS - A SUMMARY

Torbjörn Wärnheim
Institute for Surface Chemistry
P.O. Box 5607
S-114 86 STOCKHOLM
Sweden

1. INTRODUCTION

The NATO Workshop "Chemical Reactions in Organic and Inorganic Constrained Systems", held at the Château des Bézards, June 24-28 1985, offered an overview of an important, truly multidisciplinary field, which has become of increasing interest to the scientific community during the last years. The fundamental idea is simply to explore how and to what extent one can geometrically modify the immediate environment of a molecule, with subsequent effects on its reactivity. From this rather loosely outlined starting point, there may be many different ways of creating sterically or otherwise constrained systems. One may use either suitable organic molecules or polymers, often designed to mimic enzymes or enzymatic action, or one may use a two-or three-dimensional inorganic framework, like a zeolite or a layered clay, of immense importance in cracking or conversion of petrochemical raw materials, and let them affect the course of the reaction. The interest may in both cases be focused either on the constraining system and its structural features, or on the type of reactions that can be made to occur in that system; in the end, hopefully, a unified description is available which directly relates the constraints to the possible reactions.

It is perhaps presumptuous to try to summarize the lectures presented during the Workshop from this point of view and especially in an unbiased fashion when one is unfamiliar with several relevant research areas; this summary is therefore to be regarded more as a prejudiced commentary than as a scientific review.

2. CYCLODEXTRINS AND SIMILAR MOLECULES

Interesting lectures immediately related to each other were given by Prof. Breslow, Prof. Tabushi and Prof. Fujita. The cyclodextrins,which are cyclic oligomers of glucose, have found use in organic chemistry during the last fifteen years. Professor Breslow gave an overview of some of his own contributions to the field. Molecules like cyclodextrins may be used as cages to achieve a geometrical constraint for a specific reaction. The goals are, of course, to obtain as selective a reaction as possible with respect to the substrate and also with respect to regioselectivity and stereospecificity. Natural references here are the enzymatic systems themselves : how do the synthesised molecules compare ? As pointed out, in some instances it is possible to obtain systems just as efficient for a single step reaction, but not necessarily the reaction sought. Professor Breslow's talk gave a general evaluation of a field comprising not only the use of cyclodextrins but also other systems with varying geometry, creating cages and constraints for many different types of organic reactions.

A more specific discussion of the cyclodextrins was given by Professor Tabushi : by using disulfonated capped cyclodextrins as starting materials, compounds with enzymatic properties were prepared. Not only the specificity but also the total turnover of specific enzymes like racemase or aminotransferase were approached. The functionalisation of cyclodextrins by different reagents was presented by Professor Fujita.

In his presentation, Prof. Gutsche discussed the calixarenes, which are oligomers of formaldehyde and p-alkyl phenols with shape-selective properties somewhat akin to those of the cyclodextrins. The main constraints, in this case, are electrostatic and hydrogen bonding is of paramount importance in determining the course of subsequent chemical reactions carried out in the presence of these compounds.

Similarly, the concluding lecture, given by Prof. Lehn, reviewed a whole series of reactions carried out in the presence of crown ethers or cryptands (which are most efficient in imposing a molecular geometric constraint) and which can extend all the way to artificial photosynthesis.

3. ZEOLITES

A second type of constrained system to be discussed was the group of synthetic and natural zeolites, $SiO_2-Al_2O_3$ compounds with a three-dimensional framework. Their porous structure, with cavities and pores in the range of 3 to 9 Å in diameter, makes them effective and shape-selective catalysts and this, together with their thermal stability, has made them indispensable as catalysts in the conversion and cracking of petrochemical raw materials. Much research has naturally been focused on these compounds and an extensive presentation of some recent results on their structural characteristics and of the tools available for these investigations was given by Prof. J.M. Thomas who discussed solid state NMR, high resolution electron microscopy (HREM) and also the use of neutron powder profile analysis. With the advent of cross polarizing magic angle spinning NMR (CP-MAS NMR), solid-state analysis has been facilitated to a very great extent. By using mostly measurements of chemical shifts, structural features like the immediate surroundings of Si or Al atoms may be probed. Similarly, the porous structure can be accurately gauged by HREM techniques. Some technical modifications of the method, such as the averaging of microscopy images of the zeolites, can lead to a knowledge of the structure on a very detailed level.

   From the discussion that followed, it was obvious that both the
determination of the structure of various zeolites as well as the synthesis of new ones with different incorporated cations is a successful
and rapidly expanding field. As an example of the exotic possibilities
offered, the intercalation of $[Na_4]^{3+}$ was discussed.
   Zeolites and their catalytic action were the subject of the lectures by Prof. Jacobs, Prof. Ribeiro, Prof. Guisnet, Dr. Naccache, Dr.
Martens and Dr. Cartlidge. These studies aimed at clarifying the correlation between the structural and the catalytic properties of the zeolites. A central point is the evaluation of shape selectivity : does
the porous structure of the zeolite create the necessary configurational constraints needed to obtain a specific reaction product with a
given shape ? Prof. Jacobs and Dr. Martens discussed many specific examples, such as the conversion of carboxylic acids in acidic zeolites
and the hydroconversion of n-decane. The features of acidic zeolites
were also considered by Prof. Ribeiro. The actual structural classification of the zeolites can of course be obtained by monitoring their
catalytic efficiency and their shape selectivity for different reactions. Several of the contributions discussed the different reactions
that could be used for this purpose. As pointed out both in this context and the discussion of enzyme mimetics, not only must the reactants
and the products be considered but, even more so, also the effect of
the geometrical constraints imposed on the transition state by the environment.

4. LAYERED AND PILLARED CLAYS

The role of layered clays as frameworks for creating a constrained medium in a reaction were amply treated during the Workshop. The use of
pillared clays as catalysts has been an area closely coupled to the use
of zeolites ; the latter are of immense commercial importance but pillared clays provide a complementary possibility in the conversion of petrochemical products. In principle, a pillared clay is created by ion-
exchange of e.g. a 2:1 layered clay, such as montmorillonite, with poly-
hydroxyalumina compounds. By thermal treatment, the complex ion is dehydrated and a thermally stable pillared clay can be obtained. The review of the structure and synthesis of these pillared clays presented
by Prof. Pinnavaia discussed several ways of creating the stable pillars but emphasised the use of novel metal complex precursors to the
clusters, thus providing pillars containing elements which cannot be
obtained by simple ion exchange and dehydration. A point to be stressed
is the necessity of a firm knowledge of the solution chemistry of ions
of potential interest : this provides the tools for designing the suitable clustered cations and other complexes since anionic and neutral
compounds may be used as well.
   The properties of pillared clays were presented in a number of
lectures. Prof. Poncelet concentrated on the use of pillared montmorillonite and beidellite as catalysts in the cracking and isomerisation
of n-heptane : the catalytic properties of pillared beidellite are com-

parable to those of certain zeolites but the pillared beidellites are
able to withstand higher working temperatures. In his contribution, Dr.
Vaughan stressed the complementary role of zeolites and pillared clays
in the catalytic conversion of petrochemical raw materials. The very
high catalytic activity displayed by carefully designed zeolites puts
severe demands on the properties required of pillared clays yet the role
of the latter could well increase in the future, with 2:1 clays as well
as other types, like brucite, achieving technological prominence.

In his presentation, Dr. Ballantine gave an overview of the many
reactions in which layered clays have been used as catalysts. Acid pretreatment of natural clays results in ion-exchanged species which will
catalyse an extremely wide variety of acid-catalysed organic reactions
such as dimerisations, oligomerisations, Diels-Alder condensations,
additions, elimination reactions etc. Similarly, Dr. Cornelis summarised
the progress in the field of organic synthesis using clay-supported iron
or copper nitrates. Here also, it is possible to obtain selective transformation of alcohols to aldehydes or ketones, thiols to disulfides, the
nitration of phenols, Friedel-Crafts alkylations, chlorination of paraffins, and many other processes of fundamental importance in synthetic
organic chemistry.

Other contributions in the same field were those of Dr. Bram and
Prof. Foucaud. The former showed that one can obtain C- rather than
O-alkylation of bidentate molecules by using alumina-supported reagents
while the latter discussed the more specific synthesis of γ-lactones via
ene-reactions catalysed by montmorillonite. The same host structure was
shown by Dr. Ruiz-Hitzky to accept crown ethers and cryptands to form
interlayer complexes of the cations in the clays.

5. GRAPHITE

Several contributions dealt with graphite and its intercalation derivatives. Again in the field of organic chemistry, Prof. Sommer presented
the use of superacids such as SbF_5 intercalated in graphite. Even at
room temperature, the isomerisation and cracking of hydrocarbons can
easily be obtained with this catalyst, but fairly rapid deactivation
makes it ill suited for industrial use. Dr. Setton described reactions
between some alkali metal graphite intercalates and molecules as different as hydrogen and furan, and showed that charge-sharing among the
graphite, the alkali metal atoms and the third component are of primary
importance in fixing the composition of the ternary compounds which are
formed. Dr. Lagrange also described the intercalation of alkali metals
and their amalgams into graphite. The behaviour of the latter is rather
complicated: some of them intercalate as such while, with others, each
component of the amalgam reacts successively. The presentation of Prof.
Boehm discussed the variations in crystallographic parameters which were
observed during the intercalation of sulfuric acid as well as the action
of intercalated potassium on transition metal halides; this did not give
the intercalation of the transition metals as had already been claimed.
Lastly, Dr. Heinemann told of experimental water-gas reactions done with
water vapor on pure graphite and proving that the reaction can be made

truly catalytic by the adjunction of certain metal oxides which decompose the phenolic compounds shown be formed at the edge of the carbon sheets.

6. SURFACTANT ASSEMBLIES

Another type of structure which provides constraints for chemical reactions are the assemblies of amphiphilic molecules such as the micellar and microemulsion systems presented in several lectures. A feature which makes these systems somewhat different from those previously discussed is their highly dynamical state since the lifetime of a given micellar aggregate is seldom longer than a few milliseconds, in contrast to oligomeric or polymeric enzyme mimetics or crystalline inorganic frameworks.

A way of assembling such amphiphilic compounds into a more static entity was described by Dr. Büschl. Amphiphiles with two lipid-like alkyl chains, can, under certain conditions, form assemblies larger than the usual micelles. Thus, liposomes generally consist of a water core surrounded by surfactant bilayers (lamellae). They could be used to encapsulate drugs for oral intake, to provide a slow release of the active compound. By incorporating lipids with a reactive double bond (which can therefore polymerize) in the liposome, a large increase in stability is obtained. Furthermore, by mixing polymerisable molecules with fluorolipids, which form separate "isles" on the liposome surfaces, it is also possible to obtain inlets and outlets for reactants and products, with an enzyme enclosed in the liposome. The fluorolipid isles can then be opened with the cleaving agent.

Other contributions proved that one can carry out chemical reactions in normal and reverse micelles as well as in microemulsions. An overview of several types of microemulsions was given by Dr. Rivière. Among the examples mentioned were isomerisations of oxime ethers, photocycloadditions and photodimerisations. It is important to understand the correlation between the structure of a microemulsion and the course of a reaction, if one wishes to obtain high yields as well as regio- and stereospecificity. In these fluid, highly dynamic systems, often with an ill-defined interface between the hydrophilic and the hydrophobic domains, it is difficult to provide a description which includes the possibility of having a catalytic effect, pinpoints the exact location of the substrate with respect to the interface and describes the distribution of the reagents between the hydrophilic and hydrophobic domains, even though all these factors are important.

Another talk, by Dr. Sjöblom, focused on the use of acidic microemulsions as retarding systems in reactions with calcium carbonate. Here, the structural features of microemulsions, as reflected in their diffusion properties, were correlated with their reactivity. The crucial point in the practical use of microemulsions in reactive systems often seems to be a thorough knowledge of the relevant phase-equilibria and of the properties of the solution. Detailed discussions of the reaction mechanism should often be postponed until more experimental evidence has been gathered and a wider variety of reactions have been explored.

In a contribution dealing with photosensitized electron transfer reactions, Dr. Brochette compared the properties of a normal micellar system, stabilized by cationic surfactant, with those of a reverse micellar system, stabilized by an anionic surfactant. The photosensitizer was either a porphyrin or a chlorophyll, and the acceptor a viologen molecule. On comparing the reaction rates of the forward and the back electron transfer reactions, it was found that large effects occur on going from a direct to a reverse micellar system. This was interpreted in terms of the position of the acceptor with respect to the interface.

A more general discussion of the photochemistry at interfaces was given by Prof. J.K. Thomas who dealt with different processes at the interface of micellar aggregates and of colloidal particles. Studies of the pyrene-dimethylaniline electron transfer reaction showed that the course of the reaction was strongly affected e.g. by the possibility of changing the environment of the reacting species from polar to non-polar characteristics (with consequent solubilization in the micelles). Similar constraints may be imposed by the presence of cyclodextrins.

Although it does not deal with surfactant assemblies, the memoir presented by Dr. Möbius is closely connected to them since it probes the aggregation processes occuring at the air-water interface. With the help of a Langmuir-type trough and by recording the reflection spectra of different dyes, it was shown that modifications of the surface pressure resulted in changes in aggregation and stacking. The geometrical constraint here is obviously quite similar to the one found at the boundary of a micelle.

7. MISCELLANEOUS TOPICS

Another presentation is also indirectly related to the photochemical processes examined above. By measuring the lifetime of photoexcited species, Dr. Van Damme showed that this variable can, in certain cases, provide access to the fractal dimension associated with the surface of a solid ; this is an important parameter which has been suggested as a means of characterising the inner surface of pillared clays.

Prof. Schöllhorn gave a very complete presentation of solid intercalation compounds other than the ones discussed above, such as the chalcogenides, stressing the redox relationships between the guest and host lattices which result in the presence of both geometric and electronic constraints.

The technically important transition metal bronzes were discussed by Prof. Sermon. The compounds, which can house non-stoichiometric quantities of sodium or hydrogen, have interesting catalytic activities not shown by clays or zeolites and may become technologically important in the removal of nitrogen oxides pollutants.

The solid crystalline silicic acids constitute a class of compounds with many interesting properties : the strong interaction between hydrocarbons and their inner surface leads to interesting applications in gas chromatography; in a different field of action, they can, for instance, catalyse the formation of porphyrins by the condensation of benzaldehyde

and pyrrole or the formation of ethylenic compounds by the dehydration
of alcohols.

8. CONCLUDING REMARKS

A commentary on a Workshop on such a multifaceted subject must necessarily be both sketchy and somewhat complete. There will always be brea-
ches between the participants with respect to nomenclature, field of
interest, and so forth. The lectures gave an admirable impression of
scientists trying to bridge a gap between, e.g., inorganic and organic
chemists, and also between scientists whose main interest resided in the
structure of the constraining system and those whose main interest was
the reacting molecules.

Both the structural features of the systems and the reaction themselves were amply discussed during the Workshop, in a very complementary fashion. To some extent the question remains : what connections are
there between those very different systems ? From a physical chemists'
point of view, the answer will possibly be that it is the same catalogue
of intermolecular interactions that are responsible for creating the
constraints. The steric factor is always present in these considerations,
as pointed out previously ; it is a question of capturing molecules and
of the transition state in a reaction. This is of course a necessary requirement. Another is the need to activate the catalyst. The interplay
of these factors is certainly extremely complex, to make a gross under-
statement. Thus, the cooperation between scientists and the willingness
to find the common denominator of the different fields, as shown at the
Workshop, is certainly not only admirable but also necessary.

CHIRAL, SIZE AND SHAPE RECOGNITION OF GUESTS BY MODIFIED CYCLODEXTRIN

Iwao Tabushi
Kyoto University
Department of Synthetic Chemistry
Sakyo-ku Yoshida
Kyoto 606, Japan

ABSTRACT. Starting from regiospecific prim,prim A,B-disulfonate capped cyclodextrin, a series of the A,B-difunctional cyclodextrins were prepared. Typical examples are A(B)-amino-B(A)-carboxy-β-cyclodextrins and A(B)-amino-B(A)-B$_6$-cyclodextrins. These artificial compounds have remarkable characteristics of enzymes or receptors chiral specificity, guest shape and size selectivity, acceleration and turn-over. Activities of artificial racemase, artificial aminotransferase, artificial amino acid receptors will be discussed.

1. INTRODUCTION

Among many interesting and important fields of chemistry, biomimetic chemistry is now undoubtedly one of the most promising and productive fields. The increasing number of interesting reports clearly shows that biomimetic chemistry (bioorganic and bioinorganic chemistry) took off rapidly during the past several years and some successful applications of biomimetic chemistry to the synthetic, analytical, pharmaceutical, and industrial chemistries have been made. Since the biomimetic chemistry was born in the interface between organic (and inorganic) chemistry and biochemistry, it was quite natural that the first and major target of this area would be the construction of "enzyme models".

Enzyme has several unique characteristics——specific catalysis, regulation, biosynthesis, biodegradation, transport, etc. However, the "first generation" of enzyme models (or *artificial enzymes*) mostly deal with the simplest catalytic activity (see Table 1). After many successes in the preparation of artificial enzymes having satisfactory reactivity, chemists tired of continuing this rather routine trial to prepare artificial enzymes of the first generation and started to go further. Thus, more sophisticated enzyme characteristics have been investigated, such as substrate specificity or termolecular or multimolecular reaction. In the following enzymatic reactions:

1

R. Setton (ed.), Chemical Reactions in Organic and Inorganic Constrained Systems, 1–10.
© *1986 by D. Reidel Publishing Company.*

TABLE I   Monofunctionalized Cyclodextrin as Enzyme Models

Cyclodextrin	No.	Reaction	ref.
$[\beta]$–N $\overset{\frown}{}$ NH$_2$]$_2$ Cu	17	(furan)–CCH(OH)CH(furan) → (furan)–CC(furan), reaction scheme	1
$[\beta]$–X$\cdots$M, X= N N N / N N N N	18, 19	^-O_2C–(ring)–CO_2^- → (ring)–CO_2^- → (cyclohexanone)	2
M=Mg, Ni, Cu, Zn,			
$[\beta]$–OPO$_3^-$	20	(tetrahydropyranyl-O-)–(ring)–NO$_2$ → (pyranyl-OH) + HO–(ring)–NO$_2$	3
$[\beta]$–OPO$_3^-$ $[\beta]$–OPO$_3^-$	21	+(ring)–CCH$_2$OH + T$^+$ → +(ring)–CCHOH + H$^+$, T	4
$[\beta]$–OPO$_3^-$	22	(ring)–CHSCH + I$_2$ → (ring)–CH$_2$SCH + HI, O	5
$[\beta]$–S–Fe–Fe–SR RS–Fe–Fe–SR	23	—	6
$[\beta]$–S–(pyridinol OH, CH$_3$, CH$_2$NH$_2$)	24	RCCO$_2$H → RCHCO$_2$H (NH$_2$)	7
$[\beta]$–X X= NH, NH H (Ret)	25	—	8
S N=Ret H	26		
$[\alpha]$–NH$_2$(CH$_2$)–(imidazole)NH	27	AcO–(ring)–NO$_2$ → AcO$^-$ + $^-$O–(ring)–NO$_2$	9
$[\alpha]$–OCH$_2$CN–X, OH X= CH$_3$	28		
= CH$_2$–(imidazole)	29	ArOAc → ArO$^-$ + AcO$^-$	10
= (CH$_2$)$_2$NMe$_3$	30		
$[\beta]$–S(CH$_2$)$_2$CNCH OH	31	ArOAc → ArO$^-$ + AcO$^-$	11
$[\beta]$–N(CH$_3$)$_3$Cl	32	(ring)–OAc, CO$_2$ → (ring)–O$^-$, CO$_2$ + AcO$^-$	12
$[\alpha]$–OC–(ring)–C, N Ni N, N=N–O	33	(ring)–OAc, NO$_2$ → (ring)–O$^-$, NO$_2$ + AcO$^-$	13

$$E + S \rightleftarrows E{\cdot}S \rightarrow \text{Following catalytic reaction} \qquad (1)$$

where E denotes the enzyme and S denotes the substrate, formation of an E·S complex practically determines the substrate specificity. Formation of the E·S complex is controlled by the combination of various types of intermolecular interactions between the enzyme and the substrates. This process becomes a kind of "trigger" for the following processes, since the substrate molecule and the catalytic functional group(s) of the enzyme are forced to take the most favorable position for the further reaction(s) in the E·S complex. Thus, for the construction of excellent enzyme models, it is essential to provide the molecule with a binding site of known or reasonably expected "shape" and a catalytic site consisting of an appropriate combination of the functional groups. From these purposes, inclusion compounds are chosen as the enzyme model[14]. Similarly, receptors may be satisfactorily mimicked by the use of inclusion molecular systems. Here, enzyme models and receptor models are discussed, in which cyclodextrin cavity is used as artificial hydrophobic binding sites.

2. RESULT AND DISCUSSION

2.1. Regiospecific or Regioselective Capping of Cyclodextrins.

    Usually, enzymes or receptors have at least two important functional groups for guest recognition or catalysis near the hydrophobic binding sites. Often it is necessary to have two different functional groups in certain fixed spacial orientation, which may be typically described as Z (zusammen), R (rechtwinklig) and E (entgegen) (see Fig 1). This situation may be perfectly attained if regiospecific difunctionalization is performed (see Fig 2). We have been developing

Fig. 1   Typical Spacial Orientations of
Guest-Enzyme Complexes

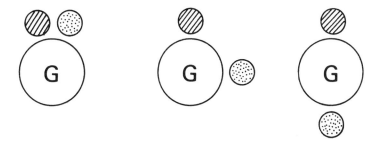

E-configuration   R-configuraton   Z-configuration

Fig 2. Three Possible Isomers of
Capped Cylodextrins

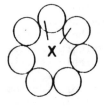 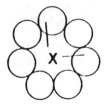 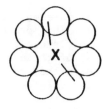

A,B-isomer   **A,C-isomer**   **A,D-isomer**

TABLE 2   Regioisomer Distribution in the
Crude Cap Mixture of β-Cyclodextrin

capping reagent	$[\beta\text{-CD}](10^{-3}M)$	T°(C)	Time(h)	Ratio(Anal.Yield%)	
				AC	: AD
	3.2	60	2	92	8
	5.9			93	7
	18			94	6
	35			92	8
	18	27	4	89(40)	11(5)
	2.9	60	2	11(3)	89(23)
	18			23(4)	77(14)
	44			26(6)	74(16)
	3.3	27	20	8(4)	92(41)
	18	60	2	66	34
	18	60	2	52	48
	18	60	2	7	93
				AB	: AC
	44	25		97	3

the "regiospecific capping" technique for the purpose of regiospecific introduction of two (different) functional groups[15f,g] From the study experimental details for A,B-, A,C, and A,D-regiospecific difunction- alization are established (Table 2).

2.2. Regiospecific Preparation of A-(pyridoxamine-5'-thio)-B-(ω-amino- ethylamino)-β-cyclodextrin.

A,B-capped cyclodextrin 1 was converted to the B_6 model 3 via di- iodide 2[16] using pyridoxamine thiol[17a,b] under the conditions listed in Chart 1. From the crude mixture, excess ethylenediamine was dis- tilled off and 6 mL of EtOH was added to the finely crushed solid. The resultant suspension was agitated, then filtered, and the powder remaining was dissolved in 2 mL of water. The pH of the solution was adjusted to 7.0 by the addition of HCl, and the solution was applied on a CM-25 Sephadex column (1.5 x 60 cm) eluted with NH_4HCO_3 (0 - 0.5 M gradient). From 0.12 - 0.15 M NH_4HCO_3 gradient via the usual work up was obtained 64 mg of 3·HCl (15 % yield).

CHART 1   Preparation of Regioisomeric Artificial B_6 Enzymes from A, B-Capped Cyclodextrin.

1 (A,B-capped β-cyclodextrin)

2 (400 mg)

3a (60%)   3b (40%)

3 total yield 15% (3a + 3b, 64mg as HCl salt)

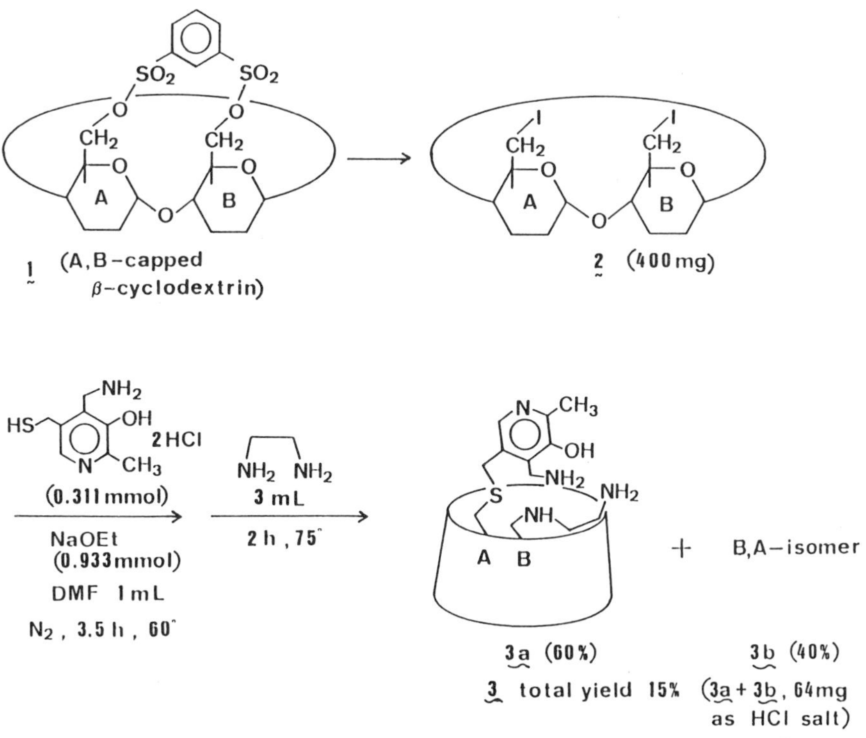

Separation of the two regioisomers, 3a and 3b, was successfully
carried out again by CM-25 column chromatography of prepurified (3a +
3b) mixture (60/40 ratio) by the use of a long, thin column (5 x 1500
mm) eluted with aq. NH_4HCO_3 (0 - 0.1 M gradient). The observed elution
diagram obtained by measuring intensities of the absorption at 324 nm
(B_6 chromophore) of each 3 mL fraction clearly shows that the two
regioisomers are satisfactorily separated. The latter eluent was
recrystallized from aqueous EtOH twice, giving pure A,B-regioisomer,
3a, while from the earlier fractions, pure B,A-regioisomer, 3b, was
similarly obtained. Both compounds showed pyridoxamine characteristic
CH_2 signals and ethylene diamine CH_2 signals in expected intensities,
but the signals of 3a SCH_2 at C_6 were found at 2.82 and 2.55 ppm while
those of 3b were at 3.01 and 2.64 ppm. Less than 1 % of the contami-
nating signal could be seen[18].

2.3. Separation of A,B-pyridoxamine Enzyme Model and B,A- Isoenzyme
Model Based on "Self-selection" Induced by α-ketoacids.

As discussed above, separation of A,B- and B,A-regioisomers of
B -artificial enzyme is not easy. As an alternative, we have studied
"self-selection" by the guest which is specific for the artificial
enzyme but not for the isoenzyme.
A regioisomer mixture of Pm·N 2 was treated with the ketoacid
substrates (4 - 6) at pH 8.0, 30°C. After 72 % conversion of total Pm
to Pl (after 3 h: determined by TLC and electronic spectrum at 320 and
420 nm) Pm and Pl were separated through a C-25 Sephadex column (see
Scheme 1). The isolated Pl·N as a mixture of A,B and B,A-isomer was

SCHEME 1

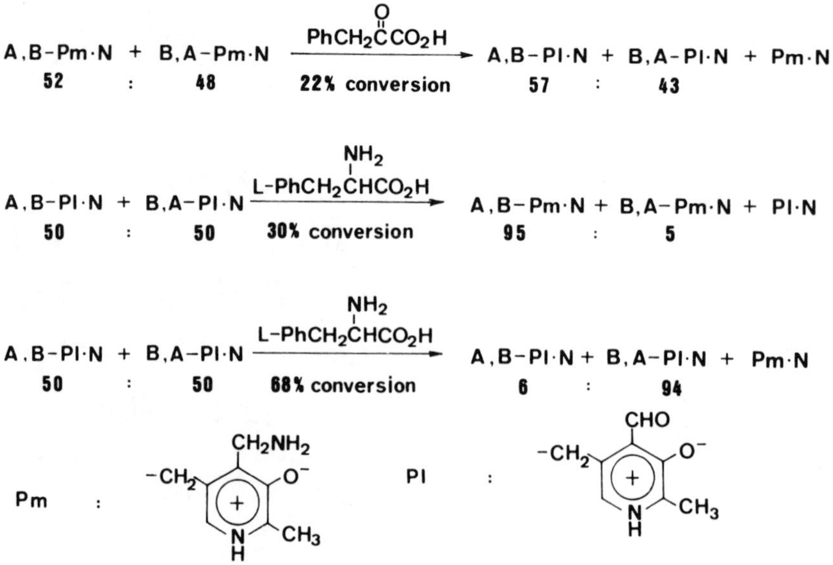

then analyzed by use of 400 MHz ^{1}H NMR, specifically at 2.33 and 2.65 ppm absorptions (C_6 protons of A,B and B,A isomers, respectively, based on the relevantly prepared A,B- and B,A-P1·N). The observed ratio of A,B to B,A isomer was 57/43 for 4, being almost nonselective.

In an independent experiment, the Pm mixture 2 was completely converted to the P1 mixture 3. The P1 mixture was then treated with one of the L-amino acids 7 - 9, the specific substrates. After 30 % conversion (48 h) of total P1 to Pm, the regioisomer ratio of A,B-Pm·N to B,A-Pm N, was determined by 400 MHz ^{1}H NMR (2.82 and 3.01 ppm for A,B- and B,A-isomer pyridoxamine CH_2, respectively). The observed ratio of A,B/B,A induced by 7 was 95/5, showing the significance of "self-selection" of artificial B_6 enzyme (A,B isomer) form the corresponding artificial isoenzyme (B,A) by host-L-amino aicd chiral recognition. At the 68 % conversion of total P1 mixture to total Pm mixture, the B,A/A,B ratio was determined for the unreacted P1 mixture to be 94/6, again showing the significance of "self-selection" made by a specific substrate[19].

2.4. Chiral Induction by Use of A,B-B_6-Artificial Enzyme.

The aminotransfer reaction smoothly proceeded between 3a and keto acids, 4a-c, under the conditions described in eqn 1, giving the corresponding amino acids, 5 a-c, respectively. The amino acids formed were further converted to the 3,5-dinitrobenazmides, 6 a-c, which were acidified (pH 3) and extracted with ether (10 mL x 3). The condensed ether layer (150 μL) was analyzed by HPLC by using a Sumipax OA-1000 chiral column eluted with a MeOH : H_2O : CH_3CO_2H mixture (95/5/1 vol). The two peaks of each racemic amino acid derivative were clearly resolved. Under the conditions, total yields of amino acids were ca 40 %[20] and the enantiomer excess observed for L-amino acid was very high, as listed in Table 3.

The observed high chiral induction for 3a may be understood on the basis of stereospecific participation[17c,20c] of the ω-NH_2 grouping toward the prochiral azomethine plane fixed nearly perpendicular toward N-lone pair during the prototropy (see eqn 1), due to the host-guest multiple recognition[15,21].

TABLE 3   Chiral Induction in Amino Acid Formed
via the Artificial B_6 Catalysis[a]

R-COCO$_2$H	L/D
R = Ph-CH$_2$-	98/2
(indolyl)-CH$_2$-	95/5 [b]
Ph-	98/2 [b]

a. at 30°C, 2 M K$_2$HPO$_4$-KH$_2$PO$_4$ buffer, pH 8.0
b. recorded ratio is lowest limit

eqn. 1

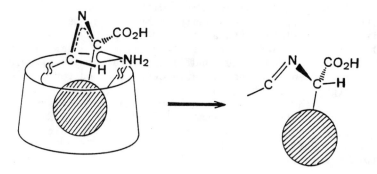

REFERENCES

1. Matsui, Y., Yokoi, T., and Mochida, K., Chem, Lett. p. 1037 (1976).
2. Tabushi, I., Shimizu, N., and Yamamura, K., J. Am. Chem. Soc. 99, 7100 (1977); Tabushi, I., Shimizu, N., and Yamamura, K., Int. Congr. Pure Appl. Chem., 26th I, 44 (1977). (Abstr.)
3. Siegel, B., Pinter, A., and Breslow, R., J. Am. Chem. Soc. 99, 2509, (1977).
4. Eiki, T., and Tagaki, W., Chem. Lett. p. 1063 (1980).
5. Siegel, B., J. Inorg. Nucl. Chem. 41, 609 (1979).
6. Breslow, R., Hammond, M., and Laner, M., J. Am. Chem. Soc. 102, 5402 (1980).
7. Tabushi, I., Kuroda, Y., and Shimokawa, K., J. Am. Chem. Soc., 101, 1614 (1979).
8. Tabushi, I., and Shimokawa, K., J. Am. Chem. Soc. 102, 5402., 1980.
9. Iwakura, Y., Uno, K., Toda, F., and Bender, M. L., J. Am. Chem. Soc. 97, 4432 (1975).
10. Kitaura, Y., and Bender, M. L., Bioorg. Chem. 4, 219, 237, 1975.
11. Tabushi, I., Kuroda, Y., and Sakata, Y., Heterocycles 15, 815, (1981).
12. Matsui, Y., and Okimoto, A., Bull. Chem. Soc. Jpn. 51, 3030 (1978).
13. Breslow, R., and Overman, L. E., J. Am. Chem. Soc. 92, 1075 (1970).
14. A Part of the description is drafted from the following article. Tabushi, I., and Kuroda, Y., 'Cyclodextrins and Cyclophanes as Enzyme Models' in 'Advances in Catalysis', Dley, D. D., Pines, H. and Weisz,P. B. Eds. 32, 417, Academic Press (1983). Academic Press (1983).
15. a) Tabushi, I., Shimokawa, K., Shimizu, N., Shirakata, H., and Fujita, K., J. Am. Chem. Soc. 98, 7855 (1978). b) Tabushi, I., Kuroda, Y., Yokota, K. and Yuan, L. C., ibid. 103, 711 (1981). c) Tabushi, I., Yuan, L. C., Shimokawa, K., Mizutani, T. and Kuroda, Y., Tet. Lett. 22, 2273 (1981). d) Tabushi, I. and Yuan, L. C., J. Am. Chem. Soc. 103, 3574 (1981). e) Tabushi, I., Kuroda, Y. and Yokota, K., Tet. Lett. 23, 4601 (1982). f) Tabushi, I., Yamamura, K., Nabeshima, T., J. Am. Chem. Soc. 106, 5267 (1984). g) Tabushi, I., Nabeshima, T., Fujita, K., Matunaga, A., and Imoto, T., J. Org. Chem., in press: For review, see Tabushi, I., Acc. Chem. Res. 15, 66 (1982).
16. Tabushi, I., Kuroda, Y., and Mochizuki, A., J. Am. Chem. Soc. 102, 1152 (1980).
17. a) Breslow, R., Hammond, M., and Lauer, M., J. Am. Chem. Soc. 102 421 (1983). b) Breslow, R., and Czarnik, A. W., ibid. 105, 1390 (1983). c) Zimmerman, S. C., and Breslow, R., ibid. 106, 1490 (1984).
18. A part of the description is drafted from the following article. Tabushi, I., Kuroda, Y., Yamada, M., Higashimura, H., and Breslow, R., submitted to J. Am. Chem. Soc.
19. Tabushi, I. and Kuroda, Y., unpublished result
20. The present yield of 6 is neither corrected for the conversion in acylation (which may not be exactly the same for D- and L-amino acid) nor the extraction efficiency.

21. a) Tabushi, I., Shimokawa, K., and Fujita, K., J. Am. Chem. Soc.
     99, 1527 (1977). b) Tabushi, I., Kuroda, Y. and Shimokawa, K.,
     ibid. 101, 1614 (1979). c) Boger, J., Brenner, P. G. and Knowls,
     J. R., ibid. 101, 7631 (1979). d) Tabushi, I., Kuroda, Y. and
     Mizutani, T., Tetrahedron 40, 545 (1984).

6A6X-DISULFONATES OF CYCLODEXTRINS

Kahee Fujita
Faculty of Pharmaceutical Sciences, Kyushu University 62,
Maidashi, Higashi-ku, Fukuoka 812
Japan

ABSTRACT. 6A6X-Disulfonates of α, β, and γ-cyclodextrins were prepared by the reactions of cyclodextrins with sulfonyl chlorides in pyridine and were purely isolated by reversed-phase column chromatography. Regiochemistries of these isomers were determined by additional sulfonation, chemical derivation to the authentic specimens, specific degradation by Taka amylase, and chemical conversion to 3,6-anhydro-derivatives followed by Taka amylolysis. Guest-binding behaviors of some modified β-cyclodextrins are shown.

1. INTRODUCTION

In the past decade, designs of enzyme (or receptor) mimics by use of cyclodextrins have attracted much attention [1]. Since enzymes (or receptors) have at least two functional groups at the active site, more sophisticated enzyme (receptor) mimics should be constructed to possess two (or more) functional groups at desirable positions of cyclodextrins. Separation of a trisubstituted α-cyclodextrin (6A6C6E-isomer) from a mixture of the positional isomers by means of silica gel column chromatography was carried out to make a model of a receptor for a phosphate anion [2], although structures of the other regioisomers were not determined. Specific activation of two primary hydroxyls of β-cyclodextrin has been attained through specific preparations of 6A6C and 6A6D-capped β-cyclodextrins from which 6A6C and 6A6D-bifunctional β-cyclodextrins are easily obtained [3]. We describe here non-selective preparation of 6A6X-disulfonated cyclodextrins followed by effective separation, determination of their regiochemistry, and their guest-binding behaviors.

2. PREPARATION, ISOLATION, STRUCTURAL DETERMINATION OF DISULFONATES

2.1. 6A6X-Disulfonates of α-Cyclodextrin

The titled compounds were non-selectively prepared by the reaction of α-cyclodextrin with mesitylenesulfonyl chloride in pyridine. A solution

11

of α-cyclodextrin (3 g, 3.1 mmol) and mesitylenesulfonyl chloride (6 g, 27 mmol) in pyridine (230 mL) was stirred for 2 h at room temperature followed by addition of water (1 mL) and concentration in vacuo. The crude concentrated mixture was applied on a reversed-phase column (Lobar column LiChroprep RP8, Merck, 25 x 310 mm). After a stepwise elution from 900 mL of 10% aqueous MeOH to 100 mL of 20% aqueous MeOH, a gradient elution with 1 L of 40% aqueous MeOH-1 L of 60% aqueous MeOH was applied to give 6A6X-dideoxy-6A6X-di-[(mesitylsulfonyl)oxyl]-α-cyclo-dextrins, 1 (298 mg, 7.8%), 2 (374 mg, 9.1%), and 3 (555 mg, 13.6%). ^{1}H NMR spectra of 1-3 and 6A6X-dideoxy-α-cyclodextrins (4-6) obtained from NaBH$_4$ reduction of 1-3 demonstrated that 1-3 were primary dimesity-lenesulfonates. Positional assignments of 1-3 were carried out as shown in Scheme I. Trimesitylenesulfonates (7-10) were prepared as standard compounds for the assignments; sulfonation of α-cyclodextrin (1 g) with 27 mol excess of mesitylenesulfonyl chloride (6 g) gave four trimesity-lenesulfonates (7-10). Separation by use of the reversed-phase column gave pure 7 (66 mg), 8 (96 mg), and a mixture of 9 and 10 from which pure 9 (58 mg) and 10 (4 mg) were isolated by a preparative reversed-phase HPLC. All of the isolated 7-10 were assigned to be regioisomeric trimesitylenesulfonates (6A6B6C, 6A6B6D, 6A6B6E, and 6A6C6E isomers) from their ^{1}H NMR spectra. Additional monomesitylenesulfonations of the dimesitylenesulfonates (1-3) were employed as criteria of assignments of 1-3 and also 7-10. The additional monosulfonation of 6A6B-, 6A6C-, or 6A6D-dimesitylenesulfonates should produce three (6A6B6C, 6A6B6D, and 6A6B6E), four (6A6B6C, 6A6B6D, 6A6B6E, and 6A6C6E), or two (6A6B6D and 6A6B6E) trimesitylenesulfonates, respectively (Scheme I). HPLC analyses showed that 1, 2, or 3 gave three (7-9), four (7-10), or two (7 and 8) products, respectively. Therefore, we assigned 1 as a 6A6B, 2 as a 6A6C, 3 as a 6A6D, 7 as a 6A6B6D (or 6A6B6E), 8 as a 6A6B6E (or 6A6B6D), 9 as a 6A6B6C, and 10 as a 6A6C6E isomer.

Scheme I.

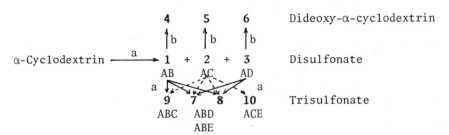

a: Mesitylenesulfonyl chloride/Pyridine
b: NaBH$_4$

2.2. 6A6X-Disulfonates of β-Cyclodextrin

A mixture of the titled compounds was prepared and each compound was purely isolated by the similar method to that described in 2.1. The

structural determinations of the ditosylates, **11** (6A6B), **12** (6A6C), and
13 (6A6D) were carried out as shown in Scheme II. The yields of **11-13**
were 6%, 11%, and 10%, respectively. ^{1}H NMR and FABMS spectra of the
dideoxy derivatives (**14-16**) which were obtained by NaBH$_4$ reduction of
11-13 showed the presence of two methyl groups (δ1.2, doublet, J = 6 Hz),
demonstrating that **11-13** were the primary ditosylates. The ditosylates
(**11-13**) were converted to the corresponding diphenylthio-derivatives
(**17-19**) by treatment with thiophenol. The authentic 6A6C and 6A6D-di-
phenylthio-derivatives were prepared by the reaction of thiophenol with
the β-cyclodextrin capped with diphenylmethanedisulfonate (**20**) [3] and
isolated by use of the reversed-phase column. By comparing the ^{1}H NMR
spectral patterns of the phenyl groups and the retention times in
reversed-phase HPLC with those of the corresponding 6A6C and 6A6D-isomers,
respectively. Therefore, **11** was reasonably assigned to the 6A6B-isomer
since there are only three kinds of disubstitutions on the primary
hydroxyls of β-cyclodextrin. Moreover, the 6A6B structure of **11** was
confirmed by an enzymatic degradation of **17** with Taka amylase followed
by NaBH$_4$ reduction to give **21**. From the FABMS spectrum of **21** showing the
presence of the molecular ion and the fragments (**22** and **23**), the tosylate
(**11**) was assigned to the 6A6B isomer. Also the assignment was independently made by the ^{13}C NMR spectrum of **21**. In conclusion, the ditosylates,
11, **12**, and **13**, were assigned as 6A6B-, 6A6C-, and 6A6D-isomers, respectively.

Scheme II.

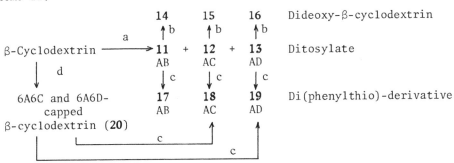

a: p-Toluenesulfonyl chloride/pyridine
b: NaBH$_4$/DMSO
c: PhSH/DMF
d: p,p'-Diphenylmethanedisulfonyl dichloride/pyridine

2.3. 6A6X-Disulfonates of γ-Cyclodextrin

Four ditosylates of γ-cyclodextrin were prepared and purely isolated by
the similar method to that described in 2.1. The yields of 6A6B (**24**),
6A6C (**25**), 6A6D (**26**), and 6A6E (**27**) isomers were 4.9%, 5.0%, 4.7%, and
4.0%, respectively. Among them, only one isomer, 6A6B-disulfonate (**24**),

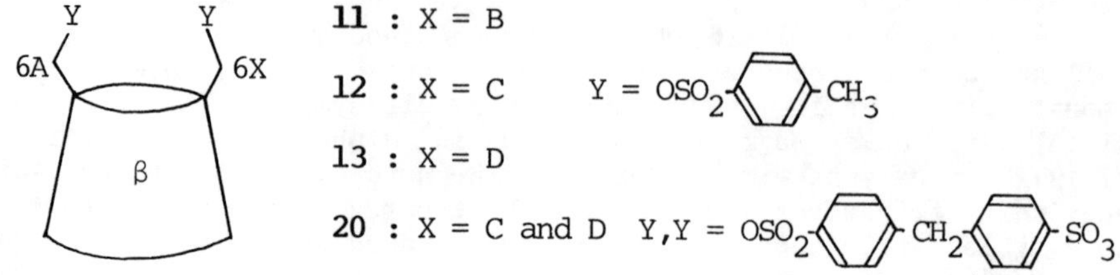

11 : X = B

12 : X = C   Y = OSO_2⟨C₆H₄⟩CH_3

13 : X = D

20 : X = C and D   Y,Y = OSO_2⟨C₆H₄⟩CH_2⟨C₆H₄⟩SO_3

21

22   23

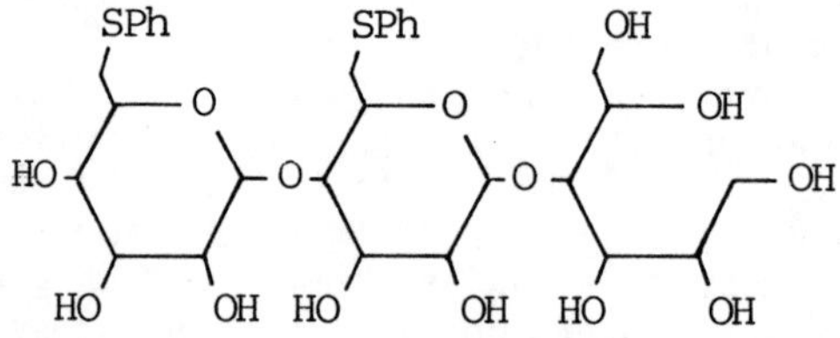

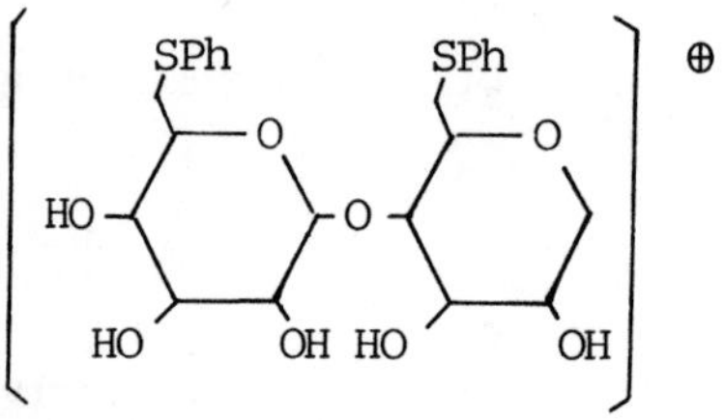

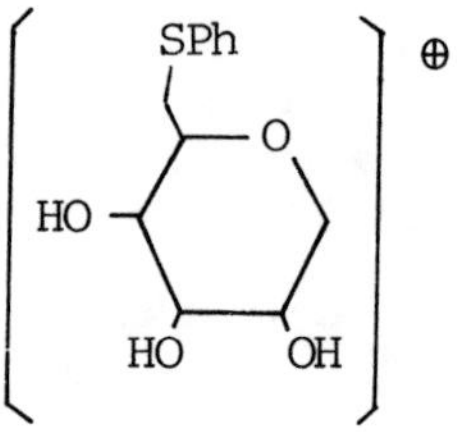

is possible to be structurally determined by the method described in 2.2 (Taka amylolysis). However, this method is not effective to the assignments of the other sulfonates (**25-27**) because they gave a same product, 6'-deoxy-6'-tosyloxyl-maltose. Moreover, there is no authentic specimens for **25-27** and the related compounds. The method which was employed in the assignment of α-cyclodextrin disulfonates was not applicable to the present determination because of the existence of too many (seven) tri-substituted isomers. To the structural determinations of **24-27**, we applied our new findings that 6-deoxy-6-sulfonyloxyl-cyclodextrins were easily converted to 3,6-anhydro-cyclodextrins and that the 3,6-anhydro-cyclodextrins gave 3",6"-anhydro-maltotetraoses in their Taka amylolyses. On the basis of these results, we expected that the Taka amylolyses of 6A6B, 6A6C, 6A6D, and 6A6E-dianhydro-γ-cyclodextrins gave 3",6";3"',6"'-dianhydro-maltopentaose, 3",6";3"",6""-dianhydro-maltohexaose, 3",6"; 3""',6""'-dianhydro-maltoheptaose, and 3",6"-anhydro-maltotetraose, respectively. The disulfonates, **24-27**, were converted to the corresponding dianhydro-γ-cyclodextrins in yields, 85%, 74%, 79%, and 74%, respectively. The anhydro-structure was confirmed by the FABMS and ^{1}H NMR spectra. Taka amylolysis of each 3A6A;3X6X-dianhydro-γ-cyclodextrin gave the oligosaccharide having the expected molecular weight (FABMS spectrum). The linear oligosaccharide was reduced to the corresponding glucitol derivative and then completely acetylated. The acetyl derivative was analyzed by EIMS and FDMS spectra to determine the positions of the 3,6-anhydro-glucose units. In conclusion, these results demonstrated that **24-27** should be assigned as 6A6B, 6A6C, 6A6D, and 6A6E-ditosylates.

3. GUEST-BINDING OF MODIFIED β-CYCLODEXTRINS

3.1. 6A6X-Ditosylates of β-Cyclodextrin

The capped cyclodextrin (**20**) (a mixture of 6A6C and 6A6D-isomers) has been found to bind 1,8-anilinonaphthalenesulfonate or 1-adamantane-carboxylate more tightly than the parent cyclodextrin by Tabushi [3]. We also found that it could bind p-nitrophenyl acetate more strongly than β-cyclodextrin (TABLE I). It is interesting to compare the guest-binding ability of the capped cyclodextrin (**20**) with those of 6A6X-disulfonated β-cyclodextrins (**11-13**), since the latters have flexible, but similar substituents on the cyclodextrin hydroxyls to that of the former. The association constants between the tosylates and p-nitrophenyl acetate were similar to that of β-cyclodextrin, although they differed from one another. This demonstrates that the flexibility of the substituents was not favorable to the strong guest-binding.

3.2. 3,6-Anhydro-β-cyclodextrins

While the glucose unit of cyclodextrins has a $^4C_1$ conformation, 3,6-anhydro-glucose is reported to possess a $^1C_4$ conformation because of its bicyclic structure [4]. Therefore, the cavity shapes of 3,6-anhydro-cyclodextrins are expected to be deformed. From the consideration on the basis of the CPK molecular models, the cavity sizes are expected to

TABLE I.   Association constant (M^{-1}) between modified cyclodextrins and some guests.

β-Cyclodextrin	PNA[a]	MO[b]	PNPA[c]
β-cyclodextrin	3.6×10^2	29.7×10^2	2.1×10^2
capped β-cyclodextrin (**20**)			8.3×10^4
β-cyclodextrin-(OTs)$_2$			
6A6B (**11**)			5.7×10^2
6A6C (**12**)			1.2×10^2
6A6D (**13**)			2.4×10^2
3,6-anhydro-β-cyclodextrin	3.9×10^2	7.3×10^2	
di-3,6-anhydro-β-cyclodextrin			
6A6B	1.0×10^2	3.3×10^2	
6A6C	-	0.8×10^2	
6A6D	-	-	

[a] p-Nitroaniline. In phosphate buffer (pH 6.86).
[b] Methyl orange. In carbonate buffer (pH 10.60).
[c] p-Nitrophenyl acetate. In carbonate buffer (pH 10.60).

become small by substitution of one 3,6-anhydro-glucose unit for one
glucose unit of cyclodextrins. This effect is shown in the decrease in
guest-binding ability (TABLE I). Also, this effect is dependent on the
relative position of the two anhydro-glucose units. Loss of free rotation of the guest molecule in the cavity of the cyclodextrin might be
partially responsible for such decreases in binding ability.

4.   REFERENCES

[1] Bender, M. L.; Komiyama, M. Cyclodextrin Chemistry, Springer
Verlag, Berlin, 1978. Tabushi, I. Acc. Chem. Res. 1982, 15, 66.
Breslow, R. Science (Washington D. C.), 1982, **218**, 532. Tabushi, I.
Tetrahedron, 1984, **40**, 269.
[2] Bojer, J.; Knowles, J. R. J. Am. Chem. Soc. 1979, **101**, 7631.
[3] Tabushi, I.; Shimokawa, K.; Shimizu, N.; Shirakata, H.; Fujita, K.
J.Am. Chem. Soc. 1976, **98**, 7855. Breslow, R.; Bovy, P.; Hersh, C. L.
Ibid. 1980, **102**, 2115. Tabushi, I.; Kuroda, Y.; Yokota,K.; Yuan, L. C.
Ibid. 1981, **103**, 711. Tabushi, I.; Yuan, L. C. Ibid. 1981, **103**, 3574.
[4] Lindberg, B.; Lindberg, B.; Svensson, S. Acta Chem. Scan. 1973,
27, 375.

ATTEMPTS TO MIMIC ENZYMES -- REPORT FROM THE BATTLE FRONT

Ronald Breslow
Department of Chemistry
Columbia University
New York, New York 10027

ABSTRACT. Combining catalytic and binding groups can lead to
multifunctional catalysts that imitate enzymes. Good progress has
been made toward bona fide enzyme mimics, but many problems remain. A
frank recognition of these problems points the way to improved
artificial enzymes.

1. INTRODUCTION

One of the great challenges of modern chemistry is to learn how to
mimic enzymes and extend their catalytic properties and style. Some
years(1) ago we coined the word "biomimetic" to describe the field
that takes its inspiration from an examination of biological systems,
and this word now seems to be widely accepted. The challenge of the
field is great: we want to make synthetic molecules very different in
chemical structure from the proteins of natural enzymes, but still
able to catalyze enzyme-like reactions with very high speeds and
selectivities. We want to do this not only for the reactions that are
already catalyzed by natural enzymes, but at least as importantly for
those reactions of interest to man for which no natural enzyme
catalysts exist.
   One of the dangers in this field seems to be a tendency to
overstate the advances that have been made. If every new system that
imitates some aspect of an enzyme catalyzed process is hailed as the
final solution, real progress will be difficult. We must not ignore
the feedback from nature, when various potential catalyst systems are
designed, synthesized and tested. If we approach the tests with a
proper critical spirit we can learn about the problems that remain and
then try to solve them.
   In this lecture I will emphasize both the advances that have been
made in our laboratory in several areas and the remaining problems. I
will also try to indicate our view of how to attack these remaining
problems. This is done deliberately, to try to discourage the
overselling that has become stylish in the field.

R. Setton (ed.), Chemical Reactions in Organic and Inorganic Constrained Systems, 17–28.
© *1986 by D. Reidel Publishing Company.*

2. PEPTIDASES

Chymotrypsin is not a particularly effective enzyme, but it is one
whose properties and mechanism have been extensively explored.
Although the natural substrates for this enzyme are peptides derived
from proteins, the enzyme will also hydrolyze esters; it has become
popular to examine phenyl esters, in particular p-nitrophenyl esters,
as substrates. With nitrophenyl esters the enzyme shows burst
kinetics, involving a rapid acylation step and a slower hydrolysis.
The acylation step is only about 100,000 faster than is hydrolysis of
these esters at the same pH in the absence of the enzyme, so the
enzymatic reaction with such nitrophenyl esters is not really typical
of most enzymatic processes. An acceleration of only 100,000, and
this only for the first step rather than the complete catalytic
hydrolysis sequence, is small compared to a number of enzyme
accelerations that attain 100,000,000 or greater catalyses compared
with reactions under the same conditions in the absence of the enzyme.
    When we started work in this area Bender had demonstrated(2) that
certain phenyl esters would react with beta-cyclodextrin at a rate
about 100 greater than the rate at which they hydrolyzed under the
same pH conditions, so even the very modest acceleration seen with
chymotrypsin and nitrophenyl esters was not attainable. It seemed to
us likely that the problem was geometric; as we have described
elsewhere,(3) we successfully modified the geometry of the starting
material so as to achieve much larger accelerations. Molecular models
suggested that the starting material could bind well into the
beta-cyclodextrin but that the transition state for acylation of the
cyclodextrin was geometrically incorrect unless the substrate was
pulled significantly out of the cavity. Of course if much of the
binding interaction is lost on proceeding from starting material to
transition state one would expect a slow reaction; we solved this
problem by a number of geometric changes that produced better-bound
transition states for acylation reactions. The most striking examples
were a series of esters, based on the ferrocene nucleus, that acylated
beta-cyclodextrin with an acceleration of more than 1,000,000 compared
with hydrolysis in the same medium in the absence of the cyclodextrin.
Studies of pressure effects on the binding constants and rate
constants for various substrates into beta-cyclodextrin(4) confirm our
picture that the fastest systems are ones that indeed retain most of
their binding on proceeding to the transition state.
    Enzyme accelerations are normally referred to reaction rates in
water. In our case the reaction without cyclodextrin was faster in
our mixed organic solvent medium than it was in water, so that the
acylation of cyclodextrin was 100,000,000 faster than was hydrolysis
of the substrate with the same buffer in a water medium.
    The chirality of cyclodextrin led to a stereospecific reaction,
with as much as a 64-fold preference for one optical isomer of the
substrate compared with the other.(3) Thus on the face of it the
achievement of such an enormous rate acceleration, exceeding that of
the enzyme, and of this very large substrate selectivity could be
advertised as a solution to the artificial chymotrypsin problem.

However, there are certainly problems remaining with this model, and in our publications we have fully and frankly discussed them.

One simple problem has to do with binding geometry. Some years ago we showed that substrates could be attacked by catalytic groups mounted on either the primary or the secondary side of a cyclodextrin,(5) and we have a number of other examples of this phenomenon since that early report.(6) In screening substrates that might be able to bind into cyclodextrin and acylate it while staying bound, we examined some derivatives of adamantane.(3) The adamantane nucleus is too large to fit fully into the cavity of beta-cyclodextrin, and when it sits occluded to the cyclodextrin face the sidechain ester group can be positioned so as to react with a cyclodextrin hydroxyl. However, we found a surprisingly large binding constant for adamantanepropiolic p-nitrophenyl ester, and a rate constant less than that expected.

It seemed likely that the problem here was non-productive binding, and this was confirmed by solving the problem. As one solution we examined a cyclodextrin with a floor that prevented the non-productive binding, and found that now the compound behaved as expected. As another solution we attached a t-butyl group to the substrate, designed so as to hold it in the correct binding geometry. Again the reaction rate and strength of binding were as expected. Thus in this case we were able to detect and solve the problem that the principal binding geometry was incorrect for reaction, but this could be a general problem in enzyme model systems. It is a problem we will return to when we discuss other models.

A second problem has not yet been solved, and is a critical one. Enzyme models that cleave phenyl esters are all very well, but the substrates of real interest are peptides and other amides. Our study of some of the best ferrocene derivatives indicated that the acceleration of hydrolysis fell off strongly as the leaving group was made poorer.(3,7) That is, with a less reactive carboxylic acid derivative not only the rate but also the relative acceleration by cyclodextrin decreased. If the system had been able to accelerate amide hydrolysis by 1,000,000 or 100,000,000 this would have been very exciting, but no such thing occurred. A consideration of the geometry indicates the likely problem. The best substrates have a good geometry for binding and also a good geometry in the tetrahedral intermediate formed in the course of the acylation reaction. However, to complete the acylation the carbonyl group must be rotated 90° from its original plane so that it can form a bond to the cyclodextrin hydroxyl group, and in the very fast substrates the binding is so rigid that this rotation is difficult.

It is an oversimplification to say that an enzyme simply must bind and stabilize the transition state of a reaction, although that is of course a critical requirement. The geometry of the enzyme-substrate complex must be such as to permit the critical freedom of motion along the reaction coordinate, so that reaction intermediates and reaction products also have an acceptable geometry. Thus in the further design of such potential catalyst systems with their associated substrates it is important to take this one type of free motion into account while restricting all of the unwanted degrees of freedom.

In chymotrypsin and other hydrolytic enzymes additional catalytic groups are required, and it would be astounding if an enzyme model could cleave a peptide or amide without an additional functional group to catalyze proton transfers. It is necessary to remove the proton from the attacking oxygen and at least as important to put a proton on the leaving group nitrogen in order to permit cleavage of a peptide. We have developed reactions that permit us to functionalize cleanly on the secondary side of beta-cyclodextrin,(8) and to convert the functionalized compound to an epoxide.(9) Some years ago we examined the use of such an epoxide to permit us to add an imidazole ring to the secondary face of the cyclodextrin. In chymotrypsin and other serine proteases, an imidazole group plays the critical role in catalyzing proton transfer. However, we found that the cyclodextrin carrying an extra imidazole ring was no better than simple cyclodextrin in cleaving phenyl esters.(10) Most important, the pH-rate profile demonstrated that the imidazole ring was not acting as a base catalyst, but that external hydroxide ion was required. Recently in another laboratory our epoxide has been used to prepare a similar compound, and again there seems to be no evidence for cooperative catalysis by the hydroxyl and imidazole groups. Thus this critical feature of the enzymes remains to be understood and added to the enzyme models.

All enzyme mimics that perform simple acylations also suffer from the flaw that one really wants to achieve turnover catalysis of the hydrolysis reaction. Thus in mimics of chymotrypsin it will also be necessary to learn how to hydrolyze the intermediate ester rapidly. If appropriate molecular design produces compounds that can cleave peptides rapidly, it seems likely that the same catalytic groups that achieved this will also be able to perform the hydrolysis of the intermediate esters as they do in chymotrypsin itself. Cram has looked at some synthetic potential catalysts derived from crown ethers and related compounds. He described one compound which seemed to show a significant rate increase of cleavage of p-nitrophenyl ester of an amino acid compared with the cleavage of the same substrate in the absence of the crown ether derivative,(11) but as we have pointed out(3d) the overall rate achieved is approximately 1,000 times slower than is the hydrolysis rate for the same ester with the same buffer in water solution. More seriously, recent work(12) demonstrates that the great bulk of the effect comes from the fact that binding of the amino acid ester to the crown ether keeps the amino acid protonated, while without the crown ether the buffer system simply neutralizes the amino acid derivative and makes it much less reactive. The potential to develop interesting catalysts that operate in nonaqueous media is great, but so far the systems prepared to mimic chymotrypsin do not work as well as simple cyclodextrin does. In particular none of the synthetic systems shows the fast turnover catalysis of peptide cleavage that is the true target of such work.

3. CATALYSIS OF THE DIELS-ALDER REACTION

The Diels-Alder reaction is one of the most useful of synthetic
processes, but there are no enzymes that are known to catalyze these
reactions. (However, there are some natural products whose structures
suggest that they may be formed by very special enzyme-catalyzed
Diels-Alder reactions.) This is just the sort of process one wants to
learn to catalyze by an enzyme-like reactant, as part of the effort to
extend the style of enzyme chemistry into new areas. As we have
reported,(13) we find that certain very simple Diels-Alder reactions
are indeed catalyzed by beta-cyclodextrin. The cavity is able to bind
both the diene and the dienophile, and promote their mutual reaction.
As expected from this mechanism, reagents that are too big to fit into
the cavity in this way do not take part in a catalyzed reaction, but
instead the reaction is inhibited by the selective binding of one of
the two reagents. In a similar fashion, the use of alpha-cyclodextrin
leads to inhibition of the simple Diels-Alder reactions catalyzed by
beta-cyclodextrin, since the smaller cavity of alpha-cyclodextrin can
bind one reagent but not both of them together.
    Although the reactions show a significant acceleration compared
to typical Diels-Alder reactions at the same concentration in various
organic solvents, it turns out that the acceleration relative to an
uncatalyzed reaction in water alone is more modest. The problem here
is that for Diels-Alder reactions that can be accelerated by
hydrophobic binding into a cyclodextrin cavity, there is also
hydrophobic association of the two reagents in water alone.(13,14)
This interesting effect has led to a number of studies(15) of the use
of water for catalyzing the Diels-Alder reaction, and it has been
found(14) that in water as a medium there are also large changes in
the exo/endo ratio in the products. However, this is clearly one of
the limitations in the biomimetic catalysis of the Diels-Alder
Reaction by cyclodextrins. Furthermore, although the Diels-Alder
reaction in a cyclodextrin cavity occurs in a chiral environment, we
have so far observed only small amounts of optical induction in the
products. If we are to achieve much better accelerations and
selectivities, in competition with the remarkable effects seen in
water solution alone, we will clearly have to use more elaborate
binding sites.

4. RIBONUCLEASE MIMICS

One of the more successful enzyme mimics up to this point is the
cyclodextrin-bis-imidazole carrying imidazole rings on the primary
side of the cyclodextrin nucleus.(16) This substance binds a
substrate into the cavity, and then performs a bifunctional catalysis
of the substrate hydrolysis. The pH-rate profile indicates that the
two imidazole groups are cooperating, one of them acting as a base
while the other one acts in its protonated form as an acid. Perhaps
even more striking, the substrate t-butylcatechol cyclic phosphate is
hydrolyzed in essentially one direction by this cyclodextrin catalyst,

while in solution without the cyclodextrin it undergoes random
hydrolysis of either of the two phosphate ester bonds. As we have
discussed elsewhere,(16) the regioselective cleavage of a single P–O
bond can be accounted for by the detailed geometry of the
enzyme–substrate complex mimic.

This is an enzyme mimic that utilizes the principal catalytic
groups of ribonuclease to perform a catalyzed hydrolysis of a cyclic
phosphate ester, just as the enzyme ribonuclease does. The model
apparently uses the same mechanism as is used by the enzyme, and it
shows a product selectivity typical of that produced by the enzyme.
The cyclodextrin–bis–imidazole performs turnover catalysis,
complicated only by the fact that the product of hydrolysis serves as
an inhibitor with a binding constant comparable to that of the
substrate itself.(17)   In all these respects the model seems to be an
excellent mimic of the enzyme ribonuclease.

However, there certainly are limitations to this model. For one
thing, the substrate on which it operates was arbitrarily selected to
fit the cavity, and the reaction itself is not of particular intrinsic
interest. The geometry of this catalyst is not such that it can fit
correctly onto RNA itself; we find that it is at best a poor catalyst
for the cleavage of RNA. Thus if a true ribonuclease mimic is to be
developed, a project on which we are actively working,(18) it will be
necessary to find compounds that bind well to RNA, preferably with
selectivity for particular structural regions, and then carry out
catalytic hydrolysis with good rate and turnover. Such a goal does
not seem inaccessible, but it has not yet been reached.

5.   ENZYMES THAT USE PYRIDOXAL AND PYRIDOXAMINE

We have done a number of studies of systems in which we have
elaborated the pyridoxal and pyridoxamine structures, adding
additional functionality to the normal coenzyme capability of these
systems. As is well known, the enzymatic reactions used in pyridoxal
and pyridoxamine, in which for instance pyridoxal phosphate is a
coenzyme, can also be imitated to some extent in simple model systems
without an enzyme. It has seemed to us attractive to elaborate these
simple models by adding binding and catalytic functionality, to try to
bring them to the velocities and selectivities typical of the enzymic
processes themselves.

Our earliest study on this(19) involved the attachment of
pyridoxamine to the primary side of beta–cyclodextrin. At the
operating pH of the medium the pyridoxamine primary amino group is
protonated, and we expected that with the resulting charge the
molecule should exist in an open form rather than in a form with the
pyridoxamine ring bound into the cyclodextrin nucleus. As expected
from this, the compound is actually comparable in reactivity to simple
pyridoxamine in transamination with small keto acids, such as pyruvic
acid, so indeed the pyridoxamine ring is fully available. More
important, the compound shows a considerable preference for reaction
with keto acids that can bind into the cyclodextrin cavity.

For instance, the conversion of phenylpyruvic acid to

phenylalanine is 50 times faster with this pyridoxamine-cyclodextrin compound than it is with pyridoxamine alone under the same conditions, and there is a 35-fold acceleration of the conversion of indolepyruvic acid to tryptophan. Thus in these cases the additional binding group contributes to the catalysis, although it is obvious that the acceleration is nowhere near the very large accelerations we have achieved in cyclodextrin acylations, for instance. We believe the problem here is that the linkage between the binding group and the catalytic group is rather flexible. It is our plan to construct more rigid molecules in which the extra degrees of freedom are frozen out, since past experience indicates that this should be very effective in increasing the rate of the reaction.

The transaminations in which substrates are inside the cyclodextrin cavity also have the interesting feature that they experience the chirality of the cyclodextrin. Thus we found that there was as much as a 5 to 1 preference for the formation of the natural L amino acid in some cases. Of course this kind of stereochemical control is rather accidental. We have also attached pyridoxamine to the secondary side of beta-cyclodextrin(9) and also to a synthetic cavity(20) based on some of the synthetic cavities studied by Koga. As expected these were similar in their ability to promote reactions of keto acids that could bind into the cavities, but of course the synthetic cavity, which is not chiral, shows no asymmetric induction in the product amino acids.

It seemed to us likely that we could achieve stereochemical control of the synthesis of amino acids during transamination if we could attach a proton transfer catalytic group to pyridoxamine, and if it would act not only to remove the proton from the CH_2 group of pyridoxamine but also to put that proton on the new asymmetric carbon of the product amino acid. As we have described,(21) kinetic studies indicate that this mechanism operates provided we use a sufficiently long flexible base. With such a basic arm attached at C-5 of pyridoxamine we observe much faster transamination, and the dependence of the rate on the length of the basic arm indicates clearly that the base is acting to deliver the proton to the new carbon, not just remove it from the CH_2 group.

Using these findings, we synthesized a pyridoxamine derivative carrying a fused ring between positions 5 and 6, and used this fused ring to carry a basic sidearm in a chiral orientation.(22) As hoped, this system was very effective in promoting the stereospecific formation of a new amino acid. Studies with related compounds made it clear that it was directing the stereochemistry by active catalysis of proton transfer, not simply passive blocking of a face of the substrate.

We have also found that we can optimize the particular reactions performed by pyridoxal or pyridoxamine depending on the positioning of a base group. The relatively long flexible basic arm is best to promote transamination, while a shorter basic arm is very effective in promoting the conversion of chloropyruvic acid to the pyridoxal Schiff base of alpha-aminoacrylic acid.(23) We have also found that a rigid long basic arm is particularly effective at promoting racemization of amino acids by pyridoxal without interference by the transamination process.(24) It is obvious that in the future such careful geometric placement of additional catalytic groups is the best approach to developing the selectivity characteristic of enzyme reactions, a selectivity that has only recently begun to appear in the model reactions involving pyridoxal and pyridoxamine.

Of course one should be able to combine these two lines of research. It would be attractive to attach both a binding group and a basic catalytic group to pyridoxamine to promote selective rapid transaminations; two approaches to this are underway. In our scheme, which has been discussed at various meetings in the past,(25) we are constructing a pyridoxamine system carrying a cyclodextrin and a basic sidearm both attached to the C-5 carbon of pyridoxamine. This synthesis has been slow to develop, but recently the synthetic problems have been overcome and the compound is essentially in hand. However, we do not yet have clear results about its behavior. The other approach to this has been examined in the laboratories of Tabushi,(26) and is described in detail elsewhere in this volume. Tabushi has been developing synthetic methods to permit the attachment of two different groups to neighboring glucose residues of a cyclodextrin; using this approach he has attached both a pyridoxamine nucleus and a basic chain. Apparently this system is quite effective in catalyzing stereospecific amino acid synthesis also.

One of the limitations of these systems that will require further improvement, besides their rather low velocities compared with the enzyme processes, is seen in our study of a mimic of the tryptophan synthetase reaction.(23) Tryptophan synthetase operates by binding two substrates to an enzyme and catalyzing their condensation. One of the substrates, serine, is bound to pyridoxal and undergoes and catalyzed dehydration to the pyridoxal Schiff base of alpha-aminoacrylic acid mentioned earlier. The second substrate, indole, is bound to a hydrophobic site of the enzyme. The two pieces are held together so that a new carbon-carbon bond can be formed, and the amino acid tryptophan is produced.

We have constructed a mimic of this enzyme by attaching pyridoxal to the primary side of beta-cyclodextrin, and have used this to condense serine with indole.(23) It was a better catalyst for the condensation than was simple pyridoxal, but with both systems the yield, based on serine, is rather small. The problem is that serine undergoes a variety of reactions catalyzed by pyridoxal. Without appropriate control, by geometric placement of catalytic groups, these other reactions compete with the dehydration to make the aminoacrylic acid derivative needed for tryptophan synthesis.

A related but more effective substrate, beta-chloroalanine, reacts with pyridoxal to generate the same aminoacrylic acid intermediate, but with fewer side reactions. This gives a better yield of tryptophan on condensation with indole catalyzed even by pyridoxal, and the pyridoxal-cyclodextrin catalyst is better yet. However, the improvement in the rate of formation of tryptophan is only about 5-fold, certainly much less than the very large improvements hoped for as a result of binding of two reacting groups. As expected, even this improvement only occurs at rather low concentrations, in which the gathering effect of the second binding

site is important. Spectroscopic measurements indicate clearly that
pyridoxal is bound into the cavity of cyclodextrin in the catalyst, so
any additional binding of substrates must be in competition with the
internal binding of a piece of the catalyst group itself. In contrast
to pyridoxamine, pyridoxal is uncharged at the operating pH of the
reaction: it has no reason not to enter the hydrophobic cavity. Thus
for an improvement in this type of system we will have to learn how to
assemble the catalytic group and the binding site in such a way that
they cannot mutually interact and "short circuit" each other.

In general for good catalysis it is important that we avoid short
circuits.(27) A basic catalytic group and an acidic catalytic group
must not be able to interact directly, or that interaction will
compete with the hoped for interaction with the substrate. In the
same way, any binding of pieces of the catalyst into the binding site
is obviously at the expense of catalytic effectiveness. While a very
good start has thus been made toward the generation of enzyme mimics
for the important reactions catalyzed by pyridoxal and pyridoxamine,
it is evident from a realistic evaluation of these systems that some
improvements are needed. Furthermore, it is clear in which direction
to move in order to achieve such improvements.

6. ENZYMES USING THIAMINE PYROPHOSPHATE AS COENZYME

Ever since our work elucidating the mechanism by which thiamine
pyrophosphate acts as a coenzyme in biochemical reactions,(28) we have
retained an interest in this system. It is an extremely attractive
coenzyme for incorporation into enzyme mimics, since in contrast to
pyridoxal and pyridoxamine it catalyzes all of its reactions without
undergoing a fundamental structural change itself. Of course
pyridoxal catalyzes amino acid racemization and tryptophan synthesis
without undergoing a permanent transformation, but in transamination
reactions pyridoxal and pyridoxamine are interconverted, and a
regeneration is required.

In our first study(29) we attached thiamine and various analogs
to beta-cyclodextrin on the primary carbon. We observed a cooperative
increase in the selectivity of the coenzyme toward some catalyzed
reactions in which the additional binding site could be used, but not
in others. For instance, thiamine and various related compounds can
catalyze the benzoin condensation of benzaldehyde; this process is not
better with the new binding site present. Reaction of the thiamine
with the first benzaldehyde is probably faster, but the intermediate
product can be bound into the cyclodextrin cavity in such a way as to
slow attack of the second benzaldehyde. To improve this system we
will need more than one binding site, so that both of the substrates
can undergo an attractive interaction with the catalyst, not merely
one of the substrates.

We did however find that reactions involving only one substrate
showed cooperative catalysis. The exchange of the aldehyde proton of
benzaldehyde and its derivatives is catalyzed by thiamine and other
thiazolium salts, since the intermediate in thiamine catalysis
activates the aldehyde hydrogen. As expected, we found that this

process was faster when the beta-cyclodextrin binding unit was also present, to help bind the substrate, but the additional catalysis was rather modest. With thiamine itself only a 4-fold acceleration was observed; it seemed to us that part of the problem might be that the pyrimidine ring of thiamine can bind into the cyclodextrin cavity, competing with substrate binding. This problem was confirmed and partially solved by using a charged benzyl group in the thiazolium salt attached to the beta-cyclodextrin. With such a charged group, which could not bind into the cavity of cyclodextrin, there was now a 10-fold improvement in the rate of reaction with this catalyst compared to the reaction with a simple thiazolium salt.

We still have a long distance to go with such systems before we have all the important catalytic groups present, and attached in such a way as to minimize the flexibility of the system and prevent nonproductive binding of pieces of the catalyst into the binding site. With such improvements one can hope to develop systems that could catalyze the very unusual reactions performed biochemically by the coenzyme thiamine pyrophosphate.

7. SUMMARY

In the preceding account we have tried to present the advances that have been achieved as we have added binding interactions and additional catalytic interactions to produce polyfunctional catalysts as mimics for enzymes. Not all such polyfunctional catalysts are effective, and it is critical that they be evaluated honestly and their shortcomings acknowledged so that improvements can be designed and incorporated. At the height of the conflict in Vietnam a U.S. senator proposed that we solve the problem by announcing that we had won the war, and then withdraw our troops. Appealing as this might be as a solution to the moral dilemma of Vietnam, it is no prescription for scientific research. Developing compounds that are really good mimics of enzymes is a tough and challenging area, but it is possible to achieve the goals provided we do not announce total victory when there are still many battles to be won.

8. REFERENCES

1. R. Breslow, Centenary Lecture, Chem. Soc. Rev., 1, 553 (1972).

2. R. L. van Etten, J. F. Sebastian, G. A. Clowes and M. C. Bender, J. Am. Chem. Soc., 89, 3242 (1967);

3.a. M.F. Czarniecki and R. Breslow, J. Amer. Chem. Soc., 100, 771 (1978); b. R. Breslow, M.F. Czarniecki, J. Emert, and H. Hamaguchi, J. Amer. Chem. Soc., 102, 762 (1980); c. G. Trainor, R. Breslow, J. Amer. Chem. Soc., 103, 154 (1981); d. R. Breslow, G. Trainor, and A. Ueno, J. Am. Chem. Soc., 105, 2739 (1983).

4. W.J. le Noble, S. Srivastava, R. Breslow and G. Trainor, J. Am. Chem. Soc., 105, 2745 (1983).

5. R. Breslow and L.E. Overman, J. Amer. Chem. Soc., 92, 1075 (1970).

6. e.g. B. Siegel, A. Pinter, and R. Breslow, J. Amer. Chem. Soc., 99, 2309 (1977).

7. G. Trainor, unpublished work.

8. A. Ueno, R. Breslow, Tetrahedron Lett., 23, 3451-3454 (1982).

9. R. Breslow, A.W. Czarnik, J. Am. Chem. Soc., 105, 1390 (1983).

10. K. Nakasuji, unpublished work.

11. D. J. Cram and H. E. Katz, J. Am. Chem. Soc., 105, 135 (1983).

12. D. Cram, International Symposium on Bioorganic Chemistry, New York, 1985.

13. D. Rideout, R. Breslow, J. Amer. Chem. Soc., 102, 7816 (1980).

14. R. Breslow, U. Maitra and D. Rideout, Tetrahedron Lett., 24, 1901-1904, (1983); R. Breslow, U. Maitra, Tetrahedron Lett., 25, 1239-1240 (1984).

15. e.g. P. A. Grieco, P. Garner and Z. He, Tetrahedron Lett., 24, 1897 (1983).

16. R. Breslow, J. Doherty, G. Guillot and C. Lipsey, J. Amer. Chem. Soc., 100, 3227 (1978).

17. C. Lipsey, Ph.D. thesis, Columbia Univ., 1979.

18. Cf. R. Corcoran, M. Labelle, A. W. Czarnik, and R. Breslow, Anal. Biochem. 144, 563-568 (1985).

19. R. Breslow, M. Hammond, and M. Lauer, J. Amer. Chem. Soc., 102, 421 (1930).

20. J. Winkler, E. Coutouli-Argyropoulou, R. Leppkes, and R. Breslow, J. Amer. Chem. Soc., 105, 7198-7199 (1983).

21. S.C. Zimmerman, A.W. Czarnik, and R. Breslow, J. Amer. Chem. Soc., 105, Tetrahedron Lett., 24, 1694-95 (1983).

22. S.C. Zimmerman and R. Breslow, J. Amer. Chem. Soc., 106, 1490-1491 (1984).

23. W. Weiner, J. Winkler, S. Zimmerman, and R. Breslow, J. Amer. Chem. Soc., in press.

24. J. Chmielewski, unpublished.

25. e.g. R. Breslow in "Biomimetic Chemistry" Kodansha, Ltd., Tokyo, 1983, p. 1-20.

26. I. Tabushi, Y. Kuroda, M. Yamada, H. Higashimura, and R. Breslow, J. Amer. Chem. Soc., in press.

27. Cf. R. Breslow, Proc. N.Y. Academy of Sci., in press.

28. R. Breslow, J. Amer. Chem. Soc., 79, 1762 (1957); R. Breslow, J. Amer. Chem. Soc., 80, 3719 (1958).

29. D. Hilvert and R. Breslow, Bioorganic Chemistry, 12, 206-220 (1984).

WEAK FORCES IN MIXED CHELATE AND MACROCYCLIC SYSTEMS

C. David Gutsche
Department of Chemistry
Washington University
St Louis, MO 63130
USA

ASBTRACT. The influences of noncovalent forces have been investigated
in two systems, one involving the operation of electrostatic
attraction in a mixed chelate and the other the operation of hydrogen
bond attraction in the synthesis and the comformational and complexing
properties of a macrocycle. The function of the mixed chelate system
is to transfer the acetyl group from acetyl phosphate to an acceptor
nucleophile via the intermediacy of a triethylenetetramine containing
a micellizing side chain and an acetyl accepting nucleophilic group.
Through the application of ^{31}P NMR measurements on a model system
it has been ascertained that a mixed chelate does form and that the
rate of exchange of the phosphate moiety with the mixed chelate is
fast enough that it is not a rate determining factor. An appropriately
substituted triethylenetetramine-type compound has been synthesized
and has been shown to be an effective acetyl phosphate hydrolysis
catalyst, thus acting as a phosphatase mimic.

Calixarenes, a class of stoma-containing macrocyclic compounds,
can be synthesized by base-catalyzed condensation of p-substituted
phenols and formaldehyde. The pathway for calixarene formation is
postulated to depend in a critical fashion on intermolecular and
intramolecular hydrogen bonding. Similarly, the conformational
behavior of the calixarenes is strongly influenced by intramolecular
hydrogen bonding. The calix[4]arenes have been shown to form complexes
with certain amines in acetonitrile solution, the forces responsible
for the complex formation probably being a combination of
electrostatic and hydrogen bond attractions.

1. INTRODUCTION

The covalent bond is the centerpiece of organic chemistry. It is
what holds atoms together in the arrays we call molecules; it
is the focus of our attention when we devise methods for transforming
molecules; it is what might be termed a "strong force". However,
weaker forces are also operative in chemical systems and may play
important roles in determining reaction pathways and chemical

R. Setton (ed.), Chemical Reactions in Organic and Inorganic Constrained Systems, 29–48.

behavior. In the work that is described in the next sections several
examples are discussed in which "weak forces" play a role, viz.
electrostatic attraction in a mixed chelate system, hydrogen bond
attraction in the synthesis of certain macrocyclic molecules, hydrogen
bond attraction in the conformations these molecules adopt, and
electrostatic attraction together with hydrogen bond attraction in the
formation of molecular complexes in solution.

2. ELECTROSTATIC FORCES IN A MIXED CHELATE SYSTEM

2.1. Overall Concept

The original goal of this project was to devise a system in which
acetyl phosphate can be used as a selective acetylating agent through
the cooperative action of mixed chelate formation, pseudointramolecular acyl transfer, and intramicellar transfer. As depicted in
Fig 1, compound 1 forms a mixed chelate (2) with a metal ion and
acetyl phosphate, the nucleophilic moiety of 2 (R = a nucleophile)
then acts as a pseudointramolecular acceptor for the acetyl group of
acetyl phosphate to form the acetylated species 3, and 3 transfers
the acetyl group to another nucleophile which has formed a co-micelle
with 2.

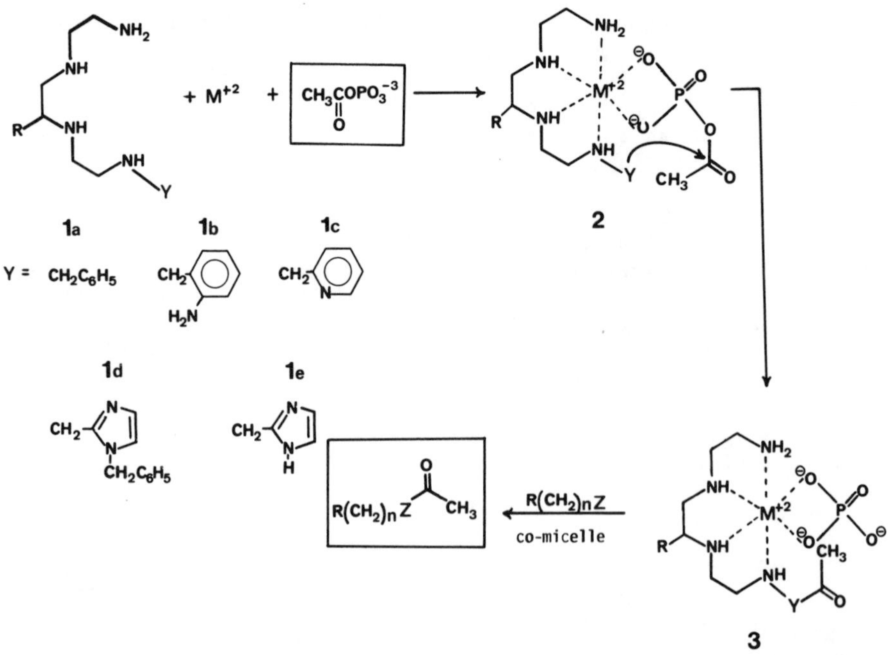

Fig 1. Selective acetyl transfer from acetyl phosphate to a
nucleophile

2.2. Initial Experiments (1,2)

The rates of decomposition of acetyl phosphate in the presence of
metal ions, with and without triethylenetetramine, are shown in Table
1. The effect of 5-dodecyltriethylenetetramine, alone and in the
presence of metal ions, on the rate of decomposition of acetyl
phosphate is shown in Table 2. The rates of decomposition of acetyl
phosphate in the presence and absence of metal ions, triethylenetetramine, and substituted triethyleneamines (1) are shown in Table 3.

		k_{obs}, $10^{-3}min^{-1}$				
[trien], M	[metal],M	Mn^{+2}	Co^{+2}	Zn^{+2}	Ni^{+2}	Cu^{+2}
none	0.005	5.4	6.2	4.6	5.9	6.3
0.005	0.005	7.4	5.7	5.5	4.5	5.0
0.025	0.025	23.0	19.0	9.1	6.3	4.8

Table 1. Rates of decomposition of acetyl phosphate at 30°in the
presence of metal ions and triethylenetetramine

		k_{obs}, $10^{-3}min^{-1}$			
[5-alkyltrien],M	no metal	Mn^{+2}	Zn^{+2}	Ni^{+2}	Cu^{+2}
.005	13.7	8.9	5.7	5.8	7.8
.010		23.0	7.4	6.4	8.6
.025	77.0	150	14.4	10.6	19.0
.050					50

Table 2. Effect of 5-dodecyltriethylenetetramine and metal ions
on the rate of acetyl phosphate decomposition at 30°

		k_{obs},$10^{-2}min^{-1}$					
metal	no tetramine	trien	1-a	1-b	1-c	1-d	1-e
none	1.7	2.51	1.84	1.71	1.84	1.77	2.01
Mn	2.4	2.83	1.93		2.42	2.40	3.60
Co	2.4	3.48	4.58	13.2	2.45	1.86	2.30
Zn	2.1	2.44	1.83	1.94	2.16	2.13	2.70
Ni	3.9	2.34	2.42	2.31	2.30	2.12	2.40
Cu		2.31	2.11	1.83	5.09	2.04	2.04

Table 3. Rates of decomposition of acetyl phosphate at 50° in the
presence of metal ions, trien, and substituted triens

2.3. NMR Studies of the Mixed Chelates of Triethylenetetramine, Phosphate, and Metal ions

The lack of significant rate enhancements in the acetyl phosphate hydrolysis, as shown by the data in Tables 1-3, might be attributed to a variety of factors. To explore the question of the extent to which mixed chelate formation occurs and the rate at which a phosphate moiety enters and leaves the mixed chelate, a system was studied that included (a) triethylenetretramine (trien) as the polyamine, (b) inorganic phosphate, and (c) the metal ions Cu^{+2}, Ni^{+2}, Mn^{+2}, or Fe^{+3}. ^{31}P NMR was used as the probe, making use of measurements of the longitudinal relaxation times (T_1) and transverse relaxation times (T_2).

2.3.1 Stability Constants of Mixed Metal Chelates via Measurement of Longitudinal Relaxation Rates. The observed longitudinal relaxation rate (R_{obs}) for the system being studied is the sum of the relaxation rate of the free phosphate (R_o) multiplied by the fraction of free phosphate ($1 - P_m$) plus the relaxation rate of the bound phosphate (R_m) multiplied by the fraction of the bound phosphate (P_m),

$$R_{obs} = R_o(1 - P_m) + R_m P_m \tag{1}$$

When $[M]_o \ll [P]_o$ essentially all of the metal ions are complexed and $[M]_o \simeq [M-trien] + [M-trien-P]$, $[trien] \simeq [trien]_o$, and $[P] \simeq [P]_o$. leading to the expression $R_{obs} = R_o + R_m P_m = R_o + R_m[M - trien-P]/[P]_o$ The equilibrium constant, $K = [M-trien-P]/[M-trien][P]$ can be simplified and rearranged to $[M-trien-P] = K[P]_o[M]_o/(1 + K[P]_o)$. The "excess" rate ($R_e$), defined as the difference between R_{obs} and R_o, is determined only by the paramagnetic relaxation mechanism;

$$R_e = R_{obs} - R_o = R_m[M-trien-P]/[P_o]. \tag{2}$$

Combining these expressions gives an expression in which R_e is proportional to the total metal ion concentration; a plot of R_e vs $[M]_o$ gives a straight line with the slope defined as

$$R_e = \alpha[M]_o \tag{3}$$

where α is defined as $R_m K/1 + K[P]_o$. A plot of $1/\alpha$ vs $[P]$ gives a straight line from which the values of R_m and K can be calculated. Measurement of R_e as a function of $[M]$ at a given $[P]$ gives a straight line, as illustrated in Fig 2. Using the values of the slopes obtained from a series of such measurements at various $[P]$ and application of expression-3 produces values for R_m and K, as illustrated in Fig 3 for Cu^{+2}.

The extent of mixed chelate formation between trien, phosphate, and the various metal ions was determined from these measurements and found to have the values shown in Table 4.

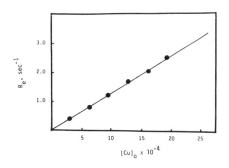

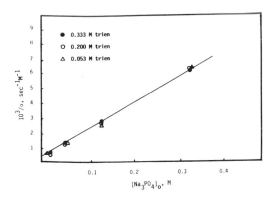

Fig 2. Relationship between R_e and Cu^{+2} at $[P]_o$ = 0.333 M, $[trien]_o$ = 0.326 M,

Fig. 3. Effect of $[P]_o$ and $[trien]_o$ on α at $25^{\circ}C$ and pH 11.6 on a system containing Cu^{+2}

Metal Ion	$[trien]_o$	R_m, sec^{-1}	K
Cu^{+2}	0.333	54	81
Ni^{+2}	0.333	360	26
	0.200	420	25
	0.053	1400	26
Fe^{+3}	0.333	1600	61
Mn^{+2}	0.333	3800	120

Table 4. Relaxation rates and stability constants for mixed chelates at $25^{\circ}C$, pH 11.6.

2.3.2. Exchange Rates of Mixed Metal Chelates via Measurement of Transverse Relaxation Rates (3). For solutions containing only one type of paramagnetic ion, the observed relaxation time $(T_{2,obs})$ is

$$\frac{1}{T_{2,obs}} = \frac{1}{T_{2,o}} + \frac{P_m}{\tau_m}\left[\frac{\frac{1}{T_{2,m}^2} + \frac{1}{\tau_m T_{2,m}} + \Delta\omega_m^2}{\left(\frac{1}{T_{2,m}} + \frac{1}{\tau_m}\right)^2 + \Delta\omega_m^2}\right] \quad (4)$$

where $T_{2,obs}$ and $T_{2,o}$ are the transverse relaxation times for nuclei in solution in the presence and absence of paramagnetic species; $T_{2,m}$ is the relaxation time in the metal complex, τ_m is the average residence time of the nuclei in the metal coordination sphere; P_m is the mole fraction of nuclei bound to the paramagnetic metal ($P_m = \tau_m/\tau_A$;

τ_A is the average lifetime of nuclei in the bulk solvent; and $\Delta\omega_m$ is the chemical shift difference between the bound and unbound nuclei. The temperature dependence of the quantity $1/T_{2,obs}$ shows a plot that can be divided into four regions to which various relaxation modes have been ascribed. Region-I, for example, is characterized by "slow exchange" in which $\Delta\omega_m{}^2 \gg 1/T_{2,m}^2$ and $1/\tau_m^2$, leading to the simplified expression

$$1/T_{2,obs} = 1/T_{2,o} + P_m/\tau_m \qquad (5)$$

By subtracting the outer sphere contribution, obtained from extrapolation of region 0, $1/\tau_m$ can be obtained from expression 5. The enthalpies and entropies of activation can be obtained by means of the Eyring expression,

$$1/\tau_m = \frac{kTe}{h} e^{-H/RT \ S/R} \qquad (6)$$

Plots of $(1/T_{2,obs} - 1/T_{2,o})/P_m$ vs $1/T$ gave graphs in which region-I could be discerned for all four of the paramagnetic ions studied, region-II could be discerned for Cu^{+2}, and region-0 could be discerned for Cu^{+2} and Mn^{+2}. The thermodynamic and kinetic data for dissociation are shown in Table 5.

Metal Ion	$\Delta H^{\ddagger}$,kcal/mole	$\Delta S^{\ddagger}$,cal/$^\circ$K/mole	$k, 10^3 s^{-1}$
Cu^{+2}	2.31	−30	15.2
Ni^{+2}	11.3	−3.7	5.0
Fe^{+3}	4.3	−17	35
Mn^{+2}	15.6	11	2.8

Table 5. Thermodynamic and kinetic data for the dissociation of mixed metal chelates

2.4. Discussion of Results of NMR Studies

The present data provide evidence that mixed chelate formation does, indeed, occur. From the data in Table 4 it can be calculated that the amount of phosphate bound in the mixed chelate with trien and metal under the conditions used in the earlier experiments (1,2) is ca 40% for Cu^{+2}, 19% for Ni^{+2}, 48% for Fe^{+3}, and 34% for Mn^{+2}. If acetyl phosphate behaves in a quantitatively similar fashion, this degree of association should be sufficient for demonstrable catalysis (a) if the other necessary features such as appropriate stereochemistry and sufficient intramolecular nucleophilicity are met and (b) if the metal mixed chelate exists for a long enough time for reaction to take place. That the brief existence of the complex is not the limiting feature in these systems is indicated by the data in Table 5 which show that its lifetime is on the order of 10^{-4} sec, a time sufficiently long for as many as 10^6 collisions to occur.

2.5. Design of a Hydrolysis Catalyst

The work summarized in Sec 2.2 shows that (a) a modest degree of micellar catalysis of acetyl phosphate decomposition is observed with a triethyltetramine carrying a long hydrocarbon chain and (b) a nucleophile attached to the polyamine (or polyamide) chelator is sometimes capable of exerting some degree of intramolecular nucleophilic assistance. The features of a long hydrocarbon chain and a polyamide chelator carrying a nucleophile are both incorporated in compound **4** which was synthesized by the reactions shown in Fig 4.

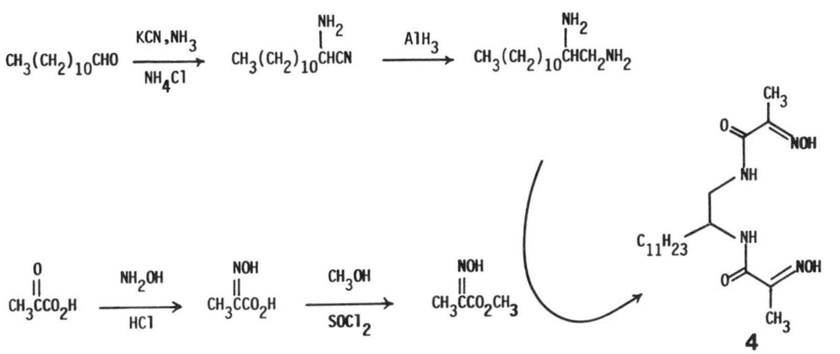

Fig. 4. Synthesis of 1-undecyl-N,N'-bis(hydroxylimino-
          propanoyl)ethylenediamine (**4**)

Experiments with **4** in neutral aqueous solution are limited by the very low solubility of **4**. Even in the presence of 0.025 M CTAB a 0.005 M solution remains cloudy. Upon adjusting the pH to 11.5, however, a clear solution is obtained as a result of ionization of the hydroxyimnino groups. When the concentration of **4** is doubled, the solution remains clear but it becomes viscous, possibly the result of a phase transition in which globular micelles change to cylindrical micelles. Addition of metal ions to these solutions results in chelate formation in some instances, as indicated by color changes and the fact that no precipitate forms. As shown in Table 6, Cu^{+2}, Zn^{+2}, and Ni^{+2} have essentially no effect on the rate of hydrolysis. These measurements were made at pH 6, since the metal hydroxides precipitate in more basic solution, and it is assumed that the effects of the metal ions would be essentially the same at pH 11.5 if the ions could remain in solution. With CTAB and compound **4** in the absence of metal ions only a modest rate enhancement is observed, while with CTAB and **4** in the presence of metals significant rate enhancements are observed, as shown in Table 6.

In all cases first order kinetics were observed. Since the initial concentrations of acetyl phosphate and metal ion (which is assumed to form the metal-**4** complex) are the same, this kinetic result indicates that the catalyst is not consumed in the reaction. A

| Compound 4 | k, $10^{-3}\,\mathrm{min}^{-1}$ | | | |
	no metal ion	Zn^{+2}	Cu^{+2}	Ni^{+2}
none		4.4	4.4	5.4
0.005 M	45	560	430	240
relative rate	11	140	108	60

Table 6. Rate of acetyl phosphate decomposition at $40^{\circ}C$ and pH 11 in the absence and presence of compound **4**

mechanism that might account for this behavior is illustrated in Fig 5-A. It is similar to one proposed by Breslow and Malmin (4) for the decomposition of acetyl phosphate by a chelated oxime in which an acyl transfer to the oxime occurs, followed by rapid hydrolysis of the oxime acetate. Another mechanism that can also accommodate these results postulates that the oxime abstracts a proton from an appropriately positioned water molecule, thereby generating hydroxide in the vicinity of the acetyl phosphate, as depicted in Fig 5-B. Compound **4**, catalyzing the transfer of the acetyl group from acetyl phosphate to water, might be considered as a phosphatase mimic.

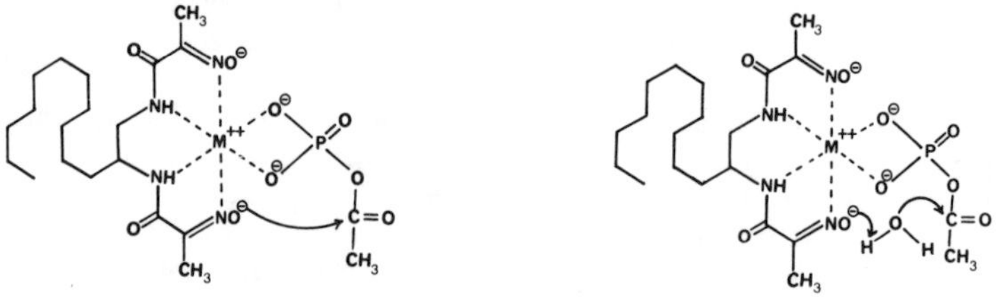

Fig 5. Catalysis of hydrolysis of acetyl phosphate by **4**

The original goal of this project, as stated in Sec 2.1, was to devise an acyl transfer mimic wherein the acetyl group becomes attached to a nucleophile other than water. Attempts to achieve this with the present system, however, have been unsuccessful. Reaction mixtures containing **4**, metal ions, CTAB, and n-hexanol, n-butylamine, or n-octylamine showed little change in rate of decomposition as the result of the addition of the alcohol or amine; no evidence for ester or amide formation was adduced.

3. HYDROGEN BOND FORCES IN A MACROCYCLIC SYSTEM

3.1. Brief History of the Calixarenes

In 1941 Zinke (5) treated various p-substituted phenols with

formaldehyde in the presence of base and obtained high melting
compounds to which cyclic tetrameric structures (5) were assigned, as
depicted in Fig 6. In 1978 Gutsche and coworkers (6) showed that

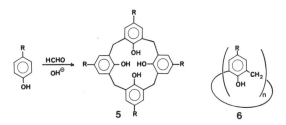

Fig 6. Synthesis of calixarenes

mixtures of products actually are obtained and that cyclic hexamers
(6, n=6) and cyclic octamers (6, n=8) can also be prepared in good
yield by the appropriate choice of reaction conditions. They named
this series of compounds "calixarenes" (Greek, calix, chalice or vase;
arene, indicating the presence of aryl rings) and designated the
number of aromatic residues in the cyclic array by a bracketed numeral
between "calix" and "arene".

3.2. Conformations of the Calixarenes

3.2.1. Conformations of the Calix[4]arenes. That the calixarenes
possess the possibility for conformational isomerism was first made
explicit by Cornforth and coworkers (7) who pointed out that four
discrete forms can exist. We have designated these as the "cone",
"partial cone", "1,2-alternate", and "1,3-alternate" conformations, as
illustrated in Fig 7. By means of temperature dependent ^{1}H NMR studies

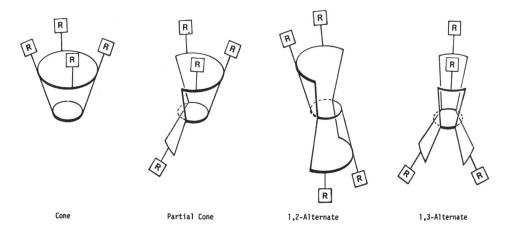

| Cone | Partial Cone | 1,2-Alternate | 1,3-Alternate |

Fig 7. Conformations of the calix[4]arenes

it has been established that the "cone" conformation is favored
(8,9,10) as the result of very strong intramolecular hydrogen bonding
(OH stretching band at 3160 cm^{-1}). The rate of inversion at the
coalescence temperature of 45° is ca 150 sec^{-1}. p-Substituents have
relatively little effect on the barrier to interconversion. For
example, in chloroform solution the greatest difference in the free
energy of activation for inversion for the p-alkylcalix[4]arenes
studied is between p-tert-butylcalix[4]arene ($\Delta G^{\dagger}$ 15.7 kcal/mole) and
p-tert-amylcalix[4]arene ($\Delta G^{\dagger}$ 14.5 kcal/mole). Calixarenes substituted
with H, isopropyl, tert-octyl, allyl, phenyl, 2-hydroxyethyl, and
benzoyl in the p-position all fall between these extremes. The
insensitivity of the inversion barrier to the steric bulk of the p-
substituent is commensurate with what is thought to be the molecular
motion involved in the process, i.e. rotation of the aryl rings in a
direction that brings the OH groups through the center of the annulus
of the calixarene ring system.

    In contrast to the small effect of p-substituents, solvents
exert a considerable influence on the rate of conformational
inversion. In nonpolar solvents (chloroform) the free energies of
inversion are somewhat higher than in semipolar solvents (acetonitrile) and considerably higher than in the basic solvent pyridine.
It is postulated that the major effect of polar solvents, especially
pyridine, is to disrupt the intramolecular hydrogen bonding.

3.2.2. Conformations of the Calix[8]arenes. Sixteen "up–down"
conformations of the calix[8]arenes can exist, along with numerous
other conformations in which one or more aryl rings project "out".
Space filling molecular models indicate that these cyclic octamers
should be much more flexible than the cyclic tetramers. Yet, in
nonpolar solvents calix[8]arenes possess temperature dependent ^{1}H NMR
spectra that are almost identical with those of the calix[4]arenes.
We originally explained this in terms of a "pinched" conformation
(10) in which the cyclic octamer puckers to create two pairs of
circular arrays of hydrogen bonds with four OH groups in each array.
An x-ray crystallographic structure of p-tert-butylcalix[8]arene (11),
however, shows that in the solid state the molecule exists in an
essentially flat form which we have named the "pleated loop"
conformation. Models show that this conformation is perfectly
constituted to facilitate "flip-flop" hydrogen bonding (12). The
similarity between the temperature dependent ^{1}H NMR behavior of the
cyclic tetramers and cyclic octamers in nonpolar solution suggests
that in both cases the interconversion may take place via a
"continuous chain" pathway in which a considerable degree of hydrogen
bonding persists throughout the transformation.

    The similarity of the temperature dependent ^{1}H NMR behavior of
the cyclic tetramers and cyclic octamers disappears in pyridine
solution. We postulate that in both cases the effect of the solvent is
to disrupt the intramolecular hydrogen bonding, leading to a weakening
of the forces maintaining the "cone" and "pleated loop" conformations,

respectively. In the cyclic tetramers, however, vestigal non bonding
forces favoring the "cone" conformation remain, which are absent in
the cyclic octamers.

3.2.3. Conformations of the Calix[6]arenes. The cyclic hexamers can
exist in eight "up-down" conformations as well as in numerous others
in which one or more of the aryl groups projects outward. Temperature
dependent [1] H NMR measurements indicate that in nonpolar solvents the
cyclic hexamers are more flexible than either the cyclic tetramers or
cyclic octamers. The pattern into which the spectrum resolves at low
temperature is commensurate with conformations in which the molecule
transannularly pinches to produce a "winged" or a "hinged"
conformation. Space filling models show that in these conformations
the six OH groups can form two clusters in which "flip-flop" hydrogen
bonding might be possible, although the H-O-H angle is very acute.

p-Substituents have little influence on the barrier to
conformational inversion for the calix[6]arenes, and the magnitude of
the "pyridine effect" is between that observed for the cyclic
tetramers and the cyclic octamers.

3.2.4. Conformations of Other Calixarenes. A calix[5]arene, a calix-
[7]arene (13), and three oxacalixarenes have also been studied. Table
7 shows the data for these compounds, along with the infrared and NMR
characteristics for the OH groups in the various calixarenes.

Calixarene	ν_{OH}	δ_{OH}	$\Delta G^{\ddagger}$, kcal/mole
Calix[4]arenes	3160	10.2	15.7
Calix[5]arenes	3280	8.0	13.2
Calix[6]arenes	3150	10.5	13.3
Calix[7]arenes	3155	10.3	12.3
Calix[8]arenes	3230	9.6	15.7
Dihomooxacalix[4]arenes	3300	9.7	12.9
Tetrahomodioxacalix[4]arenes	3370	9.0	11.9
Hexahomotrioxacalix[3]arenes	3410	8.5	<9

Table 7. Stretching frequencies, chemical shifts, and free
energies of activation for conformational inversion
of calixarenes and oxacalixarenes

3.2.5. Comparison of Shapes and Conformations in the Calixarene
Series. The conformational flexibility of calixarenes and oxacalixarenes carrying endo-annular OH groups is determined by the size
of the macrocyclic ring which, in turn, influences the nature of the
intramolecular hydrogen bonding. To achieve the most effective
intramolecular hydrogen bonding the calix[4]arenes, dihomooxacalix[4]-
arenes, and calix[5]arenes adopt a "cone" conformation; the tetrahomodioxacalix[4]arenes adopt a "flattened cone" conformation; the calix-
[6]arenes are thought to adopt a "winged" or "hinged" conformation;
the calix[7]arenes are thought to adopt a pseudo "pleated loop"

conformation; the calix[8]arenes adopt a true "pleated loop" conformation. Thus, as the ring size increases the preferred conformation becomes increasingly planar. Conformational intercon- versions in nonpolar solvents occur with decreasing free energies of activation in the order: calix[4]arenes = calix[8]arenes > calix[5]- arenes = dihomooxacalix[4]arenes = calix[6]arenes > calix[7]arenes > hexahomotrioxacalix[3]arenes. In pyridine solution the intramolecular hydrogen bonding is disrupted, and the inversion barrier becomes primarily a function of ring size; the free energies of activation decrease in the order: calix[4]arenes > calix[5]arenes > dihomooxa- calix[4]arenes > calix[6]arenes > tetrahomodioxacalix[4]arenes > calix[8]arenes > hexahomotrioxacalix[3]arenes.

3.3 Synthesis of Calixarenes

3.3.1. Recipes for Calixarene Synthesis. A variety of factors control the outcome of the condensation of p-tert-butylphenol and formaldehyde, the most important of which appear to be the amount of base that is employed as the catalyst and the temperature at which the reaction is carried out. In the "standard Petrolite procedure" (14) 1 equiv of p-tert-butylphenol, 1.5 equiv of paraformaldehyde, and 0.05 equiv of NaOH are added to xylene, and the mixture is refluxed for ca 4 hrs. The precipitate accounts for as much as 85% of the starting materials and consists mainly of cylic octamer. In the "modified Petrolite procedure" (6), 1 equiv of p-tert-butylphenol, 1.5 equiv of paraformaldehyde and 0.5 equiv (i.e. a 10-fold increase) of KOH are added to xylene, and the mixture is refluxed for ca 4 hrs. The precipitate accounts for as much as 85% of the starting material and consists mainly of cyclic hexamer. In the "modified Zinke-Cornforth procedure" (6,15) 1 equiv of p-tert-butylphenol, 1.5 equiv of aqueous HCHO, and 0.04 equiv of NaOH are heated at 120°C for 1.5 hrs; the cooled product is broken into small pieces, added to diphenyl ether and heated at 220°C under a slow stream of nitrogen for 2 hrs. Trituration of the crude product with ethyl acetate affords 50–55% of almost pure cyclic tetramer. The odd numbered calixarenes are produced in much smaller amounts by these procedures, or modifications thereof, and are much more difficult to secure.

On the basis of what we presently know about the base-catalyzed condensation of p-substituted phenols with formaldehyde it appears that (a) cyclic octamers and cyclic hexamers are generally more likely products than cyclic tetramers and (b) cyclic tetramers are obtained only from phenols carrying p-substituents very similar in structure to the tert-butyl group. For example, p-phenylphenol condenses with HCHO in the "modified Zinke-Cornforth procedure" and the "Petrolite procedure" to yield mixtures from which rather low yields of p- phenylcalix[8]arene and p-phenylcalix[6]arene can be isolated but in which no trace of p-phenylcalix[4]arene can be detected. Phenols substituted in the p-position with hetero atom groups such as halogens, NO_2, CO_2R yield either completely intractible mixtures or unreacted starting materials.

3.3.2. Mechanistic Pathway of Calixarene Formation. The determination
of the reaction pathways that are involved in the transformation of p-
substituted phenols to calixarenes provides a formidable puzzle that
has been only partially solved. It is quite certain that the first
events in the overall reaction involve the condensation of the
phenolate ion with HCHO to form hydroxymethylphenols. Subsequent
condensation of the hydroxymethylphenols with the starting phenol then
ensues to form linear dimers, trimers, tetramers, etc, and it has been
shown that this process also can take place under relatively mild
conditions. Reasonably convincing, though indirect, evidence has been
adduced (16) to suggest that this occurs via a Michael-like reaction
between phenolate anions and o-quinonemethide intermediates formed by
loss of water from the hydroxymethylphenols, as shown in Fig 8.

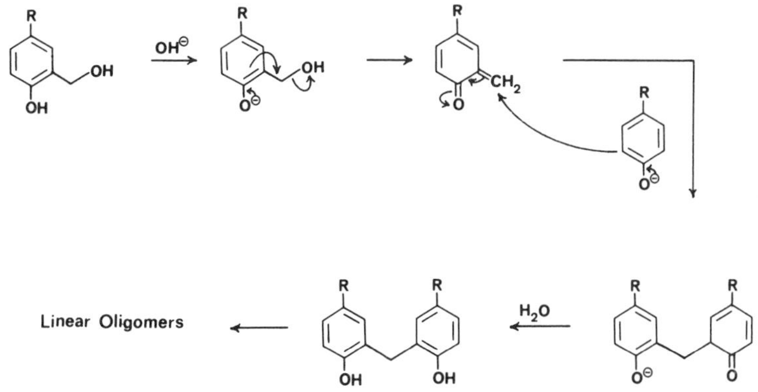

Fig 8. Formation of o-quinonemethide intermediates and their
condensation with phenolate anions.

Forming concomitantly with the diphenylmethane compounds are dibenzyl
ethers, resulting from intermolecular dehydration of the hydroxymethyl
compounds. Thus, the mixtures from which the calixarenes ultimately
emerge probably contain diphenylmethanes as well as dibenzyl ethers of
various degrees of oligomerization

     The pathway by which this mixture is transformed to the
calixarenes remains conjectural, but there is mounting evidence that
hydrogen bonding plays a key role in the process. x-Ray crystallography (17) of the linear tetramer shows that it exists in a
"staggered" conformation in the crystalline state, intramolecular
hydrogen bonding holding the phenyl groups in an almost planar, zig-
zag array. For the linear tetramer to convert to a cyclic tetramer it
must undergo a conformational inversion around an axis passing through
the center of the molecule to yield an intramolecularly bonded
species that we have designated as a "pseudocalixarene" (18). The
normal barrier to this kind of rotation, arising from nonbonded
interactions, is magnified in this case by the necessity of

breaking the hydrogen bond array. On the other hand, association of a pair of linear tetramers in the "staggered" conformation to form an intermolecularly bonded "dimer" (which we have designated as a "hemi-calixarene") requires no conformational inversion and no hydrogen bond

pseudocalix[4]arene   hemicalix[4]arene   pseudocalix[6]arene   hemicalix[6]arene

Fig 9. Schematic representations of hemicalixarenes and
       pseudocalixarenes

breaking. Not only is the staggered conformation of the linear tetramer preserved in the hemicalix[8]arene, but "flip-flop" hydrogen bonding becomes possible, thereby enhancing the stability of the system. Subsequent conversion of the hemicalix[8]arene to the calix[8]arene, which is known to exist in the solid state in the "pleated loop" conformation requires no conformational inversions and no disruption of the hydrogen bond array.

When a mixture of p-tert-butylphenol and aqueous HCHO is subjected to the "modified Zinke-Cornforth procedure" a "precursor" is formed which has been shown by HPLC analysis (19) to contain at least three dozen noncyclic components. When the "precursor" is then heated to higher temperatures cyclization occurs, primarily to cyclic octamer if the temperature is that of refluxing xylene, or cyclic tetramer if it is that of refluxing diphenyl ether. In the latter case, a precipitate of cyclic octamer forms well under the reflux temperature, this precipitate then changing to cyclic tetramer at the higher temperature. In controlled experiments we have demonstrated that p-tert-butylcalix[8]arene can be converted in ca 75% yield to p-tert-butylcalix[4]arene by treatment with base in refluxing diphenyl ether, a process that we have styled as an example of "molecular mitosis", albeit only one generation long. Therefore, we postulate that calix[4]arenes are formed by the pathway shown in Fig 10 in which the cyclic octamer is an intermediate, possibly an obligatory one.

The formation of calix[6]arenes poses additional questions, for it is possible that these compounds arise from more than one pathway. We have shown that cyclic octamer can be converted to cyclic hexamer if somewhat more base is used than for the conversion to cyclic tetramer. The requirement for larger amounts of base to drive the reaction to cyclic hexamer is also noted in the comparison of the "standard Petrolite procedure" and the "modified Petrolite procedure". To explain this we suggest that a template effect, well documented in crown ether chemistry, may be operating, because higher yields of cyclic hexamer are obtained with RbOH as the base than with CsOH, KOH,

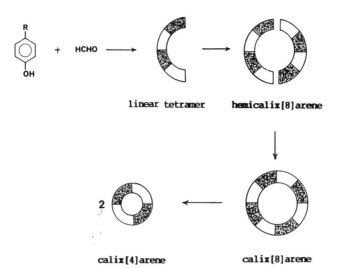

linear tetramer   hemicalix[8]arene

calix[4]arene   calix[8]arene

Fig 10. Reaction pathway in the formation of calix[8]arenes and
calix[4]arenes

or NaOH; LiOH is quite ineffective. Space filling molecular models
indicate that a conformation comparable to the "pleated loop" of the
cyclic octamer can be constructed if two of the OH hydrogens are
removed. Since the models tend to underestimate the compressibility of
the actual molecule, it may be possible that the removal of only one
hydrogen is sufficient. An important function of the base, therefore,
may be to create either (a) the monoanion of a linear hexamer which
then assumes a "pleated loop" conformation as a prelude to cyclization
or (b) the monoanion of a linear trimer which then associates with an
unionized linear trimer to form the hemicalix[6]arene anion, followed
by cyclization, as portrayed in Fig 11.

The various factors that influence the calixarene forming
reaction can be summarized as follows: (a) Solvent effect – calixarene
formation favored by nonpolar solvents (e.g. xylene, diphenyl ether)
and inhibited by polar solvents (e.g. quinoline); (b) Base
concentration effect – cyclic octamer (and cyclic tetramer) favored by
a catalytic amount of base, and cyclic hexamer favored by a
stoichiometric amount of base; (c) Temperature effect – cyclic octamer
and cyclic hexamer favored at lower temperatures, and cyclic tetramer
favored at higher temperature; (d) Cation effect – cyclic octamer and
cyclic tetramer favored by smaller cations, and cyclic hexamer favored
by larger cations (particularly Rb^+). A reaction pathway, based on the
arguments presented above, postulates that the calix[8]arene is the
product of kinetic control (and, possibly, solubility control), that
the calix[4]arene is the product of thermodynamic control, and that
the calix[6]arene is the product of template control.

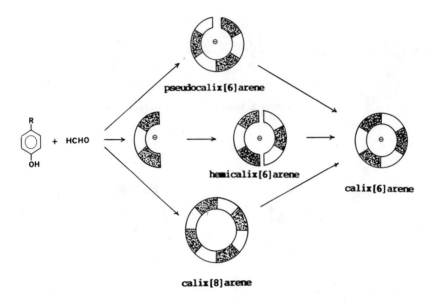

Fig 11. Reaction pathway for the formation of calix[6]arenes

3.4. Complex Formation with Calixarenes

3.4.1. Solid State Complexes with Neutral Molecules . Many of the
calixarenes form complexes in the solid state. For example, p-tert-
butylcalix[4]arene forms complexes with chloroform, benzene, toluene,
xylene, and anisole; p-tert-butylcalix[5]arene forms complexes with
isopropyl alcohol and acetone; p-tert-butylcalix[6]arene forms a
complex containing chloroform and methanol; p-tert-butylcalix[7]arene
forms a complex containing methanol; p-tert-butylcalix[8]arene forms a
complex with chloroform. The tenacity with which the guest molecule is
held by the calixarene varies widely among the calixarenes. Whereas
the cyclic octamer loses its chloroform upon standing a few minutes at
room temperature and atmospheric pressure, the cyclic tetramer and
cyclic hexamer retain their guests even after many hours under high
vacuum at high temperature. x-Ray crystallographic studies have been
carried out on many of these complexes (20), and it has been shown
that in a number of cases the guest molecule is in the calix. In some
cases, however, it is interstitially situated, and with p-tert-
octylcalix[4]arene the side chain itself is shown to be the
"guest"(21).

3.4.2. Solution Complexes with Neutral Molecules in Organic Solvents.
The strength of the embrace between certain calixarenes and guest
molecules in the solid state suggests that complexes should exist in
solution as well. However, this has been a particularly difficult
point to establish, and the available evidence is more suggestive than

convincing. To probe this question we have employed the technique of aromatic solvent induced shift (ASIS) (22), using chloroform as the reference solvent, toluene as the aromatic solvent, and the linear tetramer from p-tert-butylphenol as the noncyclic reference compound. The differences in chemical shifts in chloroform and toluene (ASIS values) for the resonances arising from the tert-butyl hydrogens, methylene hydrogens, aryl hydrogens, and OH hydrogens of the reference compound as well as the cyclic tetramer, cyclic hexamer, and cyclic octamer were measured. Significant differences, particularly for the aryl hydrogens and tert-butyl hydrogens, were noted between the linear tetramer, the cyclic tetramer, and the cyclic octamer. We have interpreted these in terms of stronger complexation between toluene and the cyclic tetramer than the cyclic octamer, consistent with the differences between the solid state complexation behavior of these molecules. The ASIS values of the tert-butyl group of the cyclic tetramer in mixtures of chloroform and toluene display an approximately linear dependence on the mole fraction of toluene and show a small but linear dependence on temperature between 20° and $100^{\circ}C$, consistent with the formation of a 1:1 complex (23) with a K_a of 1.1. A comparison of the ASIS values of p-tert-butylcalix[4]arene, p-tert-amylcalix[4]arene, and p-tert-octylcalix[4]arene) reveals that the first two are quite similar but different from the third (which is very similar to the corresponding cyclic octamer). We interpret this in terms of stronger complex formation with p-tert-butyl and p-tert-amylcalix[4]arene than with p-tert-octylcalix[4]arene, commensurate with the differences that have been noted for the solid state complexes (see Sec. 3.4.1).

p-tert-Butylcalix[4]arene appears to show little or no selectivity in complexing chloroform and toluene in solution, both of these molecules forming tight solid state endo-calix complexes. With the exception of certain amines, as discussed in Sec 3.4.3, attempts to detect complexation (in chloroform solution) with a variety of types of molecules have been unsuccessful. These failures are attributed to the inability of the putative guest molecule to "out compete" the chloroform (present in large excess) to the extent necessary to make the complex detectable by spectral measurement.

3.4.3. Solution Complexes with Amines. Chloroform solutions containing calixarenes and amines, such as tert-butylamine, fail to show any indication of complexation. However, an acetonitrile solution of p-allylcalix[4]arene and tert-butylamine shows shifts in the resonances of the tert-butyl hydrogens of the amine and the aryl hydrogens of the calixarene. Also, there is a sharpening of the methylene resonance as the ratio of amine to calixarene is increased. The effect on the amine can be duplicated by a mixture of tert-butylamine and picric acid, while that on the calixarene can be duplicated by a mixture of p-allylcalix[4]arene and NaOH. That the phenolic groups of the calixarene play an integral part is also indicated by the much smaller perturbations that are observed in the ^{1}H NMR spectrum of a mixture of the tosylate of p-allylcalix[4]arene and tert-butylamine. Thus, the

interaction appears to be initiated by a proton transfer from
calixarene to amine, forming a calixarene anion and an ammonium cation
which then unite to form an ion-pair, as illustrated in Fig 12. An

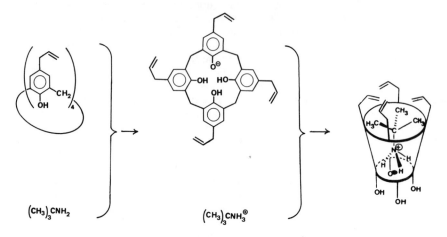

$(CH_3)_3CNH_2$

$(CH_3)_3CNH_3^{\oplus}$

Fig 12. Complex formation between tert-butylamine and p-allyl-
         calix[4]arene

important question, but a difficult one to answer, is whether the ion
pair complex is endo-calix or exo-calix. We have adduced some evidence
that it is the former by observing that (a) the relaxation time (T_1)
for the tert-butylamine hydrogens in a calixarene-amine complex is
shorter (0.79 sec) than for a mixture of tert-butylamine and the very
strongly acidic picric acid (1.21 sec) and (b) the effect of
complexation on the barrier to conformational inversion of the
calix[4]arenes is a function of the identity of the amine. For
example, a 4:1 mixture of tert-butylamine and p-allylcalix[4]arene
raises the coalescence temperature from $10^{o}C$ (calixarene alone) to
$36^{o}C$, whereas a 4:1 mixture with neopentylamine raises it only to
$19^{o}C$. On the assumption that tert-butylamine and neopentyl amine are
virtually identical in basicity, this difference can be interpreted in
terms of an endo-calix complex in which the preferred orientation of
the amine places the three ammonium hydrogens proximate to the oxygens
in the bottom of the cavity, producing a "tripod-like" association. In
the case of tert-butylamine the tert-butyl group sits quite
comfortably in the middle of the calix; in the case of neopentylamine,
however, the "bend" in the neopentyl group places its tert-butyl
portion against the side of the calixarene and makes the "tripod-like"
association less effective.

    As already noted, solvent properties play a critical role in the
calixarene-amine interaction. For example, tert-butylamine raises the
coalescence temperature of p-allylcalix[4]arene in acetone or
acetonitrile but lowers it in chloroform. In contrast to the large
shifts in the resonance of the tert-butyl hydrogens of the amine that

are observed for the calixarene-amine complexes in acetone and acetonitrile, almost no shift is observed in chloroform. However, it is interesting that the T_1 value for the tert-butyl hydrogens in chloroform solution is virtually the same (0.8 sec) as that in acetonitrile (0.79 sec), suggesting that a different kind of complex may be forming in this case.

The proton transfer that is involved in the calixarene-amine complex formation requires both a strong acid and a strong base. For example, aniline, which is a much weaker base than tert-butylamine, has virtually no effect on the coalescence temperature of the methylene resonances of the calixarene (in acetonitrile solution) even when present in a 10-fold excess. p-Nitrophenol, a weaker acid than picric acid, lowers the T_1 value of the tert-butylamine hydrogens only to 1.33 sec even when present in an 11-fold excess. That the calixarenes are stronger acids than their monomeric counterparts has been shown by Böhmer and coworkers (24) who obtained pK_1 values, in aqueous solution, of 6.0 and 4.3 for calix[4]arenes carrying a nitro group in one of the p-positions. These values are lower than that for p-nitrophenol itself, presumably the result of the stabilization of the mono-anion by intramolecular hydrogen bonding. Although pK_1 values for calixarenes in acetonitrile solution are not available, it seems probable that there is a greater difference between the acidities of simple phenols and their corresponding calixarenes in acetonitrile than in aqueous solution.

4. CONCLUSIONS

Covalent bonds are the "strong forces" that play the leading roles in the drama of organic chemistry. Non-covalent interactions such as electrostatic attractions and hydrogen bond attractions are the "weak forces" that play the supporting roles. Supporting roles, however, often have an important influence on the character of the drama, and sometimes they steal the show.

5. REFERENCES

(1)   Melhado, L.L.; Gutsche, C.D., J. Am. Chem. Soc., 1978, 100, 1850.
(2)   Lau, H-p.; Gutsche, C.D., J. Am. Chem. Soc., 1978, 100, 1857.
(3)   Swift, T.J.; Connick, R.E., J. Chem. Phys., 1962, 37, 307; Luz, Z.; Meiboom, S.,ibid., 1964;, 40, 1058, 2686; Murray, R.; Dodgen, H.W.; Hunt, J.P., Inorg. Chem., 1964, 3, 1576; Glaeser, H.H.; Dodgen, H.W.; Hunt, J.P., ibid., 1965, 4, 1061.
(4)   Malmin, J.E., Ph.D. Thesis, Columbia University, New York, N.Y., 1969; Breslow, R., Adv. Chem. Ser., 1971, No. 100, 21.
(5)   Zinke, A.; Ziegler, E., Ber., 1941, 74, 1729.
(6)   Gutsche, C.D.; Dhawan, B.; No, K.H.; Muthukrishnan, R., J. Am. Chem. Soc., 1981, 103, 3782.
(7)   Cornforth, J.W.; D'Arcy Hart, P.; Nicholls, G.A.; Rees, R.J.W.; Stock, J.A., Brit. J. Pharmacol, 1955, 10, 73.

(8) Kämmerer, H.; Happel, G.; Caesar, F., Makromol. Chem., **1972**, 162, 179; Happel, G.; Matiasch, B.; Kammerer, H., ibid, **1975**, 176, 3317.

(9) Munch, J.H., Makromol. Chem., **1977**, 178, 69.

(10) Gutsche, C.D.; Bauer, L.J., Tetrahedron Lett., **1981**, 4763.

(11) Gutsche, C.D.; Gutsche, A.E.; Karaulov, A.I., J. Inclusion Phenomena, **1985**, 3, 447.

(12) Saenger, W.; Betzel, C.; Hingerty, B.; Brown, G.M., Angew. Chem. Int. Ed. Engl., **1983**, 22, 883.

(13) Kämmerer, H.; Happel, G., "Weyerhaeuser Science Symposium on Phenolic Resins", Tacoma, Washington, **1979**, p. 143.

(14) Buriks, R.S.; Fauke, A.R.; Munch, J.H., U.S.Patent 4,259,464, filed 1976, issued 1981.

(15) Gutsche, C.D.; Iqbal, M., unpublished results

(16) Wohl, A.; Mylo, B., Ber., **1912**, 45, 2046; Hultzsch, K., ibid, **1942**, 75B, 106; v.Euler, H.; Adler, E.; Cedwall, J.O.; Torngren, O., Ark. f. Kemi Mineral. Geol., **1941**, 15A, No. 11,1.

(17) Böhmer, V.; Funk, R.; Vogt, W., Makromol. Chem., **1984**, 185, 2195.

(18) Dhawan, B.; Gutsche, C.D., J. Org. Chem., **1983**, 48, 1536.

(19) Dr. F. J. Ludwig, Petrolite Corporation, unpublished results; Ludwig, F.J.; Bailie, A.G., Anal. Chem., **1984**, 56, 2081.

(20) For recent work cf. Ungaro, R.; Pochini, A.; Andreetti, G.D.; Sangermano, V., J. Chem. Soc. Perkin Trans. II, **1984**, 1979; Ungaro, R.; Pochini, A.; Andreetti, G.D.; Domiano, P., ibid, 1985, 197. References to the earlier work of this group can be found in these papers.

(21) Andreetti, G.D.; Pochini, A.; Ungaro, R., J. Chem. Soc., Perkin II, **1983**, 1773.

(22) Williams, D.H., Tetrahedron Lett., **1965**, 2305; Jarvi, E.T.; Whitlock, H.W., J. Am. Chem. Soc., **1982**, 104, 7196.

(23) Fetizon, M.; Gore, J.; Laszlo, P.; Waegell, B., J. Org. Chem., **1966**, 31, 4047

(24) Böhmer, V.; Schade, E.; Antes, C.; Pachta, J.; Vogt, W.; Kämmerer, H., Makromol. Chem., **1983**, 184, 2361.

6. ACKNOWLEDGMENT

I am deeply indebted to my coworkers at Washington University for their tireless efforts in meeting the challenges of the chemistry that is described above. Particular acknowledgment is extended to Dr. George Mei who is responsible for the work described in Sec 2 , to Dr. Lorenz Bauer who is responsible for the work described in Secs 3.2 and 3.4, and to Drs. S. I. Chen, B. Dhawan, L-g. Lin, and M. Iqbal who are responsible for much of the work described in Sec 3.3. I wish also to express my thanks to the National Institutes of Health, the National Science Foundation, and the Petroleum Research Fund for providing financial support for this work.

ZEOLITES: THEIR CHARACTERISTIC STRUCTURAL FEATURES.

J.M. Thomas and Carol Williams,
Department of Physical Chemistry,
University of Cambridge,
Lensfield Road,
Cambridge CB2 1EP, U.K.

ABSTRACT.   In the last four years our knowledge of the structural
characteristics of both synthetic and naturally occurring
zeolites has been greatly enlarged.   This has been brought
about as a result of the development and application of three
new, powerful techniques: high resolution electron microscopy
(HREM) backed up by computer simulation; solid-state NMR,
in particular the magic-angle-spinning variety (MASNMR);
and neutron powder profile analysis (the Rietveld method).
HREM comes into its own in pinpointing structural intergrowths
that occur at the nanometre scale of spatial distribution.
It is possible to pick out a unit-cell thick sodalite sliver
in erionite for example; and ZSM-11 may be readily identified,
even when less than a unit cell thickness of it occurs in
ZSM-5.   MASNMR, coupled with rather more conventional multinuclear NMR studies using ^{13}C and ^{1}H nuclei afford dramatic
new sights into the local order that prevails in zeolitic
solids.   It is particularly good in distinguishing the possible
state of coordination in aluminium.   The neutron Rietveld
technique (and its extensions) enable us to describe in accurate
quantitative detail: (i) the disposition of certain sorbates
held within the intra-zeolite cavities; and (ii) the catalytically
active site in La-Y (cracking) catalysts.
   Apart from summarizing the salient structural features
of zeolites and highlighting those that make zeolites such
good catalysts, we outline the various methods used to convert
zeolites into active catalysts.   The role of structural
intergrowths in influencing the performance of zeolite catalysts
and ways of solving the structure of new zeolites are also
assessed.   Finally we draw attention to the analogy between
zeolite catalysts and certain types of mixed-oxide catalysts
used for the heterogeneous selective oxidation of hydrocarbons.

R. Setton (ed.), Chemical Reactions in Organic and Inorganic Constrained Systems, 49–80.

1. INTRODUCTION

The term zeolite was coined by the Swedish scientist Cronsted
in 1756 to describe the property of the mineral stilbite
which, upon heating, behaved as if it were boiling (zeo-lite:
the stone that boils).   Zeolites have been the subject of
intermittent study ever since, extensively so in recent years;
and even fifty years ago, as one may gauge from the work
of J.W. MacBain (1) and M.G. Evans (2), much was known about
their physico-chemical properties especially in regard to
the thermodynamics of reversible gain and loss of water and
other small organic and inorganic molecules.   Comprehensive
reviews of their structure, energetics, synthesis and adsorptive
behaviour have been given by Barrer (3,4) and Breck (3) whose
contributions to this subject have been considerable.
        Until quite recently it was thought that zeolites
were invariably made up of corner-sharing SiO_4 and AlO_4 tetrahedra.
Their general formula is $M_{x/n} \left[(AlO_2)_x (SiO_2)_y \right] mH_2O$ where
the cations M of valence n neutralize the charges on the
alumino-silicate framework.   Very recently, however, it
has been recognized that zeolitic behaviour — the taxononic
attribute being reversible uptake of guest species — is
also exhibited by microporous frameworks in which the tenants
of the tetrahedral sites can also be Ga, Ge B, P, As, Ti,
Zn, Fe and several other elements (see Table 1).   The term
porotectosilicates is used to describe crystalline microporous
solids consisting of cavities and channels (diameter 3 to
8Å) in which all the tetrahedral sites are tenanted by Si.
In other words these solids are new polymorphs of crystalline
silica.   One nowadays talks of faujasitic silica, mordenitic
silica, offretitic silica and so forth (5,6) — see below.
Such solids have neutral frameworks and hence   require no
exchangeable cation.   Unlike the more familiar zeolites,
which are hydrophilic, these porotectosilicates are hydrophobic.
They can be prepared either by dealumination (5-8) of a precursor
zeolite or by direct synthesis (4).   Moreover it is possible
(9,10) to insert a range of other tetrahedrally bonded elements
(notably Al and Ga) into such a porotectosilicate, thereby
generating alumino- or gallo-silicates of delicately controllable
Si/Al and Si/Ga ratios.   A large family of aluminium phosphates,
consisting of corner-sharing AlO_4 and PO_4 tetrahedra, also
exists (11,12). These microporous, crystalline materials
are analogues of the naturally occurring and synthetic alumino-
silicates.   They, too, have electrically neutral frameworks,
but can be made to accommodate exchangeable cations if by
appropriate means (12), Si atoms replace P in the frameworks.
The $AlPO_4$ family, as well as the more recently described
(13) $GaPO_4$ family, differs from the traditionally recognized
alumino silicate zeolites in that, besides 4-fold (tetrahedral)
coordination, there may also be 5-fold and 6-fold coordination.
Open framework structures containing octahedrally coordinated

transition metal ions have been prepared and characterized
by D.R. Corbin, M.M. Eddy, J.F. Whitney, W.C. Fultz, A.K. Cheetham
and G.D. Stucky (Inorg. Chem., in press). Typical stoichiometries
are $Fe_5P_4O_{20}H_{10}$, these being, respectively, the synthetic
analogues of the minerals hureaulite and alluadite.

TABLE 1 : Elements known to occupy tetrahedral sites in crystalline
microporous (open framework) structures containing cages
or channels of 3 to 8Å diameter.

STOICHIOMETRIES: TO_2, $M^+(T'_xT''_{(1-x)}O_2)$

(When M is H these are acid catalysts)

T', T''= Si, Al, B, Ga, Fe, Cr, Ge, Ti, P, V, Zn,
       Be..

These materials are low density analogues of quartz, gallium
arsenide, diamond and silicon. A recently reported[a] example
is $Zn_{0.06}Al_{0.94}PO_4$.

(a) G.C. Bond, M.R. Gelsthorpe, K.S.W. Sing and C.R. Theocharis,
    J. Chem. Soc. Chem. Comm, 1056 (1985)

Zeolites have been used for decades as water softeners and
cation exchangers, as well as vehicles for the separation
of molecules according to shape and size. In addition to
their use as molecular sieves, they have, since the mid-sixties,
been widely employed as industrial catalysts, and, of late,
high siliceous variants of zeolites have led to important
new developments in shape-selective heterogeneous catalysis
(14-18).
Since comprehensive surveys of the structural principles
of zeolites are already available (3,4,14), and since atlases
showing stereo-pairs of all alumino silicate zeolite framework
structures (17) known up until 1976 and compilations of the
coooordinates of exchangeable cations (18) in intrazeolitic
spaces are to hand, we do not dwell at length on the attributes
of the fundamental building units and relevant structural
principles in this article. When single-crystal specimens
of zeolites are available, X-ray crystallographic methods
generally yield good structural data about the location and
environment of both framework atoms and extra framework ions
or neutral guest species. But very frequently,when new
zeolites are prepared,it is found that they are not straight-
forwardly amenable to structural elucidation by conventional
X-ray based methods. Sometimes the crystals lack long-range
order because they consist of intergrowths of two or more
structures and,increasingly,it is found that phase-pure
specimens of new zeolitic solids are not available in dimensions

large enough to permit structure determination by X-ray single-crystal techniques.   We return to this topic later.

2.   THE STRUCTURAL ATTRIBUTES THAT MAKE ZEOLITES CATALYTICALLY IMPORTANT

By virtue of their structure, crystallinity and variable stoichiometry, zeolitic catalysts have:

> sharply defined pore-size distributions;
> high and adjustable acidity;
> very high surface areas (typically $600m^2g^{-1}$), the
> majority of which (ca 95 percent) is, depending
> upon crystallite size, internal and accessible
> through apertures of defined dimensions;   and
> good thermal stability (e.g. capable of surviving
> heat treatments in air up to $1000^{\circ}C$, depending
> upon the composition of the zeolite framework).

Moreover, as stated earlier, the framework composition of zeolites can be changed from the extremes of low-silica to high-silica contents, the inner walls of the channels and cages can be more or less smoothly converted from hydrophilic to hydrophobic extremes.   Other attributes, apart from their relative ease of synthesis, that contribute to the attractive features of zeolitic catalysts are that:

> the nature and siting of exchangeable cations can
> be adjusted and engineered;
> the siting and energetics of potentially reactive
> organic species housed within the catalyst pores
> can be, to some degree, also engineered;
> the catalytically active sites are uniformly distributed
> throughout the solid, being accessible at the
> inner walls of the cavities.

Finally, in view of the fact that the active sites are situated predominantly inside the zeolite, and that all these sites, which are of very high concentration — far in excess of active sites on, for example, supported metal catalysts-are, at one and the same time also in the bulk of the solid, zeolitic catalysts can be very well characterized by the powerful new tools that have recently become available for probing local environments within bulk solids.   This is particularly true of high-resolution (solid-state) multinuclear NMR, or neutron and X-ray powder profile (Rietveld) methods, of high-resolution electron microscopy and of computer graphics techniques which my colleagues and I at Cambridge, and collaborators elsewhere, have, along with others, been engaged in developing in recent years (28, 52, 63, 67, 68).

Ease of preparation of zeolite catalyst, mundane as it may seem as one of the attributes of this class of solid, is as important a factor as any in determining the widespread use of zeolites in industrial catalysts. Faujasite, for example, is a very rare mineral: it occurs in only a few locations, as   minute amounts, on earth. But its synthetic analogues zeolites X and Y (see below) are very easily prepared from solutions of silicates and aluminates. Doubtless, when reliable laboratory syntheses are evolved and subsequently scaled up in an economically attractive manner, other zeolites too will figure eminently in industrial catalysis.

A typical aperture opening in a zeolite, representing the magnitude of the effective cross-section of the channel mouth as a receptable for the guest molecule, is shown in Fig. 1. The strong O-Si or O-Al bonds that predominate in zeolitic structures are the root causes of the great thermal stability of the zeolites compared with, for example, the crown ethers or the cyclodextrins which are bound together chiefly by the weaker C-O link. It is also to be noted that the greater the proportion of Si-O bonds in a given structure, the greater is its hydrophobicity, whereas the greater the content of Al-O bonds the greater, the hydrophilicity.

3.   SALIENT STRUCTURAL FEATURES

The neutralizing and on the whole exchangeable cations present in zeolites are located at well-defined sites in the various cavities and channels within the structure, water molecules filling up the remaining voids. The water can be expelled upon heating and evacuation and may be replaced by a number of small inorganic and organic guests.

By adjusting the valency and the size of the exchangeable cation in a zeolite, the molecular sieving and hence the shape-selecting property of a zeolite may be fine-tuned. Take the Na^+ form zeolite-A, for example (Fig. 2). Replacement of Na^+ by Ca^{2+} ions results in the enlargement of the effective void space within the zeolite. Zeolite-A has four-membered, six-membered and eight-membered apertures within the structure. In each unit of Na_{12}-A all eight of the six-membered, three of the four eight-membered and one of the twelve four-membered rings are "blocked" by Na^+ ions. But in Na_4Ca_4-A, the openings of half of the six-membered rings and all the eight- and four- membered rings are vacant. It is no surprise, therefore, that ethane readily percolates through Na_4Ca_4-A, and even more readily through Ca_6-A compared with its passage through Na_{12}-A.

(a)

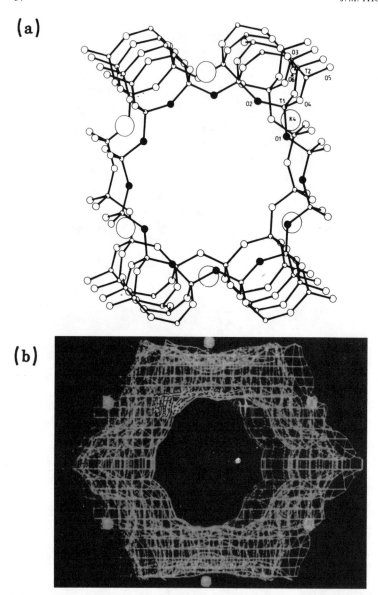

(b)

Figure 1.   View down <001> of a portion of the zeolite-L
structure.   In (a) only six of the exchangeable cations
are shown, all crystallographically equivalent.   The twelve
oxygens that line the mouth of the channel have been enlarged
(compared with analogously situated oxygens deeper in the
channel) and are shown as filled circles.   In (b) the 'network'
added to the structural drawing shows the van de Waals surface
of the framework atoms. (adapted from 68).

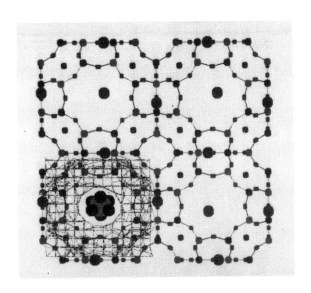

Figure 2.   Projection down ⟨001⟩ of a portion of the zeolite-A structure.   The large filled circles (●) denote position, in projection, of K^+ ions.   To facilitate adsorption, K^+ ions are replaced by divalent cations (e.g. Ca^{2+}).   At lower left, the van der Waals network associated with the framework is shown.   As indicated, a molecule of ethane may be readily sorbed by Ca^{2+}-A.

   The microporosity of a zeolite can in general be further enhanced by increasing the Si/Al ratio of the macro-anionic framework.   This is known as dealumination:

$$M_{x/n} \ (AlO_2)_x (SiO_2)_y \ mH_2O \ \xrightarrow{-Al, -M, -H_2O} \ SiO_2$$

(Hydrophilic)   (Hydrophobic)

Indeed when the NH_4^+-exchanged form of zeolite-Y is heated under hydrothermal conditions, the process of dealumination is effected, and the resulting structure is said to be ultra-stabilized in view of the fact that it subsequently with-stands high-temperature treatment at ca 1000°C without loss of structural integrity.   During the course of stabilization, the Si/Al ratio of the framework changes from an initial value of around 2.4 (depending upon the preparation of the original zeolite) to a final one of beyond 10 or 100.   Of late, it has been found possible (19-21) readily to dealuminate certain zeolites (e.g. those based on faujasite (zeolite-Y)) simply by exposure to the vapour of $SiCl_4$ at elevated temperature

and to achieve Si/Al ratios of greater than 1000.   Hydrothermal methods work well for the dealumination of zeolites
such as mordenite and offretite and ZSM-5 (to silicalite)
and even acid leaching suffices to dealuminate clinoptilolite.

Zeolites X and Y are structurally analagous to
the mineral faujasite.   The building units are truncated
octahedra, also known as sodalite cages, β-cages or tetrakaidecahedra —
all three terms are synonymous.   These cages, seen in the
centre of Fig. 2 and further represented in Fig. 3, are linked
(in zeolites X and Y) to adjacent ones via hexagonal biprisms,
thereby yielding larger supercages.   In zeolite-A and zeolite
ZK4 the β-cages are linked via cubes or so-called double-
four rings (Fig. 4).

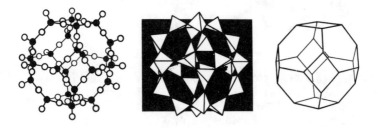

Figure 3.   Schematic representation of how a β-cage (sodalite
cage) is built up by a process of corner-sharing of twenty
four TO_4 tetrahedra (T ≡ Si or Al).   The β-cage is a truncated
octahedron.

In zeolites X and Y the principal sites for the extra-framework
cations are designated as S(I), S(I'), S(II) and S(II") sites.
The Si-O-Al framework and the locations of the principal
cation sites are shown in Fig. 5, from which we can expect
variations in the occupancy factors of the respective exchangeable
cation sites.   From Fig. 5 we see that cations preferring
higher coordination numbers usually occupy S(I) sites;   we
also see that adjacent S(I) and S(I') sites are not simultaneously occupied by cations, and that almost all of the
S(II) sites, situated as they are on the wall of the supercage,
tend to be occupied.

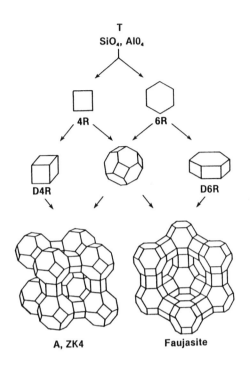

Figure 4.   Illustration of how zeolites X and Y and zeolites
A and ZK-4 may be pictured as being assembled from primary
(i.e. TO$_4$) and secondary building units (cubes or double-
four (D4) rings; hexagonal prisms or double-six (D6) rings,
etc).

    More than fifty distinct structures have been identified
in the zeolite kingdom for some time.   The reader is referred
to other sources for further details (22-26).   Suffice it
to say that there are many well recognized secondary building
units, as shown in Fig.6, that are utilized in the various
architectural patterns adopted by zeolites.   Zeolite-rho
consists of the α-cages (that are present in zeolite-A) joined
together in a cubic structure via octagonal prisms (i.e. double
eight (D8) units).

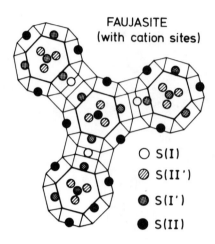

FAUJASITE
(with cation sites)

○ S(I)

⊘ S(II')

◉ S(I')

● S(II)

Figure 5.   The principal (idealized) cation sites in faujasite
(zeolites X and Y).   S(I) is at the centre of the hexagonal
biprisms (D6R) which connect the β-cages, (forming a diamond
lattice of β-cages).   S(II) sites are in the super-cages,
but sites S(I') and S(II'') are within the β-cages.   For
a specific cation-exchanged faujasitic zeolite, the precise
cation positions differ somewhat from the idealised positions
shown here.

     ZSM-5 and ZSM-11 are closely related to one another
(25). The former consists of sheets, themselves made up of
connected chains of 5-membered rings, joined through centres
of inversion (designated 'i').   The latter has the same
sheets joined at mirror planes (designated 'o') - see Fig. 7
The Si/Al ratios of ZSM-5 and ZSM-11 can vary from about
10 to 100. At the very high silica extreme (Si/Al ratio beyond
a 1000 or so — the precise ratio is arbitrarily defined),
these structures are termed silicalite I and silicalite II,
respectively.   But offretite and erionite, which tend to
coexist and intergrow — a property they share with ZSM-5
and ZSM11 — are members of a large family known as ABC-6
zeolites, details of the structure of which have been described
elsewhere (27). Individual members of this family differ
from one another according to the stacking sequences adopted

by puckered sheets, at every vertex of which there is a TO_4 group (T = Si,Al) corner-sharing via oxygen with four other TO_4 units.

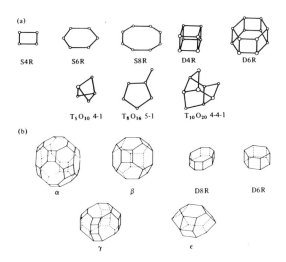

Figure 6. A selection of the secondary building units (double four, D4; double six, D6; double eight, D8; sodalite, cancrinite, gmelinite cages, etc) from which the structures of zeolites are derived.   The gmelinite cage is shown at bottom left, the cancrinite cage bottom right.

It is important to emphasize that although X-ray crystallography has yielded the structure of several zeolites (synthetic and natural), there are very many new zeolites recently isolated, which because of a combination of factors, have so far not been structurally determined.   First there is the difficulty of growing zeolite crystals of adequate size for conventional X-ray analysis.   A further obstacle is the tendency for one zeolite to form an intergrowth with another.   And even though elegant use has been made (28) of synchrotron radiation as an X-ray source (so that smaller dimensions of single crystals can be tolerated for four-circle diffractometry), many unconventional methods of structural analysis have had to be invoked (29,30).   Good progress has been achieved in correlating the performance of zeolite catalysts with the structural characteristics that have been retrieved by the combined use of solid-state NMR, neutron and X-ray (powder) methods, computer modelling and high-resolution electron microscopy.

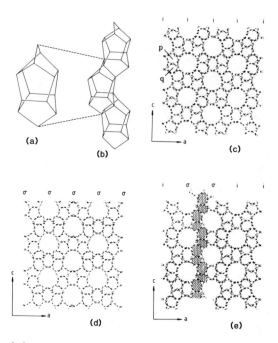

Figure 7. (a) Segment of the structure of ZSM-5 and ZSM-11
showing connected 5-membered rings composed of linked tetrahedra
(SiO_4) and (AlO_4).   Each connecting line represents an oxygen
bridge.   (b) The chains from which the ZSM-5, and ZSM-11
structures are built are themselves made up by linking the
units shown in (a).   (c)   In ZSM-5 chains are linked such
that (100) slabs are related by inversion (i).   (d)   In
ZSM-11, chains are linked such that (100) slabs are mirror
images (σ) of one another.   (e) Representation of intergrowth
of ZSM-5 and ZSM-11.   In (c) p and q refer, respectively,
to the larger and smaller 5-membered rings.   (The rings
are in reality of equal size, but do not appear so in projection).

4.   WHY ARE ZEOLITES SUCH GOOD CATALYSTS?

Although there are several reactions catalyzed by zeolites
in which acidity plays a minor role (see ref. 31), in the
main, zeolitic catalysts exert their influence because of
their high acidity, this being usually of the Brønsted kind.
In essence, we may interpret the behaviour of zeolitic (acid)
catalysts in terms of reactions involving carbonium ions
(32), just as we may interpret the behaviour of clay catalysts
in like terms (34-38).   Leaving aside for the moment precisely
where the "free" proton is situated, and its provenance,
within the zeolite (see below), we utilize the broad mechanistic
principles outlined by Whitmore (39) and others (40).   First

carbonium (alkylcarbonium) ions are formed from alkenes by
reversible protonation:

$$H^+ \quad + \quad >C = C< \longrightarrow R_1^+$$

(from catalyst)   (Alkyl intermediate)

Subsequently we may have:

$$R_1^+ \longrightarrow R_2^+$$

$$R_1^+ \longrightarrow R_2^+ \quad + \quad >C = C <$$

$$R_1^+ \quad + \quad >C = C< \longrightarrow R_2^+$$

Within this broad set of principles we can also account
for the alkylation of benzene by ethylene to yield ethyl
benzene over a ZSM-5 catalyst.

Here, as with several other comparable reactions (41) including
isomerization and hydrogenation, the shape-selectivity of
the ZSM-5 comes into play, and very little diethyl (or other
polyalkylated) benzenes are formed.   The reality of the
shape-selective quality of zeolite catalysts is highlighted
when we compare the relative diffusion coefficients of simple
and substituted aromatics covering a range of molecular cross-
sections (Fig. 8).   These data are readily interpretable
when we recall that the diameters of the pores in ZSM-5 are
ca 5.5 Å.   It is not surprising, therefore, that the diffusivity
of benzene far surpasses that of its trisubstituted analogue.
We may also readily interpret catalytic data such as those
shown (42) in Table 2 where the results of competitive isomerizations of hexene-1 on the one hand are set alongside the
isomerization of its progressively more highly branched isomers.

TABLE   2. Competitive isomerization of alkenes (over an
acid form of ZSM-5).

1.	2.	Selectivity[a] (k_2/k_1)	Temp (°C)
6-methyl-1-heptene	Hexene-1	1.8	150
3-ethyl-1-pentene	Hexene-1	35	175
4,4-Dimethyl-1-hexene	Hexene-1	120	175

a The selectivity is defined as the ratio of the percentage
  conversions of 2 and 1 for a given time of contact. (After
  R.M. Dessau). J.Cat., 77, 304 (1982)

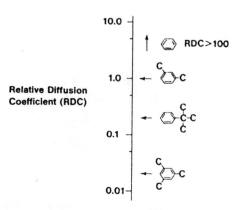

Figure 8.   Relative diffusion coefficients (RDC) of simple aromatics in ZSM-5.

As to the source of the acidity (i.e. the locus of the detachable proton required for the initial act of Brønsted catalysis),   the results of recent investigations leave little doubt as to its precise origin.   There are two distinct ways in which detachable protons can be generated in a zeolitic catalyst:

(a)   by hydrolysis of a strongly polarizing cation; and

(b)   by the generation of neutralizing entities to compensate for Al^{3+} ions housed within the framework.

The first of these is represented typically thus:

$$La(H_2O)^{3+} \longrightarrow La(OH)^{2+} + H^{+}$$

The second by:

It is the hydrogen bound to the oxygen adjacent to the tetrahedrally incorporated Al that has a propensity to free itself as a proton – the active catalytic agent – thereby leaving a macroanionic framework. Infra-red evidence substantiates this picture in many convincing ways.

First, experiments on the protonated form of zeolite-Y (H^+-Y) reveal that there are two O-H stretching frequencies (Fig. 9) that are of especial significance since they refer to the bonds that are ruptured on ionization, thereby yielding detached protons. The fact that NH_3 gas, when introduced to H^+-Y, wipes out these two frequencies shows the corresponding OH groups to be both accessible to small (basic) molecules. Only the higher frequency peak is eliminated when pyridine is introduced, indicating that the O-H group responsible for the ca 3650 cm^{-1} frequency protrudes into the supercage. X-ray structural analysis by Olson (43) on H^+-Y shows where the two O-H bonds are situated (Fig. 9). One of these, associated with the lower frequency vibration of ca 3550 cm^{-1}, points inwards to the D6R and is therefore inaccessible to molecules as bulky as pyridine. When a molecule such as cyclopropane, thought (44,45) to be capable of forming stable carbonium ions by picking up a proton, is introduced to H^+-Y, the higher frequency peak is diminished (Fig. 9); but on thorough evacuation it is reestablished when the bound cyclopropane is desorbed.

Another illustration of the value of infra-red spectroscopy as a tool in clarifying the catalytic performance of zeolites is shown in Fig. 10 (46). The fact that protons are detached from the catalyst, become part of a carbonium ion and are subsequently re-incorporated into the catalyst, is proven by the predicted isotope $\sqrt{2}$ shifts in the frequencies of the acid O-H groups upon deuteration. These spectra also show that the C-H (and C-D) bonds can be activated to the extent of rupture by zeolite catalysts, a fact which is not generally appreciated by those (47) concerned with novel ways of converting relatively unreactive hydrocarbons into more useful products. Perdeuterocyclohexane, for example, can partially convert a H^+-Y catalyst into its D^+-Y form, just as cyclohexane converts D^+-Y into H^+-Y.

Further support for the view that the acid sites in H^+- exchanged zeolites are intimately associated with Al tetrahedrally incorporated into the framework, as schematized above, comes from the results of Haag et al. (48) who showed (Fig.11) good correspondence between catalytic activity and the concentration of tetrahedrally bound Al as determined by ^{27}Al solid-state NMR. The recent results of Chu and Chang (49) as well as of Pfeiffer et al. (50) and Veeman et al (51) lend further credence to this view.

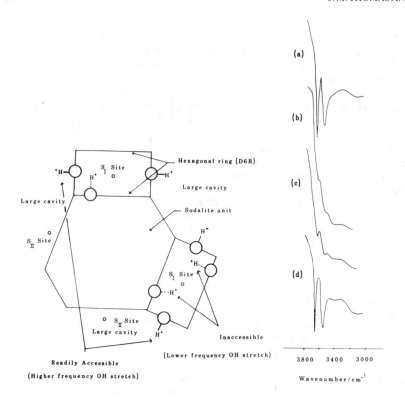

Figure 9.   Illustration of the two types of OH groups present
in the protonated form of zeolite HY (LHS) and Fourier transform
infrared spectra of the acidic OH groups of zeolite HY (RHS):
(a) after heating NH$_4$Y overnight in vacuo at 400°C;   (b) after
addition of cyclopropane at room temperature.   The band
at 3650cm^{-1} is reduced in intensity and another appears in
the region 3400–3200 cm^{-1}, suggesting that some adsorbed
cyclopropane is hydrogen-bonded to the 3650cm^{-1} OH groups;
(c) after partial removal of cyclopropane, showing the gradual
reappearance of the 3650cm^{-1} OH band; (d) after further
evacuation at room temperature the original OH bands reappear
while the broad hydrogen-bonded OH band has almost completely
disappeared.

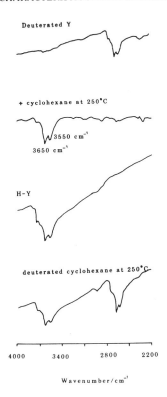

Figure 10. Infrared spectra of the OH and OD groups of zeolite H$^+$ (or D$^+$)-Y. Addition of cyclohexane at 250°C to D$^+$-Y converts the zeolite into H$^+$-Y. Similarly, addition of perdeuterocyclohexane to H$^+$-Y partially converts the zeolite into the D$^+$-Y form, by protonation of the cyclohexane to form a carbonium ion, with subsequent re-incorporation of the H$^+$ or D$^+$ into the zeolite.

It was stated earlier that there was another way in which detachable protons can be generated in a zeolitic catalyst: by hydrolysis of a strongly polarizing cation. That this is important (and real!) is confirmed by neutron powder Rietveld analysis of La^{3+}-exchanged zeolite Y. The presence of (LaOH)$^{2+}$ species as well as the locus of the released proton, held at very low temperatures (10 K) to a framework oxygen is seen (52) in Fig. 12.

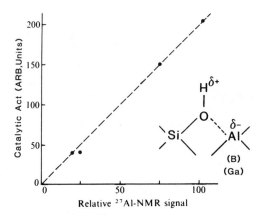

The activity of HZSM-5 plotted against the tetrahedral
aluminium NMR signal.

Figure 11.   The Catalytic activity of ZSM-5 shows a linear
dependence on the concentration of tetrahedrally coordinated
Al since this is the root cause of the acidity.   A similar
situation exists when either B or Ga replace the Al.
(After Haag et al (48))

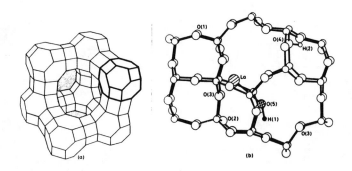

Figure 12.   (a) a view of the framework structure of zeolite Y:
the truncated octahedron shown in (b) is indicated by the thicker
lines: (b) the truncated octrahedral element in La-Y, showing
one of the hexagonal prismatic linkages to the adjacent element.
The $(LaOH)^{2+}$ species is shown with La in site SI$'$ and H(2) is
the proton released by hydrolysis of the cation.

5. CONVERTING A ZEOLITE TO ITS CATALYTICALLY ACTIVE FORM

For a given zeolite, with a well-defined Si/Al ratio, greater catalytic activity will ensue if polyvalent rather than monovalent ions occupy the sites of the exchangeable cations.   Other things being equal, therefore, for the reasons given in the preceding section, a La^{3+} or Ca^{2+} exchanged zeolite exhibits greater (Brønsted) catalytic activity than the Na^{+} exchanged analogue.   The logical extension of this argument is for the best activity to be achieved when H^{+} cations constitute the exchangeable ions, and indeed this is so.   One way in which the H^{+} form of a zeolite can be prepared is simply by washing with mineral acid.   In general, however, this procedure is not satisfactory because the zeolite framework is broken down in the process.   To circumvent this difficulty the NH_4^{+} exchanged form is first prepared and this is then heated under hydrothermal conditions to yield the H^{+} zeolite after liberation of NH_3.   During the conversion there is much structural reorganization; but prolonged annealing generally succeeds to heal many but not all of the local defects.   In zeolite-L, for example, coincidence boundaries are formed (53,54) as a result of rotation about an axis parallel to <001>, of one part of the crystal with respect to the other.   The $\sqrt{13} \times \sqrt{13}$ R x 32.2° coincidence lattice formed in zeolite L results in a marked diminution of the diffusion of reactant and product molecules along the channels which run parallel to <001> in zeolite-L.
     It is important to appreciate that both the thermal stability and the catalytic activity of a zeolite are, in general, improved by dealumination.   Until recently it was very difficult to monitor the precise degree of dealumination. In zeolites X and Y, for example, the unit cell dimension of the cubic structure was used as a criterion.   The inherent cause of the small diminution of unit cell parameter upon dealumination is the replacement of a Al–O bond (ca 1.76 Å) by a Si–O (ca 1.60 Å).   X-ray methods are used (55) for this purpose.   Other procedures are based on chemical analysis of Si and Al contents either by wet methods or by X-ray fluorescence or atomic emission spectroscopy.   These methods are, however, fraught with difficulties as they record <u>total</u> Si and Al content.   There are framework and non-framework locations.   But ^{29}Si MASNMR (56,57) is readily capable of determining the framework Si/Al ratio in a non-destructive fashion,

$$(Si/Al)_{NMR} = \sum_{n=0}^{n=4} I_{Si(nAl)} \Big/ \sum_{n=0}^{n=4} \frac{n}{4} I_{Si(nAl)}$$

where $I_{Si(nAl)}$ is the intensity of the peak arising from

Si surrounded by n AlO_4 tetrahedra.   A striking example of the way
in which ^{29}Si MASNMR surpasses X-ray emission for the purposes
of determining framework composition is illustrated in Fig. 13,
taken from the work of Wright (58) on zeolite-rho.

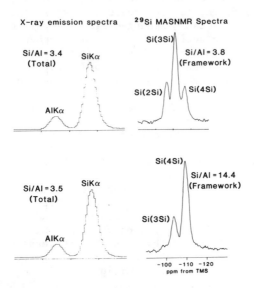

Figure 13.   When zeolite-rho is dealuminated by hydrothermal
treatment of its NH_4^+ exchanged form the X-ray emission spectrum
(LHS) remains essentially unchanged as this method of chemical
analysis does not discriminate between framework and non-
framework elements. ^{29}Si MASNMR (RHS), on the other hand,
is sensitive (14) to changes in framework composition.

     Other techniques, besides MASNMR, are of value in charac-
terizing a zeolite converted into its catalytically active
state: infra-red spectroscopy, X-ray diffraction, neutron
diffraction, thermogravimetric analysis, temperature programmed
reduction, EXAFS, and acidity gauges such as the use of Hammett
indicators.   In the final analysis actual catalytic test
reactions, such as the decomposition of cumene or the isomerization of xylenes or the cracking and hydrocracking of a
straight-chain hydrocarbon, are used as realistic yardsticks.
     Temperature programmed reduction (TPR) merits some
elaboration as it is particularly useful in characterizing
the state of zeolitic catalysts that have within them finely
divided metal particles, nominally designated as Ni^0, Pt^0
or Pd^0, which convert the acid-catalyst into a bi-functional
one by conferring the additional property of hydrogenation,
dehydrogenation and hydrogen transfer.   (At high enough
temperatures, and with the right zeolite, hydrocracking can
be effected even without incorporation of the finely divided,

classic transition or noble metal catalysts: ferrierite, for
example, succeeds in efficiently hydrocracking gas oil without
addition of Pt, Ni or Pd). More often than not, the metal
in question is first introduced in its cationic form by ion
exchange. Thereafter reduction, either prior to partial
or full dehydration, precipitates out the cominuted metal
catalyst. Fig. 14 shows typical TPR curves for Ni-containing
zeolite catalysts of various kinds (45). The significant
point here is that quite different catalysts are prepared
depending upon the particular zeolite used and the pretreatment
to which it is subjected. X-ray diffractograms (Fig. 14(c))
are useful in revealing the phases present, especially the
development of well-formed metallic crystallites. But it
is already recognized that EXAFS and XANES are superior as
tools to most others in detecting whether, for minute assemblies
of atomic species (< 10 Å diameter), metal-metal bonds are
formed. Denley's work (59) on Rh-exchanged zeolites after
reduction illustrates this point nicely.

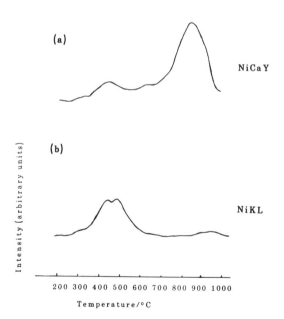

Figure 14. (a) and (b): Typical temperature-programmed
reduction (TPR) profiles for (a) NiCaY and (b) NiKL. The
hydrated zeolite samples are heated in a H2/N2 mixture
(5% vol/vol), at a heating rate of 10°C/min. H2 consumption
is measured by means of a thermal conductivity detector,
the area under the TPR curve being proportional to the amount
of H2 consumed during reduction.

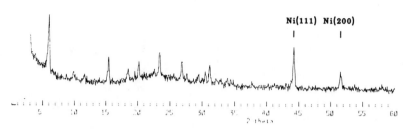

Figure 14. (c): X-ray diffractogram of the NiCaY sample, after
TPR, showing the partial breakdown of the zeolite framework
and the presence of metallic nickel.

6. THE INFLUENCE OF INTERGROWTHS UPON CATALYTIC PERFORMANCE
   OF ZSM-5

Reference has been made earlier to the fact that the diffusivity
of organic molecules in samples of zeolite-L can be adversely
affected if there are certain kinds of boundary contained
within the crystals of the zeolite. In effect, at the boundary
a new (local) structure exists. We also made earlier reference
to the tendency for ZSM-5 and ZSM-11 to intergrow (see Fig. 7).
We shall now illustrate, with specific reference to ZSM-5
and ZSM-11, what the catalytic repercussions of such intergrowths
amount to.
     First it is relevant to recall that when methanol is
converted to gasoline a distribution of aromatic products,
typified by that shown in Fig. 15, is observed. In practice,
however, product distribution curves of this kind are found
to vary quite significantly from one preparation of ZSM-5
to another. Moreover, with ZSM-11 the carbon number for
the peak, as well as the distribution curve itself, shifts
to higher values, compared with the results for ZSM-5.
It is not difficult to appreciate why this is so. There are
in ZSM-11 two kinds of cavities created by the intersecting
channels. One cavity has a volume about 1.6 times as large
as that of the other, the volume and shape of which is exactly
the same as the single type of cavity found (32) in ZSM-5
(see Fig. 16). The shape and size of these cavities crucially
govern (25) the maximum dimension attainable in the transition
state when the akylcarbenium ions, formed from methanol,
are established inside the zeolite catalyst (60). The shift
of the distribution curve (Fig. 15) to higher values is therefore
explicable in terms of the subtleties of the mode of operation
of shape-selective catalysts. And the reason why preparations
of ZSM-5 exhibit variability in catalytic performance
vis à vis product distribution is attributable to the fact
that intergrowths of ZSM-11, present to a variable degree,
occur within the ZSM-5. Proof of the occurrence of intergrowths

has been presented previously (29) and is seen in the high resolution micrograph that constitutes Fig. 17.

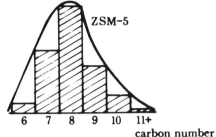

carbon number

Aromatics distribution from methanol.

Figure 15.   A Typical distribution of aromatic products obtained in the conversion of methanol to petrol over a ZSM-5 catalyst.

Figure 16.   Photograph of a scalar model showing an intergrowth of ZSM-5 (right) and ZSM-11 (left). Inflated balloons have been placed at the channel intersections to show that in the ZSM-11 structure, one of the two types of   intersections has a larger volume (about 60 percent more) than   the intersections in the ZSM-5 structure (See also Fig. 17).

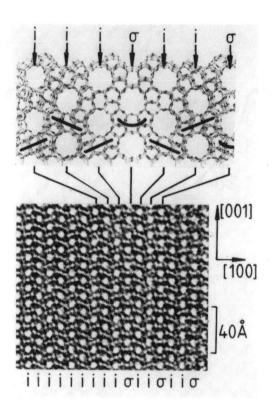

Figure 17.   High resolution micrograph of a typical ZSM-5/ZSM-11
catalyst which contains some intergrowths as structural defects.
In perfectly ordered ZSM-5 one sheet is related to its predecessor
by inversion (i), whereas in ZSM-11 successive sheets are
in mirror (σ) relation.

Intergrowths in which one particular zeolitic structure
exists locally within another have been found, by HREM, to be
quite common (33,53,61,62).   In erionite, a catalyst that
is widely used (41) in the shape-selective process so-called
selectoforming, we have found direct evidence for the coexistence
within the structure of offretite and of sodalite.

7.   ZEOLITES WITH UNSOLVED STRUCTURES

Earlier we noted that there are many zeolitic materials now
extant for which no structural models are yet available.
We also remarked that X-ray structural analysis is not a
feasible proposition for these microcrystalline materials,
even recognizing that synchrotron radiation (as an X-ray

source) enables much smaller crystals than hitherto to be
examined.   At present new strategies are being evolved to
determine the required structures.   A combination of high-
resolution electron microscopy, electron diffraction, solid-
state NMR, catalytic testing (63), neutron and X-ray powder
diffractometry, infra red spectroscopy, gas adsorption studies,
together with model building of one kind or another, is now the
"method" of proceeding.   Items of information garnered from
all these disparate, individual methods of attack are then
synthesized into the final model, but progress so far has not
been dramatic.   Nevertheless, there have been real successes
as the recently determined structures of ZSM-23 (64) and
Theta-1 (29) testify.   The point to note about such a multi-
pronged structural analysis is that vital clues emerge from several
different sources.   Thus electron diffraction may yield the
space-group and the unit cell dimensions;   IR may reveal whether
or not there are 5-membered rings (as in ZSM-5, ZSM-11 and
ferrierite); ^{29}Si MASNMR on the dealuminated zeolite and the
space-group tell (65) us the maximum number of distinct crystallographic sites there are in the material; the catalytic
testing reveals the size of the channel or cavity apertures
and so on.   If a plausible model can then be constructed,
its reliability can be tested in two distinct ways.   First
idealized coordinates are assigned to the atoms (working on
the assumption that Si-O bond lengths are close to 1.60Å
and Al-O to 1.76Å) and an X-ray powder refinement procedure
is undertaken.   This can be improved using the distance-least-
squares (DLS) approach pioneered by Meier and Baerlocher.
Second, on the basis of the model, HREM images are calculated
(77,78)as a function of specimen thickness and defect of focus
A good correspondence between computed and observed images
signifies the trustworthiness of the structural model.
(Le Roy Eyring and A. Rae Smith have evolved quantitative
procedures for assessing the degree of correspondence of
computed and observed images)
     Figure 18, taken from our recent work (66), summarizes
information garnered in the manner described in the preceding
paragraph.   We see that ZSM-23 is effectively a twinned
version of Theta-1 (29) as is exemplified in Figure 19.
Also included in Figure 18 is a proposed model (in projection)
for the structure of ZSM-12.   Doubtless many zeolites will
yield their structures according to the stratagems outlined
in this and other kinds of approach.

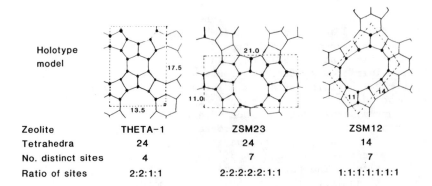

Zeolite	THETA-1	ZSM23	ZSM12
Tetrahedra	24	24	14
No. distinct sites	4	7	7
Ratio of sites	2:2:1:1	2:2:2:2:2:1:1	1:1:1:1:1:1:1

Figure 18.   Models of the projected structures of Theta 1, ZSM-23 and ZSM-12 (the unit meshes are indicated with the dimensions given in Å)

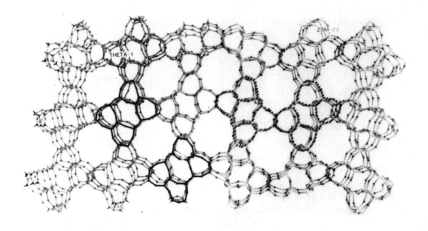

Figure 19.   Schematic illustration of how the structure of ZSM-23, hitherto unsolved, may be regarded (64) as having been derived from Theta-1 by recurrent twinning.   The left half is Theta-1, and the strip of structure immediately to the right of the halfway boundary is the twinned (mirror related) version of the strip to the immediate left.   In the ZSM-23 structure each strip is a twinned version of its predecessor.   The translationally related units in Theta-1 and the mirror related ones in ZSM-23, have been delineated for clarity.

8.   THE SITING AND ENERGETICS OF GUEST SPECIES INSIDE
     ZEOLITE CATALYSTS

To determine the siting of a guest one would nowadays use
neutron powder Rietveld methods (67,68), these procedures
being at present superior to all other experimental methods
since they can, in principle, be used under conditions that
are very close to those used in catalytic practice.   X-ray
powder methods are also promising (69,70) in this regard.
There are, however, theoretical approaches which can also
be employed.   Using atom-atom pairwise evaluation procedures
such as those that have proved eminently successful in organic
solid-state chemistry (71,72), very considerable progress
can be made in pinpointing the siting of the guest within
the zeolite cavity and channel (67).   Early on Kiselev and
coworkers (73,74) demonstrated that computational procedures
yielded values for the enthalpy of adsorption of organic
and inorganic molecules inside zeolites that were very close
to those determined experimentally from adsorption isotherms.
     In a recent study by Wright, Thomas, Cheetham and Nowak
(68), encouraging results have been obtained in a joint experimental
(neutron Rietveld) and computational determination of the
siting of pyridine molecule inside the channels of zeolite-L.
Zeolite-L is a potentially important catalyst for a wide
range of cracking, hydrocracking and other reactions. And
although pyridine is not the typical reactant consumed in
heterogeneous catalysis involving zeolites, it is the archetypal
'acid poisoner'.   Knowledge of its siting within zeolite-
L is, therefore, required.

9.   ZEOLITE CATALYSTS IN A WIDER CONTEXT

During the course of exerting their influence, zeolitic (and
indeed clay-based (35)) catalysts entail removal, transfer
and reincorporation of H^+ ions already present in the solid
catalyst and distributed more or less uniformly throughout
their bulk.   There is another class of catalyst, of growing
importance, that functions in a similar fashion but with
the important difference that oxygen ions, rather than protons,
are now the entities that are removed, transferred and
re-incorporated.   This class embraces several mixed-oxide
systems which are widely used for the selective catalytic
oxidation of hydrocarbons. Bi_2MoO_6, and other so-called
bismuth molybdates, $Bi_2Mo_2O_9$ and $Bi_2Mo_3O_{12}$ with or without
substitutional additives, are, for example, employed (75)
in the inconversion of propylene to acrolein.   And sub-
stoichiometric $Ca\,Mn\,O_{3-x}$ (0 < x < 0.5) can be used to convert
(76) propylene to benzene and isobutene to paraxylene (Fig. 20).
As with the zeolites, this class of catalyst has its active
sites (or potential active sites) distributed uniformly throughout the bulk solid.   These solid catalysts (in marked contrast

to say Pt or Ag as selective oxidation catalysts) can release
oxygen from incorporation of gaseous O_2 into the depleted
structure.   Experiments using $^{18}O_2$ and secondary ion mass
spectrometry leave little doubt that the solid oxide releases
its structural oxygen in the crucial act of catalysis, and
that incorporation of O_2 is facilitated by the electrons freed
in the step: $O^{2-} \longrightarrow O + 2e$.

This analogy between selective oxidation (mixed-oxide)
catalysts and zeolites could well be of value in designing
new heterogeneous catalysts from the established principles
of solid-state chemistry.

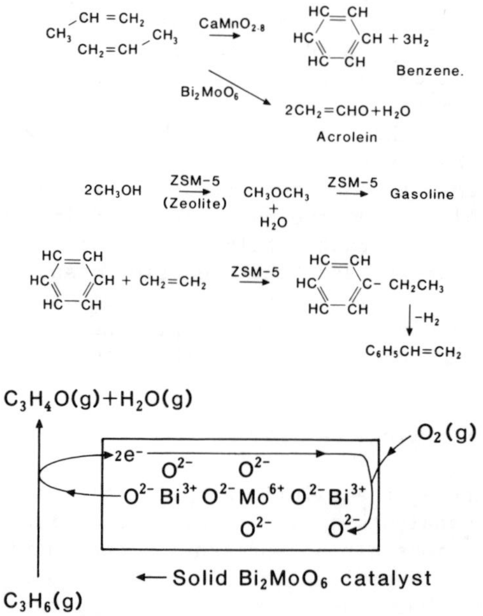

Figure 20.   The analogy between zeolitic and certain mixed
oxide (selective oxidation) catalysts.   Both in catalysts
such as Bi_2MoO_6 (or $CaMnO_{3-x}$) and in the acidic zeolites,
the active sites, from which ions are detached ($O^{=}$ and H^{+}
respectively) are distributed uniformly throughout the solids.
Ions are removed, incorporated into the reactants, which
are then converted to products, and are subsequently re-incorporated
into the solid catalyst. The detailed steps in the conversion
of propylene to acrolein are schematized here.

10.   ACKNOWLEDGEMENTS.

We thank our colleagues, mentioned in the text, for their
stimulus and help.   We are also indebted to the following
for clarifying discussions: Drs. C.J. Adams, W.J. Ball,
S.A.I. Barri, A.K. Cheetham, Christine Marsden, J.M. Newsam,
M.S. Spencer, D.E.W. Vaughan and P.B. Weisz.

REFERENCES

1.   J.W. McBain, "The Sorption of Gases and Vapours by Solids"
     Routledge, London 1932, Chapt. 5.
2.   M.G. Evans, Proc. Roy. Soc., A134, 96, (1931).
3.   (a)   R.M. Barrer, "Zeolites", Academic Press, 1978.
     (b)   D.W. Break, "Molecular Sieves", J. Wiley, 1974
4.   R.M. Barrer, "Hydrothermal Synthesis of Zeolites",
     Academic Press, 1982.
5.   J.M. Thomas, Inorganic Chemistry Towards 21st Century
     (ed. M.H. Chisholm) ACS Series No. 211, p.446 (1983).
6.   J. Klinowski, J.M. Thomas, M. Audier, C.A. Fyfe,
     J.S. Hartman and S. Vasudevan, J. Chem. Soc. Chem. Comm.,
     570, (1981).
7.   M.W. Anderson, Ph.D. Thesis, University of Cambridge 1984.
8.   J.M. Thomas, J. Klinowski, C.A. Fyfe, G.C. Gobbi,
     S. Ramdas and M.W. Anderson, ACS Series No. 218, 159 (1983)
9.   Liu Xin Sheng and J.M. Thomas, to be published.
10. Liu Xin Sheng, Ph.D. Thesis, University of Cambridge,
     1986.
11. S.T. Wilson, B.M. Lok, C.A. Messina, T.R. Cannon and
     E.M. Flanigen, J. Amer. Chem. Soc., 104, 1146 (1982).
12. J.M. Bennett, J.P. Cohen, G. Artioli, J.J. Pluth and
     J.V. Smith, Inorg. Chem., 24, 188 (1985).
13. J.B. Parise, J. Chem. Soc. Chem. Comm., 1449 (1984).
14. J.A. Rabo, "Zeolite Chemistry and Catalysis",
     ACS Monograph 171, (1976) and references therein.
15   P.B. Weisz, Pure Appl. Chem., 52, 2091 (1980).
16. J.M. Thomas, "Proc. 8th Intl. Congress on Catalysis",
     (Berlin 1984), Verlag Chemie, Vol. 1, p.31, (1984).
17. D.H. Olson, and W.M. Meier "Atlas of Zeolite Structures"
     Structure Commission of International Zeolite Assocn.,
     1978 (Zurich).
18. W.J. Mortier "Compilation of extra-framework sites in
     zeolites", Structure Commission of International Zeolite
     Assocn., (1982).
19. J. Klinowski, J.M. Thomas, M. Audier, C.A. Fyfe,
     L.A. Bursill and S. Vasuden, J. Chem. Soc. Chem. Comm.,
     570 (1981)
20. J.M. Thomas, J. Molec Catalysis, 27, 59 (1984).
21. M.W. Anderson, Ph.D. Thesis, University of Cambridge 1984.
22. R.M. Barrer, "Zeolites", Academic Press (1978).
23. D.W. Breck, "Zeolites Molecular Sieves", Wiley, New York, 1974.

24.   D.H. Olson and W.M. Meier, "Atlas of Zeolite Structures",
       Structure Commissional of International Zeolite Assocn.,
       1978 (Zurich)
25.   J.M. Thomas and G.R. Millward, J. Chem. Soc. Chem.
       Comm., 1380 (1982).
26.   J. Klinowski, Progress in NMR Spectroscopy, 16, 237
       (1984).
27.   R. Rinaldi and H.R. Werk, Acta Cryst., A35, 825 (1979).
28.   P. Eisenberger, M. Leonowiz, D.E.W. Vaughan and J.A. Newsom,
       Nature (1984).
29.   S.A.I. Barri, G.W. Smith, D. White and D. Young, Nature,
       312, 533 (1984).
30.   J.M. Thomas and J. Klinowski, Adv. Catalysis, 33 (in
       press) (1985).
31.   G.K. Boreskov and M. Minachev (editors), "Applications
       of Zeolites in Catalysis", Akademiac Kiado, Budapest, 1979.
32.   P.A. Jacobs, "Carboniogenic Activity of Zeolites",
       Elsevier, Amsterdam 1977.
33.   O. Terasaki, S. Ramdas, J.M. Thomas, J.C.S. Chem. Comm.,
       216 (1984).
34.   D.T.B. Tennakoon, W. Jones, J.M. Thomas, L.J. Williamson,
       J.A. Ball and J.H. Purcell, Proc. Ind. Acad. (Chemical
       Sci.), 92, 27 (1983)
35.   J.M. Thomas in "Intercalation Chemistry" (ed. M.S.
       Whittingham and A.J. Jacobson), Academic Press, p.56
       (1982).
36.   J.A. Ballantine, J.H. Purnell and J.M. Thomas, Clay
       Minerals, 18, 347 (1983).
37.   J.A. Ballantine, J.H. Purnell and J.M. Thomas, J. Molec.
       Catalysis, 27, 157 (1984).
38.   J.M. Thomas, Phil. Trans. Roy. Soc., A311, 271 (1984)
39.   F.C. Whitmore, J. Amer. Chem. Soc., (1932).
40.   D.M. Brouwer in "Chemistry and Chemical Engineering
       of Catalytic Processes" (R. Prins and G.C.A. Schmit,
       editors) P137, NATO ASI Series 39 (1980).
41.   P.B. Weisz, Pure and Appl. Chem., 52, 2091 (1980).
42.   S.M. Caiszery, Zeolites, 4, 202 (1984).
43.   D.H. Olson and E. Dempsey, J. Catalysis 13, 221 (1969).
44.   N.T. Tam, R.P. Cooney, G.J. Curthoys, J. Catal. 44,
       81, (1976)
45.   C. Williams, Ph.D. Thesis, University of Cambridge
       1986.
46.   M.W. Anderson and J.M. Thomas (unpublished)
47.   J.M. Thomas, Nature, 314, 669 (1985).
48.   W.O. Haag, R.M. Lago and P.B. Weisz, Nature, 309, 589
       (1984).
49.   C.T-W Chu and C.D. Chang, J. Phys. Chem. 89, 1589 (1985).
50.   D. Freude, T. Foehlich, H. Pfeiffer, G. Scheler,
       Zeolites 3, 171 (1983)
51.   K.F. Scholle, A.P.M. Kertzens, W.S. Veeman, P. Frenken
       and G.P.M. Vander Veldan, J. Phys. Chem., 88, 5 (1984).
52.   A.K. Cheetham, M.M. Eddy and J.M. Thomas,
       J. Chem. Soc. Chem. Comm, 1337 (1984).

53. O. Terasaki, J.M. Thomas, G.R. Millward, Proc. Roy.
Soc., A394, 223 (1984).
54. O. Terasaki, S. Ramdas, J.M. Thomas, J. Chem. Soc.
Chem. Comm., 216 (1984).
55. R.H.Jarman, Zeolites, 5, 213 (1985).
56. J. Klinowski, S. Ramdas, J.M. Thomas, C.A. Fyfe, J.S. Hartman,
J. Chem. Soc., Faraday Trans. II, 78, 1025 (1982).
57. G. Engelhardt, U. Kohse, E. Lippma, M. Tarmak, M. Magi,
Z. Anorg. Allg. Chem., 482 (1981)
58. P.A. Wright, Ph.D. Thesis, Univ. of Cambridge, 1985.
59. D.R. Denley, R.H. Raymond, S.C. Tang, J. Catalysis,
87, 414 (1984).
60. N.Y. Cheng, W.W. Kaeding and F.G. Dwyer, J. Amer. Chem.
Soc., 101, 6783 (1979); see also E.G. Derouane,
J. Catalysis, 72, 177 (1981) and J.R. Anderson, K. Foger,
T. Mole, R.A. Rajadhyaksha, and J.V. Sanders,
J. Catalysis, 58, 114 (1979).
61. G.R. Millward, J.M. Thomas, O. Terasaki and D. Watanabe,
submitted.
62. J.M. Thomas, G.R. Millward, J.C.S. Chem. Comm., 1380
(1982).
63. P.A. Jacobs and P. Martens, Proc. 8th Intl. Congr.
Catalysis, Berlin, 4, 357 (1984).
64. P.A. Wright, J.M. Thomas, G.R. Millward, S. Ramdas
and S.A.I. Barri, J. Chem. Soc. Chem. Comm., in press
(1985).
65. J.M. Thomas, J. Klinowski, S. Ramdas, B.K.T. Hunter
and D.T.B. Tennakoon, Chem. Phys. Lett., 102, 158 (1983).
66. P.A. Wright and J.M. Thomas, in preparation.
67. P.A. Wright, J.M. Thomas, S. Ramdas and A.K. Cheetham,
J. Chem. Soc. Chem. Comm., 1338 (1984).
68. P.A. Wright, J.M. Thomas, A.K. Cheetham and A. Nowak
(submitted).
69 P.A. Wright, I. Gameson, T. Rayment and J.M. Thomas,
in preparation.
70. I. Gameson and T. Rayment, in preparation.
71. A.I. Kitaigorodskii, Chem. Soc., Rev., 7, 133, (1978)
72. S. Ramdas and J.M. Thomas in "Chemical Physics of Solids
and their Surfaces", Vol 7, p31 (The Chemical Soc.,
London, 1978)
73. A.V. Kiselev and P.G. Du, J. Chem. Soc. Faraday Trans. II,
77, 17 (1981).
74. A.V. Kiselev, V.I. Lygin, R.V. Stardubceva, J. Chem.
Soc. Faraday Trans. II, 68, 1793 (1979).
75. R.K. Grasselli and J.D. Burrington, Adv. Catalysis,
30, 133 (1981); see also A.W. Sleight in "Advanced
Materials in Catalysis" (ed.J.J. Burton and R.L. Garten)
p191, Acad. Press (1977).
76. A. Reller, J.M. Thomas, D.A. Jefferson and M.K. Uppal,
Proc. Roy. Soc., A394, 223 (1984).

77. J.G. Allpress, E.A. Hewat and A.F. Moodie, Acta Cryst.,
     A28, 528 (1972).
78. P. Goodman and A.F. Moodie, Acta Cryst., A30, 280 (1974).

RHODIUM AND IRIDIUM CARBONYL COMPLEXES ENTRAPPED IN ZEOLITES.
STRUCTURAL AND CATALYTIC PROPERTIES

C. Naccache, F. Lefebvre, P. Gelin, Y. Ben Taarit
Institut de Recherches sur la Catalyse,
Laboratoire Propre du C.N.R.S.,
conventionné à l'Université Claude Bernard, LYON 1
2, avenue Albert Einstein - 69626 - Villeurbanne Cédex -

ABSTRACT. Zeolites have been used to anchor soluble transition metal complexes. The synthesis of mono and polynuclear carbonyl compounds was performed within the zeolite cavities. Infrared spectroscopy was employed to identify the carbonyl compounds, and the catalytic performances of these materials studied in a gas flow microreactor. Rhodium (and iridium) trivalent cations were found to react with carbon monoxide with resultant formation of monovalent rhodium (and iridium) dicarbonyl complexes. These complexes anchored in the zeolite cavities showed catalytic activity in the carbonylation of methanol in the presence of methyliodide. The mechanisms of the heterogeneous catalytic reaction parallel the scheme proposed for the methanol carbonylation in solution. The zeolite was considered as a solid solvent. In the presence of a mixture of $CO-H_2$ (or $CO + H_2O$) Rh (or Ir) exchanged NaY zeolite reacts to produce $Rh_6(CO)_{16}$ (or $Ir_4(CO)_{12}$) metal carbonyl clusters stabilized towards aggregation and thus maintaining their structural integrity over a wide range of temperature. Furthermore the interaction of the cluster with the zeolite framework affected the bonding of CO with probably the subsequent increase of CO lability.

INTRODUCTION. Soluble transition metal complexes were found active for a variety of reactions under mild conditions. To easily recover the soluble catalyst several authors have attached the soluble transition metal ion complex to a non soluble solid. For this purpose the solid was functionalized, and the metal ion compound attached through the surface ligands. An alternative way to convert soluble catalyst into non soluble material was to introduce the active complex in the intercrystal space of layer lattice silicate, zeolites. The use of zeolites for anchoring metal complexes appeared very promising. However one of the major disadvantages of zeolites is that in several cases the desirable organometallic complexes are too bulky to enter the zeolite pore mouth, but small enough to fit the dimension of the zeolite cavities. This latter difficulty may be overcome if one synthesizes the metal complexes directly in the zeolite cavities. In the present research we have attempted to synthesize in the 1.3 nm diameter

81

R. Setton (ed.), Chemical Reactions in Organic and Inorganic Constrained Systems, 81–86.
© 1986 by D. Reidel Publishing Company.

cavities of NaY zeolite , faujasite-like structure, mono and
polynuclear rhodium and iridium carbonyl compounds. Their ability to
catalyze methanol carbonylation into acetic acid was tested.

Materials and Methods.

NaY zeolite supplied by Union Carbide Linde Division was used as the
starting material. A series of samples Rh-NaY and Ir-NaY were prepared
by exchanging the sodium form with solutions containing the desired
amount of $[Rh^{3+}(NH_3)_5Cl]$ or $[Ir^{3+}(NH_3)_5Cl]$ cations. The exchanged
zeolites were then washed with distilled water, filtered and dried at
323 K. Metal contents were determined by atomic adsorption
spectrophotometry. The infrared spectra of the carbonyl compounds were
recorded on a Perkin-Elmer 580 type spectrophotometer. An IR cell was
used which allowed in situ reaction and treatment. The zeolite powder
was pressed into then wafers and mounted in the IR cell. The catalytic
reactions were carried out in a differential flow reactor at
atmospheric pressure. The reaction mixture was analyzed by gas
chromatography. The methanol carbonylation reaction was carried out in
a fixed-bed flow reactor at atmospheric pressure. CH_3OH and CH_3I
pressures were controlled by the temperature of the saturator. Feed and
products were analyzed by gas chromatography.

Results and discussion

Mono nuclear carbonyl complexes. Recent experiments have demonstrated
that the interaction of rhodium (III) exchanged NaY zeolite with CO at
293 K gives Rh(I) $(CO)_2$ species (1). This assumption was supported by
the IR spectra of the resulting mononuclear rhodium carbonyl complex
exhibiting two infrared bands at 2100-2020 cm^{-1} and attributed to the
symmetric and antisymmetric C-O vibration in Rh(I)$(CO)_2$. The reduction
of Rh(III) to Rh(I) was further supported by XPS studies which
indicated that the Rh 3d 5/2 binding energy was shifted from 310.8 eV
to 308.1 eV as expected for reduction of Rh(III) to Rh(I). Finally it
was shown by IR that CO_2 is formed upon reaction of Rh-NaY with CO. The
general equation for the formation of mononuclear dicarbonyl monovalent
rhodium is the following

$$Rh^{3+} + 3(AlO_4)^- + 3 CO \xrightarrow{H_2O} Rh^+(CO)_2 + 2H^+ + 3(AlO_4)^- + CO_2$$

According to the above equation, in the presence of H_2O, reduction
of Rh(III) to Rh(I) occurs with the subsequent formation of protons.
The type of bonding of $Rh^+(CO)_2$ with the zeolite framework was further
investigated by infrared technique (2). It was found that the IR
frequencies of the gem-CO ligands changed with the hydrated state of
the zeolite. The results were interpreted in terms of variation in
Rh(I) environment. In the absence of H_2O, Rh(I) $(CO)_2$ is bonded to the
zeolite framework through two lattice oxygen, the Rh(I)$(CO)_2$ complex
being characterized by two CO bands at 2101 and 2022 cm^{-1}. Addition of
H_2O led to the replacement of one O^{2-} ligand by H_2O molecule, ν CO
were found at 2116 and 2048 cm^{-1}. Finally fully hydrated species gave

an IR spectrum with two CO bands at 2090 and 2030 cm^{-1} attributed to the formation of the dimer $(Rh_{\frac{1}{2}}(CO)_4(OH)_2)$. In the latter case, excess water prevented the dicarbonyl rhodium complex to interact with the zeolite. Similarly iridium exchanged NaY zeolite reacts with CO with the subsequent reduction of Ir(III) to Ir(I) and the formation of Ir(I)(CO)$_3$ complex characterized by two IR CO band at 2086 and 2001 cm^{-1} (3). More recently the interaction of carbon monoxide with Ir-NaY was further investigated and it was concluded that CO reacts with Ir(III) cations to produce Ir(I)(CO)$_2$ (4).

Carbonylation of methanol

The carbonylation of CH_3OH is of great industrial importance and has found application in the production in large scale of acetic acid. The use of rhodium based catalyst in the presence of methyl iodide as promotor for the liquid phase methanol carbonylation has been developped by Monsanto scientists (5). The mechanism of the reaction

$$CH_3OH + CO \xrightarrow{ICH_3} CH_3COOH$$

has been established by IR study (6). In solution the detailed proposal for the mechanism involved in the formation of acetic acid or methyl acetate was the following : products formation occured with the catalytic active species changing between Rh(I) and Rh(III) states. In the presence of ICH_3, oxidative addition occured on Rh(I)(CO)$_2$ complex with the formation of Rh(III)(CH$_3$)(I)(CO)$_2$ followed by a rapid methyl shift to produce the Rh(III) acetyl complex Rh(III)(CH$_3$CO)(I)(CO) . Subsequent hydrolysis or methanolysis of the acetyl ligand would form acetic acid or methyl-acetate. The carbonylation of methanol was carried out in the gas phase in the presence of zeolite-supported Rh(I) or Ir(I) dicarbonyl materials as catalysts (9). The reaction was carried out at 423-453 K and atmospheric pressure, while in the liquid phase the reaction was generally carried out at high pressure. The kinetic studies have revealed that the rate of the reaction was first order in CH$_3$I, zero order in CH$_3$OH and CO for Rh-NaY catalyst similarly to what observed in liquid phase (7) and first order in CH$_3$OH and zero order in CO and CH$_3$I for Ir-NaY also identically to the kinetic observed for iridium based catalyst in solution (8). Finally the IR study of the interaction of CH$_3$I with Rh(I)(CO)$_2$-NaY indicated that rhodium dicarbonyl complex experienced an oxidative addition of CH$_3$I with the subsequent formation of the rhodium acetyl species, showing an IR band at 1710 cm^{-1}. It is concluded that in the heterogneous phase the methanol carbonylation followed the same path than in solution. The results parallel those obtained with the homogeneous rhodium or iridium catalyst. The zeolite appeared to act as a solid solvent where the active species are atomically dispersed.

Polynuclear carbonyl clusters

A large number of solids have been used to "heterogeneize molecular" metal carbonyl clusters. The aim of these studies was to attach the metal carbonyl cluster onto functional groups of functionallized surface of the support. $Rh_6(CO)_{16}$, $Ir_4(CO)_{12}$ $Os_3(CO)_{12}$ have been attached to polymer, SiO_2, Al_2O_3 and it was stated that the molecular clusters have retained their integrity upon adsorption on solids. The adsorption of $Rh_6(CO)_{16}$ onto NaY zeolite has been followed by IR spectroscopy (10). It is known that $Rh_6(CO)_{16}$ consists of an octahedral arrangement of rhodium atoms with two linearly bonded CO showing IR CO bands at 2073 and 2026 cm^{-1} and one CO bonded to three Rh atoms showing a sharp band at 1800 cm^{-1}. When $Rh_6(CO)_{16}$ was adsorbed on NaY from the vapor phase IR bands developped at 2088 and 1830 cm^{-1}. It was concluded that $Rh_6(CO)_{16}$ was adsorbed on the external zeolite surface near the pore mouth. The decarbonylation of $Rh_6(CO)_{16}$ by reaction with O_2 at 373 K followed by recarbonylation with CO at 373 K resulted in the appearance of two intense narrow CO bands at 2095 and 1764 cm^{-1}. The sharp bands at 2095 and 1764 cm^{-1} were ascribed to the terminal and bridging CO in the entrapped Rh_6 cluster. The Rh_6 carbonyl cluster entrapped within the zeolite cavities appeared to be easily decarbonylated and recarbonylated without lost of its integrity. However the retention of the cluster integrity was observed only for low Rh loading, less than 1 wt %. Indeed NaY zeolite loaded with 2-3 wt % Rh, showed after decarbonylation and recarbonylation a more complex IR spectrum analyzed as resulting from the presence on the zeolite of several rhodium carbonyl species, one of these species being identified to $Rh_6(CO)_{16}$. It was concluded that excess $Rh_6(CO)_{16}$ which could not enter the zeolite cavities, deposited on the zeolite external surface. Upon decarbonylation and recarbonylation $Rh_6(CO)_{16}$ on external surface was no more stabilized by the zeolite matrix and thus lost its molecular integrity.

Another elegant mean to stabilize molecular metal clusters in zeolite was recently developped. This goal was reached by synthesizing directly within the zeolite cavities the desired metal carbonyl clusters. In this paragraph we report the results concerning the synthesis of rhodium and iridium clusters. Reaction of Rh(III)-NaY zeolite with $CO-H_2$ (or $CO-H_2O$) mixture at temperature lower than 353 K and at atmospheric pressure gave a coloured solid. The IR spectrum of this solid exhibited intense and narrow bands at 2090-2080 cm^{-1} due to singly bonded CO and at 1832 and 1760 cm^{-1} due to bridged CO. The bridged CO at 1832 cm^{-1} was relatively unstable under CO at higher temperature yielding the band at 1760 cm^{-1} comparison of the observed IR spectra with the data of the literature has led the authors (11) to conclude that $Rh_4(CO)_{12}$ and $Rh_6(CO)_{16}$ clusters were formed within the zeolite cavities. The IR band at 1832 is attributed to the carbonyl bridged between two rhodium atoms in $Rh_4(CO)_{12}$. This assignement was also in agreement with the known relative stability of Rh_4 and Rh_6 carbonyl clusters, Rh_4 being easily converted into Rh_6 clusters by raising the temperature

$Ir(NH_3)_5Cl$ -NaY exchanged zeolite has been used as the starting material to synthesize the iridium carbonyl cluster in the cavity. As previously shown for rhodium exchanged zeolite, the reaction of Ir-NaY with a $CO-H_2O$ mixture at 443 K and atmospheric pressure produced a yellow solid (11). During the reaction CO_2 was evolved. The IR spectrum of the hydrated yellow sample was characterized by sharp bands at 2086-2040 cm^{-1}. Upon dehydration at room temperature a shift of this doublet to 2073-2030 cm^{-1} was observed with the subsequent appearance of a strong and sharp IR band at 1813 cm^{-1}. Quantitative measurement of the CO consummed during the reaction and isotopic exchange $^{12}CO-^{13}CO$ on the iridium cluster formed upon $CO-H_2O$ reaction have led the authors to conclude that the yellow iridium carbonyl cluster synthesized with the zeolite cavity was characterized by a CO/Ir ratio of 3. The general formula of this cluster was $Ir(CO)_{3\ n}$. The IR spectrum of the hydrated sample is similar to the one observed by impregnating NaY with a cyclohexane solution of $Ir_4(CO)_{12}$. Furthermore $Ir_4(CO)_{12}$ is yellow and exhibits only linearly bonded CO around 2070-2020 cm^{-1}. It was concluded that the iridium cluster produced in the zeolite cavitiy should be the dodecacarbonyl-tetrairidium $Ir_4(CO)_{12}$. The configuration of the CO ligands in this cluster depends strongly on the substitution of CO by other ligands. Hence it has been shown that as CO is replaced by a phosphine or an anion X^- in $Ir_4(CO)_{12}$ to give $Ir_4(CO)_{11}X^-$ several linear CO are converted into bridged CO with the subsequent appearance of an IR band around 1800 cm^{-1} (12). This was attributed to an increase of the electron density around the metal nucleus brought by the ligand, the extra densitiy being released by converting linear CO into bridged CO. The analysis of the IR results obtained on zeolite entrapped $Ir_4(CO)_{12}$ demonstrated that the iridium cluster experienced a strong interaction with the zeolite framework. Indeed $Ir_4(CO)_{12}$ in the dehydrated NaY sample showed an IR band at 1813 indicating that part of CO ligands are bridging two Ir atoms. This configuration in $Ir_4(CO)_{12}$ should be interpreted as the results of the strong electrostatic field of the zeolite acting on the iridium cluster. Addition of water to the sample results in the partial filling of the zeolite cavities by H_2O. In this state $Ir_4(CO)_{12}$ would be surrounded by water molecules in the zeolite cavities and thus protected from the electrostatic field of the zeolite. Thus no electron charge excess around iridium atoms exists and the structure of the Ir_4 cluster in the zeolite will be simular to that encountered in $Ir_4(CO)_{12}$ crystal, that is all CO are linearly bonded to the metal. Such zeolite-cluster interaction affects the C-O bonding as well the M-CO thus would affect the reactivity of the CO ligands.

In conclusion these results have demonstrated that the zeolite matrix could act as a "solid solvent" in which active metal centres are atomically dispersed and coordinatively bonded to the zeolite framework. These metal centres form with various molecules, trapped metal ion complexes and thus can be used as catalysts for a variety of reactions, these materials being the homologous of the soluble catalysts. Furthermore the electrostatic field of the zeolite matrix could help the oxido-reduction process occuring at the central metal atom during the catalytic reaction steps and also may increase the lability of the ligands coordinated to the metal atoms.

C. NACCACHE ET AL.

References

1. Primet M., Védrine J.C., Naccache C., J. Mol. Cat., 1978, 4, 411.
2. Lefebvre F., Ben Taarit Y., Nouv. J. Chim., 1984, 6, 387.
3. Gelin P., Coudurier G., Ben Taarit Y., Naccache C., J. Catal., 1981, 70, 32.
4. Lefebvre F., Auroux A., Ben Taarit Y., Proc. Int. Symp. Zeolites, (Yougoslavia 1984) to be published in "Studies in Surface Science and Catalysis" Elsevier.
5. Roth J.F., Craddock J.H., Hersham A., Paulik F.E., Chem. Technol., 1971, 600.
6. Foster D., J.A.C.S., 1976, 98, 846.
7. Robinson K.K., Hersham A., Craddock J.H., Roth J.F., J. Catal., 1972, 27, 389.
8. Matsumoto T., Mizoroki T., Osaki A., J. Catal., 1978, 51, 96.
9. Gelin P., Ben Taarit Y., Naccache C., "New Horizon in Catalysis" (Seryama and Tanabe eds.) Elsevier 1980, 7, 398.
10. Gelin P., Ben Taarit Y., Naccache C., J. Catal., 1979, 59, 357.
11. Lefebvre F., Gelin P., Naccache C., Ben Taarit Y., Proc. 6th Int. Zeolite Conference (Olson D. and Bisio A., eds.) Butterworths 1984, 435.
12. Gelin P., Naccache C., Ben Taarit Y., Diab Y., Nouv. J. Chim., 1984, 8, 675.
13. Raithby "Transition metal clusters" (Johnson B.J.G. ed), Wiley and Sons 1980, 66.

SHAPE SELECTIVE ALKYLATION REACTIONS OVER MEDIUM PORE HIGH
SILICA ZEOLITES

A.G. Ashton[a], S.A.I. Barri[a], S. Cartlidge[a], J. Dwyer[b]

(a) BP Research Centre   (b) Dept. of Chemistry
     Sunbury on Thames   UMIST
     Middlesex   Manchester
     TW16 7LN   England
     England

ABSTRACT. The medium pore high silica zeolites Theta-1,
ZSM-5 and ZSM-11 have been synthesised. Intrinsic differences
in shape selectivity were determined by monitoring the formation
of diisopropylbenzenes and xylenes from alkylation reactions.
Theta-1 shows enhanced selectivity towards para-substitution
which can be attributed to increased transition-state selectivity.
Coke laydown in ZSM-5 is more severe when nonframework species,
eg M^+, Al_2O_3, are present which leads to improved selectivity
and decreased conversions.

1.   INTRODUCTION

The alkylation of aromatic molecules using solid acid catalysts is
often the first step in the production of raw materials for the chemical
industry. Fundamental research on alkylation reactions has led to the
study of shape selective catalysts. Molecular shape selectivity has
been observed when the spatial dimensions of reactant or product
molecules, or transition state intermediates, approach those of the
catalyst pores. Zeolites form a major group of shape selective catalysts
and reviews have been published (1,2,3). It is the structure of the
channel system that gives a zeolite shape selective properties.

     Recently the structure of Theta-1 a novel medium pore high silica
zeolite has been published (4). The channels in Theta-1 are
unidimensional (Figure 1) with elliptical 10-T-ring openings. Zeolites
ZSM-5 and ZSM-11 have orthogonally intersecting channels with 10-T-ring
openings (5). The volumes created at the channel intersections are
nearly 0.9 nm in diameter (6) and are slightly larger in ZSM-11 compared
to ZSM-5.

     The formation of coke deposits has been found to improve the
shape selective properties of a given zeolite structure type at the
expense of reactant conversions (7). This paper presents information

87

R. Setton (ed.), Chemical Reactions in Organic and Inorganic Constrained Systems, 87–94.
© *1986 by D. Reidel Publishing Company.*

on the role of non-framework species in the production of coke and its
effect on the formation of diisopropylbenzenes. In addition the different
selectivities of Theta-1 and ZSM-5 are compared for the alkylation of
toluene with methanol.

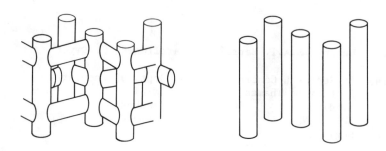

FIGURE 1:   CHANNEL SYSTEMS IN ZSM-5 (LEFT) AND THETA-1

2.   EXPERIMENTAL

Theta-1 was synthesised in a pure form using diethanolamine or a sodium,
ammonium system (9).   Tetrapropylammonium hydroxide and tetrabutyl-
ammonium hydroxide were used as templates for the formation of
gallosilicate MFI and ZSM-11 respectively.   The aluminosilicate ZSM-5
crystallised from a gel containing sodium and ammonium cations (8).
The purity of each zeolite was determined by X-ray powder diffraction
and scanning electron microscopy.

     The conditions of activation and catalytic experiments have been
reported (8,9).   Diisopropylbenzenes were formed by passing propene and
benzene over the catalysts listed in Table 1 at 270°C and atmospheric
pressure.   For the alkylation of toluene with methanol zeolite H-Theta-1
(Si/Al = 32) and H-ZSM-5 (Si/Al = 19) were precalcined at 650°C.
Products were analysed by GLC techniques.

3.   RESULTS AND DISCUSSION

3.1   The Formation of Diisopropylbenzenes

Isopropylbenzene is the major product from the alkylation of benzene
with propene and less than 2 per cent weight diisopropylbenzenes and
1 per cent coke are formed.   Only small quantities of o-diisopropyl-
benzene were detected over each catalyst (Table 1) and in most cases
none of this isomer was formed.   The changes observed in the ratio of
p-/m-diisopropylbenzene with increasing time-on-stream are presented
in Table 1 with the respective molar conversions of propene.   For some
catalysts there exists a direct correlation between propene conversion
and increase in the p-/m-ratio.   Propene is primarily converted at active
sites within the zeolite channels by polymerisation and cracking reactions.

All catalysts show either no change or an increase in shape selectivity with time-on-stream in favour of the p-isomer. The most marked increase is seen for ZSM-5 containing sodium cations (Figure 2) where after ca. 2 hours-on-stream only p-diisopropylbenzene is produced. The concomitant decrease in propene conversion over Na,H-ZSM-5 activated at 400°C, indicates coke laydown on active sites. Since no such decrease in propene conversion is observed for H-ZSM-5 activated at 400°C it can be concluded that sodium cations are responsible for coke-forming reactions.

TABLE 1: PROPENE CONVERSIONS AND RATIO OF p-/m-DIISOPROPYLBENZENES OVER THETA-1, ZSM-5 AND ZSM-11

Zeolite	Propene Conversion % (molar)[a]			p-/m-DIIPB Ratio			
	Hours-on-Stream			Hours-on-Stream			
	0.5	1.5	4.5	0.5	1.5	2.5	4.5
Na,H-ZSM-5 (400, 14)[b]	75	40	37	0.4	6.3		
H-ZSM-5 (400, 14)	99	99	97	0.8	1.1	1.3	1.6
H-ZSM-5 (700, 14)	99	94	85	0.4	2.0	2.7	3.2
H-ZSM-5 (400, 19)	97	97	95	0.4	0.8	1.0	1.1
H-ZSM-5 (400, 19)[c]	99	98	98	0.3	0.2	0.2	0.1
H-Theta-1 (400, 35)	91	79	72	1.4	1.5	1.5	1.5
H-ZSM-11 (400, 18)	99	98	98	0.3	0.1	0.3	0.2

a - Reaction temperature 270°C; benzene/propene molar feed ratio 3:1; benzene weight hourly space velocity 11 h^{-1}.

b - Activation temperature, °C, Si/Al ratio.

c - Si/Ga ratio.

By activating H-ZSM-5 with a Si/Al ratio of 14 at 700°C, hydroxy-alumina is deposited in the channels (8). The presence of non-framework alumina seems to promote coke formation since the propene conversion over H-ZSM-5 activated at 700°C (Table 1) decreases with increasing reaction time. In addition p-diisopropylbenzene is produced with

increasing selectivity. These observations indicate that coke is
deposited around non-framework species to create a severe pore blockage.

The H-ZSM-5 zeolites with Si/Al ratios of 14 and 19 after activation
at 400°C do not contain metal cations or alumina deposits. Only marginal
decreases in propene conversions are observed over these catalysts
(Table 1, Figure 2). The increases in p-/m-diisopropylbenzene ratios
with time-on-stream are small but significant. This result indicates
the importance of non-framework species on the shape selective properties
of medium pore zeolites.

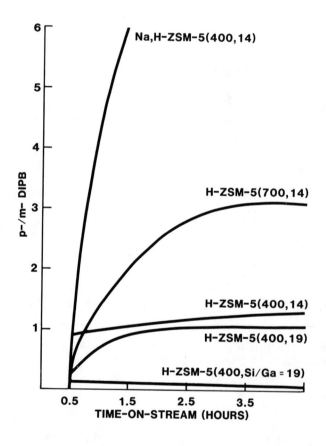

FIGURE 2:   RATIOS OF p-/m-DIISOPROPYLBENZENES FORMED OVER ZSM-5
            TYPE ZEOLITES

In comparison the gallosilicate H-ZSM-5 with a Si/Ga ratio of 19 converts close to 100% of the propene fed over 4.5 hours-on-stream. No change in the ratio of p-/m-diisopropylbenzene is observed over this period (Table 1). If gallia was present within the channels of the gallosilicate, coke formation and changes in shape selectivity might be expected. However, the Bronsted acid sites of gallium containing frameworks are reported to be stronger than those of the aluminosilicate counterparts (10). Therefore coke precursors may be cracked and the intrinsic shape selective properties of the MFI structure type preserved.

The effect of channel topology on shape selectivity is obtained by comparing H-Theta-1, H-ZSM-5 and H-ZSM-11 (Table 1). Larger channel volumes in ZSM-11 reduce the effect of coke laydown and no change in the p-/m-ratio is observed. In the case of Theta-1 the invariant ratio of p-/m-diisopropylbenzene indicates an intrinsic stability in the shape selective properties of this catalyst. The marked decrease in propene conversion over H-Theta-1 means that sites on the external crystal surfaces are active for coke forming reactions.

3.2   The Formation of Xylenes

Zeolite H-Theta-1 is a highly selective catalyst for the production of p-xylene (Table 2). The selective para-alkylation of monoalkylbenzenes can be predicted using a simple model based on restricted transition state selectivity (11). For a near-cylindrical channel with lengthwise orientation of toluene, para-alkylation will predominate, if in Figure 3 a > c > b. In this case the approach of methanol to the ortho- and meta-positions of toluene is sterically hindered. Para-xylene will then be formed when a molecule of toluene approaches a molecule of methanol polarised along the $HO-CH_3$ bond to induce electrophilic substitution.

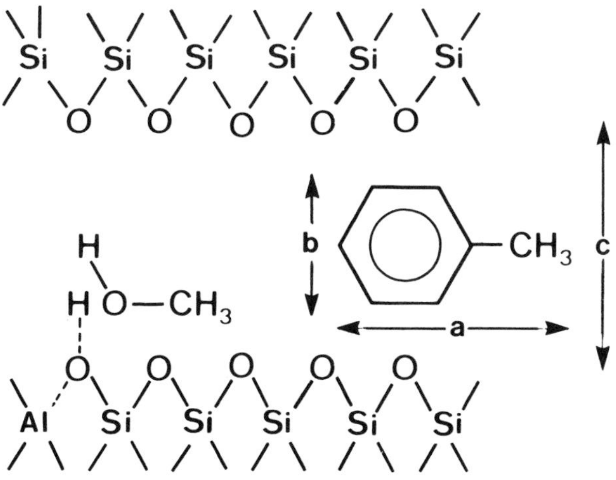

FIGURE 3:   MODEL FOR THE PREDICTION OF PARA-ALKYLATION (11)

TABLE 2: PRODUCT COMPOSITIONS FROM THE ALKYLATION OF TOLUENE WITH
METHANOL OVER H-THETA-1 AND H-ZSM-5

	THETA-1	ZSM-5	THETA-1	ZSM-5	ZSM-5
Conditions					
Temperature °C	550	550	650	650	600
TOL/MeOH molar	2/1	2/1	2/1	2/1	2/1
LHSV h^{-1}	1	1.2	1.2	1.2	4.8
Time-on-Stream hours	0.75	0.75	0.75	0.75	1.5
Conversion %					
Toluene	23.9	37.9	46.5	46.5	31.8
Methanol	100	100	100	100	100
Liquid Product % wt					
Benzene	0.7	9.4	1.3	13.8	3.4
Toluene	70.8	56.0	67.3	54.2	67.0
Para-xylene	10.2	7.3	12.1	6.7	11.1
Meta-xylene	6.5	15.5	8.4	13.9	11.1
Ortho-xylene	6.0	7.1	7.4	6.1	4.6
1,3,5-trimethylbenzene	0.4	1.1	3.0	1.3	0.2
1,2,4-trimethylbenzene	4.3	3.3	3.2	3.0	2.4
1,2,3-trimethylbenzene	N/D	0.1	N/D	0.5	N/D
Total	98.9	99.8	100.0	99.5	99.8
Xylene Composition %					
Para	44.9	24.4	43.4	25.2	41.4
Meta	28.6	51.8	30.1	52.0	41.4
Ortho	26.5	23.8	26.5	22.8	18.2

Clearly the channel systems of ZSM-5 and ZSM-11 are not compatible with the above model. The cage-like channel intersections provide adequate space for all the xylene isomers to be formed. Thus, zeolite H-ZSM-5 does not show initial shape selectivity for the production of xylenes from toluene and methanol (Table 2). Xylene isomers are produced in a ratio close to that calculated for thermodynamic equilibrium ie, p:m:o = 24:52:24 per cent weight (7).

The deposition of coke inside the channels of ZSM-5 increases the selectivity to p-xylene. A sample of H-ZSM-5 containing 9 per cent weight of coke produces 41.4% of the p-isomer compared to 25.2% for the clean catalyst. The ratio of 1,2,4-trimethylbenzene/1,3,5-trimethyl-benzene is about 11 over H-Theta-1 but for H-ZSM-5 the ratio is 3 (Table 2). Once again coke formation enhances the production of the isomer with smallest kinetic diameter and the ratio increases to 12 for the coked ZSM-5.

Benzene is produced (Table 2) by dealkyaltion and transalkylation reactions. The rate of dealkyaltion which will be greatest at strong acid sites (12) decreases over H-ZSM-5 as these sites are fouled by coke laydown. Transalkyaltion reactions require sufficient space for bulky bimolecular transition states to be formed. The channel inter-sections in ZSM-5 provide the necessary space to allow the 1,1-diphenyl-methyl transition state to form. Coke laydown in ZSM-5 increases tortuosity and reduces the accessible free space to that available in Theta-1 and the disproportionation activities of the catalysts are then similar.

4.   CONCLUSIONS

Non-framework species eg, M^+, Al_2O_3, accelerate the increase in shape selectivity of ZSM-5 due to coke laydown. Zeolites which do not contain such deposits exhibit the intrinsic selectivities determined by the constraints of their channel systems. Zeolite Theta-1 has a higher intrinsic selectivity than ZSM-5 and ZSM-11 for the production of para-disubstituted benzenes. The enhanced selectivity of Theta-1 is primarily due to space restrictions imposed on transition-state intermediates.

5.   ACKNOWLEDGEMENTS

We thank Dr D.J.H. Smith and Dr W.J. Ball for constructive comments, Analytical Services Research and Development for valuable contributions and BP International plc for permission to publish.

6.   REFERENCES

(1) Csicsery, S.M., Zeolites, 4, 202, 1984.

(2) Weisz, P.B., Pure and Appl.Chem., 52, 2091, 1980.

(3) Derouane, E.G., in Catalysis by Zeolites, (Eds. Imelik, B.
     et al) 5 (Elsevier, Amsterdam, 1980).

(4) Barri, S.A.I., Smith, G.W., White, D., Young. D, Nature (London),
     312(5994), 1984.

(5) Kokotailo, G.T., Meier, W.M., in The Properties and Applications
     of Zeolites, (Ed. Townsend, R.P.) 133 (Chemical Society Special
     Publication No. 33, London, 1980).

(6) Derouane, E.G., Gabelica, Z., J.Catal., 65, 486, 1980.

(7) Kaeding, W.W., Chu, C.C., Young, L.B., Weinstein, B.,
     Butter, S.A., J.Catal., 67, 159, 1981.

(8) Aukett, P.N., Cartlidge, S., Poplett, I.J.F., Zeolites
     (accepted for publication).

(9) Ashton, A.G., Barri, S.A.I., Dwyer, J., Proceedings of
     the International Symposium on Zeolite Catalysis, 25
     (Siofok, Hungary, 1985).

(10) Tielen, M., Greelen, M., Jacobs, P.A., Proceedings of the
     International Symposium on Zeolite Catalysis, (Siofok, Hungary,
     1985).

(11) Shabtai, J., La Chimica El'Industria, 61(10), 735, 1979.

(12) Jacobs, P., in Carboniogenic Activity of Zeolites, 169,
     (Elsevier, Amsterdam, 1977).

ZEOLITE-INDUCED SELECTIVITY IN THE CONVERSION OF THE LOWER ALIPHATIC CARBOXYLIC ACIDS

M. Vervecken, Y. Servotte, M. Wydoodt, L. Jacobs, J.A. Martens and P.A. Jacobs
Katholieke Universiteit Leuven, Laboratorium voor Oppervlaktechemie, Kardinaal Mercierlaan 92, B-3030 Leuven (Heverlee), Belgium

ABSTRACT

The selectivity of the acid catalyzed conversion of acetic, propionic and butyric acid was determined on zeolites with discrete structures and different Si/Al ratios. The reactions were done in the vapor phase, using a fixed bed continuous flow reactor at different temperatures and for different contact times.

With acetic acid only pentasil zeolites with intermediate acid strength were inducing zeolite-dependent selectivities, giving in this way xylenol isomers. With large pore zeolites, the selectivities obtained in mineral acids are found. The phenolic-type products cannot be formed with butyric acid as feed. Propionic acid is an intermediate case. With butyric acid a highly selective ketonization reaction into 4-heptanon occurs on zeolites containing an erionite cage in their structure.

I. INTRODUCTION

Shape-selectivity in zeolites for hydrocarbon reactions

The active protons or Brønsted acid sites in zeolites are for the major part located in the interior of the crystals. Consequently, the products of an acid-catalyzed reaction can to some extent be determined by the size of the zeolite pores, by their structure (dimensionality of the pore system) as well as by their shape. For acid catalyzed conversions of hydrocarbons, this subject is extremely well documented and this area of research has been reviewed several times (1-9).

R. Setton (ed.), Chemical Reactions in Organic and Inorganic Constrained Systems, 95–114.

Matching the sizes and shapes of the reactants, products and/or transition state in a chemical reaction with the size, shape and tortuosity of the zeolite channels and pores, is the basis for reaction selectivity. The first paper on shape-selective zeolite catalysis appeared in 1960 (10) and dealt with an example of size exclusion of branched alcohols from the cages of zeolite CaA. Since then other forms of shape-selectivity have emerged. Csicsery (2) has categorized these shape-selective effects as follows :

- Reactant selectivity takes place when the zeolite acts as a molecular sieve and excludes a molecule or a class of reactants from the intracrystalline void space, while others with smaller size can enter freely this space. A whole variety of zeolites is now available so that this critical exclusion limit can be varied over a wide range of sizes;

- Product selectivity occurs in the intracrystalline volume and is the result of different diffusivities of product molecules in the zeolite pores;

- Restricted transition state selectivity takes place if the spatial configuration around an active site is restricted and consequently bulky transition states cannot be formed.

Other less common expressions for molecular selectivity in the zeolite crystal have been reported :

- the concentration effect (11a) describes the enhanced concentration of hydrocarbons in faujasite-type zeolites, thereby increasing the rate of bimolecular over monomolecular reactions. A special case of this concentration effect is the so-called cage effect (11b, 16). On zeolites containing the erionite cage as structural element (such as erionite, ZSM-34 and zeolite T), the mobility of n-heptane and n-octane is lowered drastically since they fit perfectly in this cage. As a consequence the diffusivity of n-paraffins goes through a minimum at these carbon numbers, while the reactivity goes through a maximum;

- molecular traffic control (12-14) assumes that in a zeolite with discrete sets of channels (or pores) reactants may enter preferentially through one set and products leave the crystal by another set thus minimizing counter-diffusion;

- molecular circulation (15) affects the way by which molecules have access to pores and is therefore related to product selectivity.

Most recently the concept of "energy gradient selectivity" (17) was introduced. It was shown that the smaller and more tortuous the intracrystalline environment is, in which the reacting molecules move, the higher is the field and field gradient and the more secondary cracking reactions of hydrocarbons are favoured.

These shape-selective effects have a more than academic importance, since they are at the basis of several large scale industrial processes as (7) : i, selectoforming and dewaxing, which represent selective n-paraffin cracking out of specific petroleum cuts; ii, xylene isomerization, ethylbenzene production and toluene disproportionation, in which product selectivity is the major selectivity enhancing parameter; iii, methanol-to-gasoline. On the other hand, the enhanced selectivity in selected hydrocarbon conversion reactions has been used to estimate the void size and structure of newly synthesized or modified zeolites. The hydroconversion of n-decane (18), the isomerization of m-xylene (19,20), the disproportionation of alkyl benzenes (20,21) seem to be very promising candidates in this respect.

From the organo-chemical point of view, however, the potential of acid zeolites remains largely unexplored. Moreover, it remains uncertain how useful the concepts derived for hydrocarbons can be to predict the reaction behaviour of molecules containing specific functional groups. Therefore, it has been the aim of the present study to determine how different zeolites can direct the selectivity in the conversion reactions of short chain aliphatic carboxylic acids and of their primary decomposition products. Reactions were performed in vapor phase continuous flow reactors at atmospheric pressure, in conditions free of inter- and intracrystalline diffusion effects. The information is withdrawn from product distributions, which are compared at identical conversions. The zeolites used have been extensively characterized by common physico-chemical techniques.

Chemistry of short chain carboxylic acids and ketones over non-zeolitic catalysts

The bimolecular "ketonization" reaction of short chain aliphatic carboxylic acids has been reported to occur over the oxides of cerium and thorium as catalysts (22) :

$$2\ R{-}COOH \xrightarrow[\substack{CeO_2 \\ ThO_2}]{673K} R{-}\overset{\displaystyle O}{\underset{\displaystyle \|}{C}}{-}R\ +CO_2 +H_2O$$

In this reaction oxygen is eliminated via decarboxylation and dehydration reactions. The primary formed ketones can further undergo self-condensation reactions. Aldol condensations which occur readily with acetone in mineral acids can be schematically represented as follows :

The main product of self-condensation of acetone is phorone (PO), formed from three molecules of feed (pathway A-C) (22). The reaction proceeds via mesityl oxide (MO) as intermediate (reaction A). In concentrated mineral acid, however, also three acetone molecules condense but 1,3,5-trimethylbenzene (1,35 MBz) (pathway A-B-D) is the main product (22). When PO reacts in the presence of sulfuric acid cyclisation into isophorone (iPO) occurs (pathway E) (23). On heterogeous catalysts of the mixed oxide type in the vapor phase, the synthesis of variable amounts of MO and iPO out of acetone have been reported (24).

Some of the intermediates are susceptible to thermal cracking, MO being a notable exception (25). iPO in this way can be converted into xylenols, with 3,5 dimethylphenol (3,5 MPh) as most abundant isomer (26) :

(3,5MPh)

Only upon heating of acetone itself above 773 K, it undergoes pyrolysis into ketene ($CH_2 = C = O$) and methane (27).

II. CONVERSION OF ACETIC ACID ON ACID ZEOLITES

Reactions of acetic acid on zeolite H-ZSM-5

Carboxylic acids in severe reaction conditions are readily converted into aromatic hydrocarbons over H-ZSM-5 zeolites (28). At lower reaction temperatures and for H-ZSM-5 catalysts with intermediate acid site density (when 1 to 2 % of the lattice T-atoms are Al), high selectivities towards xylenols have been reported (29).

TABLE 1
Reactions of acetic acid[1] on HZSM-5 (100)[2]

	593	633
Reaction temp. (K)	593	633
Conversion (%)	47	97
Product distribution (wt %)		
CO	0.6	1.0
CO_2	10.8	12.9
acetone	22.5	15.1
p-cresol	1.0	2.5
2,4-xylenol	39.5	13.8
3,4-xylenol	6.6	4.8
C9-phenol	2.3	6.6
C10-phenol	3.0	8.9
hydrocarbons	13.7	34.5
Hydrocarbon distribution (wt %)		
$C1_=$	1.0	0.8
$C2_=$	1.7	3.0
$C3_=$	5.4	11.3
$C4_=$	51.3	20.7
aromatics	39.0	61.1
% iso in $C4_=$	78.8	53.9

1, WHSV = 0.1; 2, Si/Al = 100

Typical product distributions from acetic acid on such a zeolite are shown in Table 1 and illustrate this point. Even on such a zeolite, the high yields of xylenols disappear at high reaction temperatures at the expense of hydrocarbons. Acetone, the expected primary intermediate is observed. All other intermediates from acetone such as MO, PO and iPO do not desorb from the pore system of the ZSM-5 zeolite.

The consecutive formation of the individual reaction products is illustrated in Figs. 1 and 2 with acetic acid and acetone as feeds, respectively.

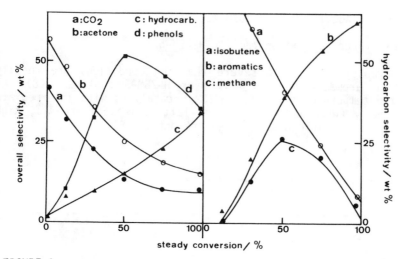

FIGURE 1
Overall reaction selectivity at steady conversion of acetic acid on
H-ZSM-5 (100) at 593 K.

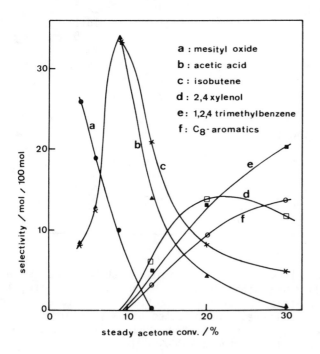

FIGURE 2
Overall reaction selectivity at steady conversion of acetone on H-ZSM-5
(100) at 620 K.

From both figures, the following relatively detailed reaction network can be derived :

$$CH_3-COOH$$
$$\downarrow 2\times$$
$$CH_3-CO-CH_3+CO_2+H_2O$$
$$\downarrow 2\times$$
$$\left[CH_3-\overset{O}{\overset{\|}{C}}-CH_2-\overset{OH}{\overset{|}{C}}-(CH_3)_2\right](DAA)$$

$$CH_3-COOH+\text{(isobutene)} \qquad (CH_3)_2C=CH-\overset{O}{\overset{\|}{C}}-CH_3+H_2O$$
$$\downarrow 2\times \qquad\qquad\qquad\qquad (MO)$$
$$\qquad\qquad\qquad\qquad +CH_3-CO-CH_3$$
$$\qquad\qquad\qquad\qquad -H_2O$$
$$\left[(CH_3)_2\underset{(PO)}{C=CH}-\overset{O}{\overset{\|}{C}}-CH=C(CH_3)_2\right]$$

$$CH_4 + (3,5\,MPh) \leftarrow (iPO) \rightarrow (OH)$$

$$(3,5\,MPh) \quad (iPO)$$

$$(2,4\,MPh)$$

$$-H_2O$$

$$(1,2,4\,MBz)$$

The carboxylic acid on the H-ZSM-5 (100) zeolite just as in a mineral acid is "ketonized" and acetone is formed. The machanistic aspects of this reaction step have been discussed earlier (29) and are not relevant for the present purpose. The formation of C8-aromatics, and 1,2,4-trimethylbenzene (1,2,4 MBz) can be explained by assuming that the classical intermediates are formed. However, none of these intermediates is able to desorb from the ZSM-5 matrix. Diacetone alcohol (DAA) which is the common acetone condensation product (22), is either cracked or dehydrated. In the former case, isobutene is formed together with acetic acid. The former molecule is dimerized, while the latter reenters the reaction cycle. The DAA cracking reaction also occurs over supported H$_3$PO$_4$ (25). This reaction has also been invoked to explain the similarity of the acetone and methanol cracking products on very acidic ZSM-5 at high reaction temperatures (672 K)(30). In the latter case, the existence of the classical intermediates MO, PO and iPO are postulated. The formation of 1,2,4 MBz (1,2,4-trimethylbenzene) out of iPO can be understood via classical carbenium ion rearrangements and dehydration (29). A dehydrodemethylation reaction of iPO gives 3,5-dimethylphenol (3,5 MPh) as primary xylenol. The high yields of the 2,4-isomer (Table 1) can be explained by a secondary isomerization reaction which representes then an example of product or restricted transition state selectivity with an oxygen-containing molecule. The simultaneous formation of methane and phenols, which is strong evidence for this dehydrodemethylation reaction is clearly visible in Fig. 1.

The yield of phenolics out of acetic acid on H-ZSM-5 (100) is further dependent on the impurity iron content of this zeolite (Table 2). On an iron-free catalyst distinctly less phenolics are fomed. In view of this, the dehydrodemethylation reaction was considered to be catalyzed by occluded iron oxide (29), in disagreement with an earlier proposal (31). A potential reaction mechanism is therefore the following :

TABLE 2
Acetic acid conversion on H-ZSM-5 (100)[1]

	iron-free	iron-containing[2]
conv./%	50	76
product distribution/%		
isobutene	13.8	15.8
acetone	60.0	37.0
aromatics	3.8	12.8
phenolics	22.4	34.4

1, Si/Al = 100; 2, 400 ppm of Fe

Influence of the site density of H-ZSM-5 on the product selectivity from acetic acid and acetone

Since the acid-catalyzed "ketonization" of acetic acid is a bimolecular reaction, and the acetone condensation involves at least three feed molecules, it can be anticipated that the reaction selectivity in H-ZSM-5 will, _inter alia_, be determined by the site density and thus by the Al content of the zeolite. Fig. 3 shows for H-ZSM-5 zeolites with low _site density_ that mainly acetone and hydrocarbons are formed. Over ZSM-5 zeolites with a high proton concentration mainly acetone is formed. Since these selectivities refer to 50 % steady acetic acid conversion, it follows that the latter materials have rapidly deactivated and only the primary reaction product

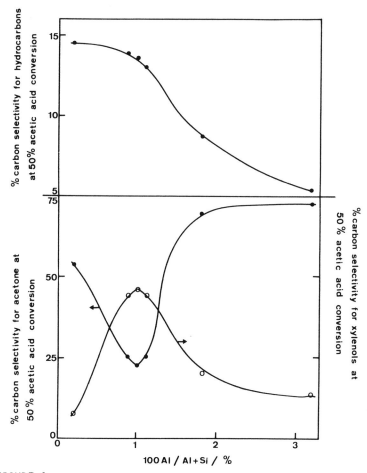

FIGURE 3
Product selectivity from acetic acid over H-ZSM-5 with different site densities or amount of aluminium in the framework.

is preferentially desorbed. At intermediate site densities (corresponding roughly to 1 proton for every two intersections) the selectivity for phenolics is highest. In terms of the proposed reaction network, this suggests that for high alumina ZSM-5, the DAA formed is mainly cracked or further polymerized into coke. On very low alumina ZSM-5, acetone oligomers are probably formed which are then catalytically cracked into hydrocarbons. Only at intermediate Al content of the zeolite (1 site for 2 pore intersections), the site density is such that the proposed sequence of events (DAA → MO → PO) can occur neatly, free of secondary reactions. After cyclization of PO into iPO, either xylenols or 1,2,4 MBz are formed depending on the reaction temperature. Since per acid site the rate of disappearance of acetic acid or acetone when both are used as feed, is identical on a given zeolite, this indicates that the rate determining event is not located in the ketonization reaction. The present data do not allow to be more specific on the nature of this step but it results from this section that H-ZSM-5 zeolites with intermediate degree of dilution of the Brønsted sites direct the conversion of acetic acid into the formation of specific products which are not formed on other non-zeolitic catalysts.

Acetic acid decomposition on acid zeolites

        To determine how far the formation of phenolics out of acetic acid is dependent on the nature of the zeolite pores, an extensive screening study using zeolites with different structure and degree of

TABLE 3
Steady state conversion of acetone on acid zeolites[1]

Zeolite[2]	% conversion at	
	520 K	608 K
H-L	7	10
US-Y[4]	35	40
H-X-30[3]	5	28
H-BETA	15	45
H-PHI	3	9
H-MOR	1	3
H-OFF	3	9
H-ZSM-35	4	10
H-ZSM-11	20	35
H-ERI	3	4
H-ZSM-34	6	8

1, WHSV = 0.7 h^{-1}; 2, materials described in ref. 18; 3, degree of H^+ exchange is 30 %; 4, ultrastable Y.

dilution of the Brønsted sites was performed. The steady activity for several zeolite structures is given in Table 3. Most of these materials deactivate very rapidly as shown by their temperature independent conversions at steady state. As far as the formation of phenolics is considered, Fig. 4 shows that only significant amounts are formed on H-ZSM-5 and H-ZSM-11 zeolites. Both materials are end-members of the so-called pentasil-family of zeolites (32). The amount of xylenols is even higher on H-ZSM-11 (100) than on the H-ZSM-5 with the same Si/Al ratio. This enhanced selectivity may reflect the presence of a set of pore intersections which are slightly larger than those in ZSM-5. Thus it seems that another key step in the reaction is the cyclisation of phorone into isophoron.

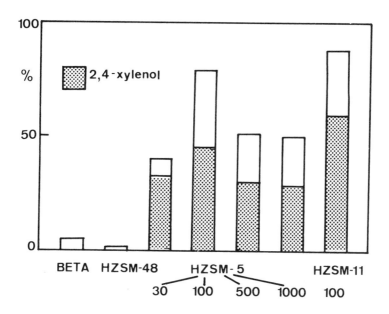

FIGURE 4
Maximum selectivity for phenolics from acetic acid on different zeolite structures (with Si/Al ratios ranging from 30 to 1000 for H-ZSM-5).

The reaction selectivity on the most active zeolites of Table 3 is shown in Fig. 5. Typical product distributions are shown for an highly acidic large pore zeolite (USY), a large pore zeolite with low acidity (H-X-30), a high-silica large pore size zeolite (H-Beta) and a highly acidic pentasil zeolite (H-ZSM-11(30)). On the low acidity large pore material, mesityl oxide and isophorone are the major products, their selectivity being only conversion dependent. On the same structure, but with higher acid strength, mesitylene (1,3,5-trimehylbenzene) is formed mainly at the expense of isophorone.

This effect of acid strength on product selectivity is in no way different from that observed in mineral acids (see higher and ref. 22). On the acidic H-ZSM-5(30), cracking of diacetone alcohol into isobutene is important and is seen to compete with the formation of phenolics <u>via</u> a series of intermediates which do not desorb. At increasing conversion, the selectivity shifts from phenolics to 1,2,4-trimethylbenzene as explained earlier. On H-Beta, the selectivity is intermediate between USY and H-ZSM-11, as could have been expected from its intermediate pore size.

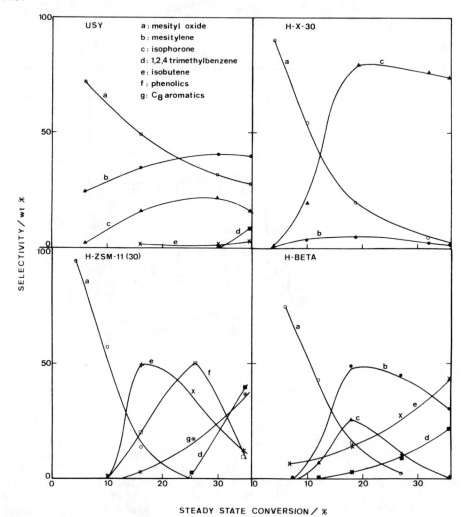

STEADY STATE CONVERSION / %

FIGURE 5
Conversion of acetone over H-zeolites with different structure and/or acid strength and site density (WHSV = 0.7 h^{-1}; reaction temperature between 520 and 610 K)

The results of acetic acid conversion collected using different zeolites can be summerized as follows :

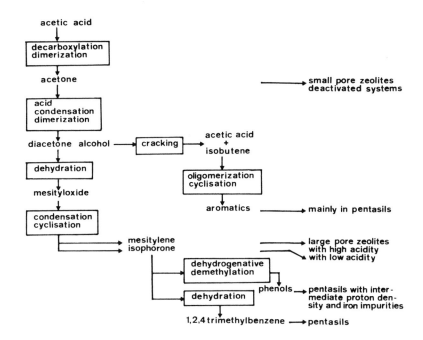

Thus it seems that the nature of the zeolite, its acid strength and/or site density as well as its impurity content in the conversion of acetic acid are important selectivity directing parameters. Only the medium pore zeolites impose constraints on the selectivity of this reaction.

III. ACID CATALYZED CONVERSION OF PROPIONIC ACID ON ZEOLITES

In analogy to the reactions of acetic acid, similar sequences can be written starting from propionic acid :

$$CH_3CH_2COOH \xrightarrow{H^+} CH_3CH_2\overset{+}{C}O + H_2O$$
$$\searrow CH_3CH_2COO^- + H^+$$

$$CH_3CH_2COCH_2CH_3 + CO_2$$

$$CH_3CH_2\underset{\overset{|}{+}}{\overset{OH}{C}}CH_2CH_3 \quad \downarrow H^+$$

$$CH_3-CH_2-\underset{\overset{||}{O}}{C}-\underset{CH_3}{CH_2}$$

$$\rightarrow CH_3CH_2\underset{\overset{||}{O}}{C}\overset{+}{C}-\overset{CH_2CH_3}{\underset{CH_2CH_3}{\overset{|}{C}}} + H_2O$$

$$\downarrow -H^+$$

Thus, it is seen that the expected primary ketonization product is 3-pentanone and as phenolics 2,3- or 2,4-methylethylphenols should be formed. It is also seen that the intermediates in the formation of these phenols are substantially more bulky than those proposed for acetic acid. On most zeolites a fast deactivation due to ketone polymerization is therefore expected. This is in fact observed and only on a few zeolites (pentasils and offretite-erionite materials) can an appreciable conversion be observed. Pertinent selectivity data for these families of zeolites are shown in Fig. 6. On zeolite T, a member of the offretite-erionite family of zeolites (18), the decomposition stops right after the ketonization event. The origin of ethylene at high conversions is not unequivocally established at this stage of the study. On the H-ZSM-5 (100) catalyst which was most selective for phenols in the acetic acid conversion, the expected phenols are observed mainly at high conversions. The selectivity for these products has, however, significantly decreased at the expenses of aromatics. Although the origin of these aromatics is difficult to trace, the present results show that the cyclisation of a branched C10-ketone is indeed difficult to perform and the effect of site dilution has vanished completely. Indeed, in Fig. 7, it is shown that the maximum in the rate of acetic acid decomposition over H-ZSM-5 zeolites with various Al contents has vanished completely when propionic acid is used as feed. This illustrates the obvious phenomenon that shape-selectivity is only for a given feed determined by the zeolite, while for a given zeolite the selectivity is also substrate-dependent. The latter effect is much less pronounced for hydrocarbons, since it has been demonstrated for at least two cases that irrespective of the nature of the feed hydrocarbons, the product selectivity remains virtually unchanged (28,33).

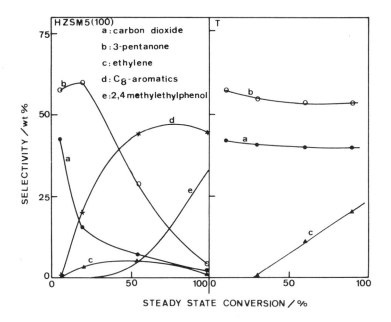

FIGURE 6
Product selectivities from propionic acid on two acid zeolites at different temperatures and for WHSV = 0.2 h^{-1}

IV. CONVERSION OF BUTYRIC ACID OVER ACID ZEOLITES

The addition of one more carbon atom to the aliphatic chain of the carboxylic acid feedstock changes the selectivity pattern completely. Most acid zeolites including the pentasils show an initial activity which declines very rapidly.

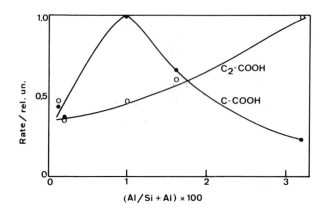

FIGURE 7
Comparison of the rate of decomposition of acetic and propionic acid (in relative units) over H-ZSM-5 with different Al contents.

From an extensive screening study, it results that only on zeolites containing the erionite cage as structural element, a reasonably high steady state activity remains. Moreover, such zeolites, including ZSM-34, zeolite T and erionite can be very easily regenerated by treatment with air at reaction temperature (600 K).

The selectivity for a typical zeolite of the offretite-erionite family (zeolite T) in the butyric acid conversion reaction is given in Fig. 8 at different times on stream.

Initially, a complex mixture of hydrocarbons appears but at steady state 4-heptanone is the major product together with non-negligeable amounts of butyric anhydride. The latter is a dehydration product from butyric acid and is formed at the external surface of the crystals of zeolite T. Indeed, when this zeolite is treated with triphenylchlorosilane so as to poison in a selective way the hydroxyls present at the outer surface of the crystals, the amount of anhydride formed decreases considerably. The ketone observed is definitely the primary product of a classical ketonization reaction.

It remains to be explained why this C7-ketone is only formed in significant yields on zeolites containing the erionite cage. On such zeolites the cage effect (11b, 16) has been demonstrated with normal paraffins. In a similar way as for C7 and C8 paraffins a low-energy trap may exist for two butyric acid molecules on such zeolites. Given the dimensions of the erionite cages, both molecules will have to be adsorbed in an end-to-end configuration. When the reaction mechanistically occurs as with acetic acid (29), an acyl carbenium ion has to interact with butyric acid. Both species when trapped in an

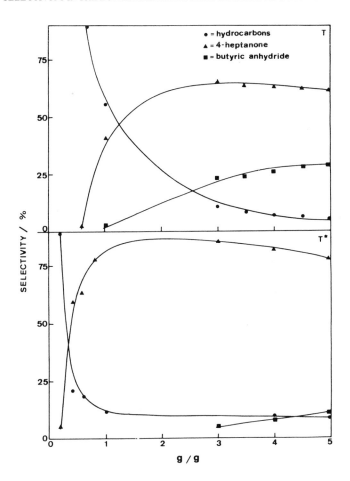

FIGURE 8
Product selectivity from butyric acid over H-T zeolite at 600 K at incrasing time on stream (expressed as g butyric acid fed to the reactor per g of catalyst). T* represents a H-T zeolite treated with triphenyl-chlorosilane.

erionite cage should then be in an ideal position to react. This can be schematized as follows :

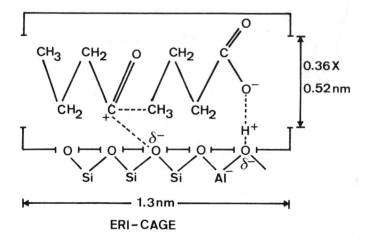

ERI-CAGE

Zeolites devoid of this specific structural element catalyse this reaction far beyond the dimerization stage of the butyric acid. Under mild conditions the catalyst rapidly deactivates since aldol condensations of 4-heptanone probably occur in an uncontrolled way. Under severe reaction conditions, this sorbed polymer is then cracked into a mixture of water and hydrocarbons.

V. CONCLUSIONS

The present paper illustrates the selectivity-directing role of a zeolite matrix in the acid catalyzed conversions of short chain aliphatic carboxylic acids.

Acetic acid first undergoes a ketonization reaction with acetone formation, just as on any other catalyst for this reaction. On large pore zeolites, as e.g. faujasite, an acid catalyzed "aldolization" reaction of the ketone formed occurs, and depending on its acid strength either 1,3,5-trimethylbenzene or mesityl oxide and phorone are main products. This chemistry is identical to that occuring in mineral acids of different strength. On medium pore zeolites, and more specifically on pentasil zeolites with intermediate Brønsted site concentration (1 site for every 2 pore intersections) containing also iron at the impurity level, high yields of xylenols are obtained. The key step in such environment seems to be the preferential formation of isophorone followed by its dehydrogenative demethylation. In severe reaction conditions water-elimination from the isophorone intermediate gives rise to the formation of 1,2,4-trimethylbenzene. In zeolites with pore size intermediate between these two extremes, also intermediate selectivities are observed.

With propionic acid a similar chemistry occurs, but since the transition state of the key intermediate is bulkier than in case of acetic acid, the absolute yields of phenolic products are much lower.

With butyric acid, this transition state can not be formed any longer and no phenolics are found. A selective ketonization reaction with 4-heptanone formation occurs on zeolites that contain the erionite cage. This selectivity can be rationalized with the help of the so-called cage-effect, found for n-paraffin cracking.

ACKNOWLEDGMENT

P.A.J. and J.A.M. acknowledge N.F.W.O. for a research position as "Onderzoeksleider" and "Aspirant", respectively. This research was sponsored by the Belgian Government in the frame of a concerted action on catalysis.

REFERENCES

1. Derouane, E.G., "Intercalation Chemistry", M.S. Whittingham and A.J. Jackson, Eds., Academic Press, New York, 1982, p. 101.
2. Csicsery, S.M., ACS Monogr. 171, "Zeolite Chemistry and Catalysis", J.A. Rabo, Ed., A.C.S., Washington, 1976, p. 680.
3. Weisz, P.B., Pure Appl. Chem., 52, 1980, 2091.
4. Derouane, E.G., "Catalysis by Zeolites", B. Imelik et al., Eds., Stud. Surf. Sci. Catal., 4, Elsevier, Amsterdam, 1980, p. 5.
5. Derouane, E.G., "Catalysis on the Energy Scene, S. kaliaguine and A. Mahay, Eds., Stud. Surf. Sci. Catal., 19, Elsevier, Amsterdam, 1984, p. 1.
6. Csicsery, S.M., Zeolites, 4, 1984, 202.
7. Haag, W.D., Lago, R.M. and Weisz, P.B., Faraday Disc. Chem. Soc., 72, 1982, 317.
8. Dwyer, J., Chem. Ind., 7, 1984, 229.
9. Derouane, E.G., "Zeolites : Science and Technology", F.R. Ribeiro et al., Eds., Nato ASI Ser. E, 80, Martinus Nijhoff Publ., 1984, p. 347.
10. Weisz, P.B. and Frilette, V.J., J. Phys. Chem., 64, 1960, 382.
11a.Rabo, J.A., Bezman, R. and Poutsma, M.L., Acta Phys. Chem., 24, 1978, 39.
11b.Chen, N.Y., Lucki, S.J. and Mower, E.G., J. Catal., 13, 1969, 329.
12. Derouane, E.G. and Gabelica, Z., J. Catal., 65, 1980, 486.
13. Derouane, E.G., Gabelica, Z. and Jacobs, P.A., J. Catal., 70, 1981, 238.
14. Derouane, E.G., Dejaifve, P., Gabelica, Z. and Vedrine, J.C., Faraday Disc. Chem. Soc., 72, 1981, 331.
15. Mirodatos, C. and Barthomeuf, D., J. Catal., 57, 1979, 136.
16. Gorring, R.L., J. Catal., 31, 1973, 13.
17. Mirodatos, C. and Barthomeuf, D., J. Catal. 93, 1985, 246.

18. Martens, J.A., Tielen, M., Jacobs P.A. and Weitkamp, J., Zeolites, 4, 1984, 98.
19. Dewing, J., J. Molec. Catal., 27, 1984, 25.
20. Guisnet, M., "Catalysis by Acids and Bases", B. Imelik et al., Eds., Stud. Surf. Sci. Catal. 20, Elsevier, Amsterdam, 1985, p. 283.
21. Karge, H.G., Sarbak, Z., Hatada, K., Weitkamp, J. and Jacobs, P.A., J. Catal., 82, 1983, 236.
22. Goldstein, R.F. and Waddams, A.L., "The Petroleum Chemicals Industry", Spon, London, 1967, p. 354-358.
23. Szabo, D., Acta Chim. Acad. Sci. Hung., 33, 1962, 925.
24. see e.g., U.S.P. 4,165,339 and 4,458,026.
25. McAllister, S.H., Bailey, W.A. and Bouton, C.M., J. Am. Chem. Soc., 62, 1940, 3210.
26. B.P. 584,256, assigned to Schell Dev.
27. Goldstein, R.F. and Waddams, A.L., "The Petroleum Chemicals Industry", Spon, London, 1967, p. 361.
28. Chang, C.D. and Silvestri, A.J., J. Catal. 47, 1977, 249.
29. Servotte, Y., Jacobs, J. and Jacobs, P.A., Proceed. Int. Symp. Zeolite Catalysis, Siofok (Hungary), 1985, Act. Phys. Chem. Szegedensis, p. 609.
30. Chang, C.D., Lang, W.H. and Bell, W.K., "Catalysis of Organic Reactions", W.R. Moser, Ed., Chem. Ind., 5, Marcel Dekker, New York, 1981, p. 73.
31. Chang, C.D., Chen, N.Y., Kolnig, L.R. and Walsh, D.E., Prepr. ACS Div. Fuel Chem., 28, 1983.
32. Kokotailo, G.T. and Meier W.M., "Properties and Applications of Zeolites", R.P. Townsend, Ed., Chem. Soc., London, 1979, p. 133.
33. Martens, J.A., Weitkamp, J. and Jacobs, P.A., "Catalysis by Acids and Bases", B. Imelik et al., Eds., Stud. Surf. Sci. Catal. 20, Elsevier, Amsterdam, 1985, p. 427.

SELECTIVITY INDUCED BY THE VOID STRUCTURE OF ZEOLITE BETA AND FERRIERITE IN THE HYDROCONVERSION REACTION OF N-DECANE

J.A. Martens[1], J. Perez-Pariente[2] and P.A. Jacobs[1]
1. Katholieke Universiteit Leuven, Laboratorium voor Oppervlaktechemie, Kardinaal Mercierlaan 92, B-3030 Leuven (Heverlee), Belgium.
2. on leave from : Instituto de Catalysis y Petroleoquimica, Madrid, Spain.

ABSTRACT

Ferrierite, BETA and ultrastable Y zeolites with restricted differences in chemical composition, crystallite size, platinum content and prepared via discrete methods were transformed into bifunctional catalysts and catalytically tested in the hydroconversion reaction of decane.
From a detailed inspection of the product distribution, criteria were selected which enabled to characterize each structure. The Ferrierite zeolites were easily distinguishable from the other large pore zeolites. The discrimination of zeolite ultrastable Y from zeolite BETA required the simultaneous consideration of all selected criteria. The limited variation of the physical and physico-chemical properties of a given zeolite structure was not found to influence the product distribution in a significant way.

INTRODUCTION

Presently, the discovery of a new zeolitic material is claimed very often in the patent literature. These new materials are in many instances described only by their chemical composition, their more or less specific X-ray diffraction pattern and their adsorption capacity for probe molecules such as n-hexane, cyclohexane and water. The crystallographic structure is however never known at the time the new material is patented. Indeed, the resolution of the structure with X-ray diffraction techniques is time-consuming and requires pure, highly crystalline large, single crystals of the material.

Catalytically, zeolites more than any other heterogeneous catalyst are able to improve the selectivity of a reaction towards the desired products due to their specific channel structure. This effect is called "molecular shape-selectivity" of the zeolite catalyst. In case of acid catalysis the active protons are for the major part located in the intracrystalline void volume. The size and shape of the

R. Setton (ed.), Chemical Reactions in Organic and Inorganic Constrained Systems, 115–129.
© 1986 by D. Reidel Publishing Company.

diffusing reactant and product molecules match closely the pores of the
zeolite, which may give rise to several types of selectivity
enhancement. Molecular shape-selective catalysis in zeolites was
reported first by Weisz and Frilette (1) in 1960. Since then the
subject has been reviewed several times (2-8).

Clearly, the knowledge of the dimensions and the structure of
the pores, cages and windows of the zeolite is essential in searching
for catalytic applications for a zeolitic material and in designing
tailor-made materials for a given reaction. One approach to get insight
in the void structure of a zeolite is by using a catalytic test
reaction that is very sensitive to structural changes in the zeolite
cavities. The determination of the "constraint index" (CI) is such a
method (9). A mixture of equal weight of normal hexane and
3-methylpentane is cracked over the zeolite sample in well defined
experimental conditions. The constraint index is then obtained as :

$$CI = \frac{\log \text{(fraction of n-hexane remaining)}}{\log \text{(fraction of 3-methylpentane remaining)}}$$

The constraint index of medium-pore zeolites such as ZSM-5 is high (CI
= 8.3) (9). This is attributed to steric constraints exerted by the
intracrystalline cages or pores on the bulky transition states which
are required for the occurrence of a hydride transfer reaction (10).
The constraint index of a zeolite was found to correlate well with
certain catalytical properties such as its para-selectivity in xylene
isomerization.

The isomerization of m-xylene was also advanced as a
potential reaction to probe the size of the pores of new zeolites with
unknown structure or of modified known zeolites (11). In this method
the ratio of the relative rates of formation of p-xylene and o-xylene
is used as a numerical value that is characteristic for a given pore
size. An important drawback of this method is the occurrence of side
reactions such as disproportionation and coke formation. These
reactions are suppressed by the favourable pore size and structure of
ZSM-5 and this is the reason why this method has exclusively been used
for ZSM-5-type materials.

Gnep et al., (12) extended this method to larger pore
zeolites and suggested that the relative rate of coke formation and
disproportionation with respect to isomerization of m-xylene is a
measure of void space availability in the zeolite. Another proposal by
the same authors is to use the disproportionation of
1,2,4-trimethylbenzene as a test reaction. The formation of the bulkier
isomer 1,3,5-trimethylbenzene is shown to be restricted in the pores of
mordenite, but not in the cages of Y-type zeolites.

Recently, the hydroconversion reaction of n-decane for
probing the zeolite channels was also proposed (13). This choice was
based on several arguments. First, a high number of feed isomers with

different degree of branching and with different sizes of the branching element can be obtained by isomerization of a long chain paraffin. Normal decane, e.g., has 73 branched isomers, containing methyl-, ethyl- or propyl-side chains. Also numerous cracked products can be obtained, varying in carbon number and degree of branching. Hence the distribution of the reaction products can be very sensitive to void volume of the catalyst used. Another argument in favour of this reaction is that it is performed with a bifunctional catalyst, containing next to the acidic centers, a hydrogenation- dehydrogenation function, which prevents the deposition of heavy coke molecules on the catalyst. An additional advantage of the use of normal paraffins as probe molecules is that they can diffuse through 8-membered ring zeolites (14). Consequently, all molecular sieves with potential application in hydrocarbon reactions can be screened with this reaction since the intracrystalline void volume of such zeolites consists either of cages with windows consisting of 8- or 12-membered rings of oxygen ions or of channels, intersecting or not, and circumscribed by 10- or 12-membered rings of oxygens ions.

In the first report (13) on the n-decane hydroconversion reaction for the exploration of zeolite voids, it was shown that when isomerization and hydrocracking of n-decane was performed in zeolites with different pore sizes and structures, pronounced differences in the composition of the reaction products were observed indeed. Based on the compositions of the isomerization and hydrocracking products obtained over 12 zeolite catalysts with different pore structures, five independent criteria were selected, which correlated well with the information available on the structure and dimensions of the intracrystalline void space of the respective zeolites :
  - the selectivity for multibranched feed isomers at the maximum isomerization conversion
  - the content of ethyloctanes in the monobranched isomers at 5 % conversion
  - the ratio of the yield of 3- to 4-ethyloctanes as a function of reaction temperature
  - the amount of 2- and 5-methylnonane present in the methyl-branched isomers at 5 % isomerization conversion
  - the absolute yield of isopentane formed in the hydrocracking reaction.
These five criteria could then give evidence on the structure of 7 zeolites with unknown crystallographic structures : ZSM-12, ZSM-34, ZSM-47, ZSM-48, zeolite BETA and zeolite PHI.

For a given zeolite framework, however, the composition of the reaction products from long chain paraffins such as n-decane is also sensitive to the number and strength of the Brønsted acid sites, the platinum loading and its degree of dispersion, or in other words to the balance between the metal and the acid function (15). On the other hand, the composition of the reaction products might also be influenced by the zeolite crystal size (10). Furthermore, changes in reaction temperature may also amplify or weaken the observed shape-selectivity (2).

If all these parameters would be influencing strongly the composition of the reaction products, the interpretation of the results obtained with the n-decane test reaction would be not straightforward. In the present study, the influence of chemical composition, synthesis method, minor changes in crystal size and reaction temperature on the product distribution is evaluated for the following zeolite frameworks : ferrierite (FER), zeolite BETA and ultrastable Y (USY).

The crystal structure of ferrierite contains 10-membered ring channels parallel to the crystallographic c-axis and intersecting with 8-membered ring channels which are parallel to the b-axis (16,17). The crystal structure of zeolite BETA remains unresolved, although it was synthesized as early as 1967 by Mobil researchers (18). According to the n-decane test reaction, it consists of pores with 12-membered rings and lobes (13). Ultrastable zeolite Y is a faujasite structure with mesopores throughout the crystals, created during the hydrothermal treatments (19).

EXPERIMENTAL

Table 1 contains information on the samples of ferrierite, zeolite BETA and ultrastable zeolite Y. The materials ZSM-35 and ferrierite have very similar XRD-patterns (20). Samples with this XRD-pattern are arbitrarily denoted as ferrierites (FER), although FER/1 and FER/2 were obtained according to a procedure for the preparation of ZSM-35. The three ferrierite samples are synthesized with different templates and the Si/Al_2 ratio of the gel of FER/3 was about three times higher than for FER/1 and FER/2.

All zeolite BETA samples were synthesized with tetraethylorthosilicate as silica source. BETA/1 is a material synthesized as reported recently (22). BETA/2 to 5 samples were obtained according to a method described for the synthesis of ZSM-20. All samples showed the same degree of crystallinity, based on XRD.

Zeolite Y is made ultrastable according to a previously reported procedure (24). USY/1 and USY/2 are different batches, ultrastabilised separately by the same procedure.

The FER and BETA samples were calcined at 873 K and 773 K, respectively. All samples were exchanged with NH_4^+ by contacting the zeolites during 4 h under reflux conditions with an aqueous solution, 0.5 M in NH_4Cl. Pt is introduced in the zeolites either by impregnation or exchange of the $Pt(NH_3)_4^{2+}$ complex in aqueous solution. The samples were loaded with 0.5 or 1.0 % by weight of Pt. BETA/4 and /5 denote the same sample, impregnated and exchanged with the Pt-complex, respectively.

TABLE 1
Synthesis and characteristics of the zeolite catalysts used

Sample notation	method ref.	Synthesis template	Si/Al$_2$ ratio of gel	Si/Al$_2$ ratio of zeolite	Crystal size (μm)	Pt-loading weight %	Pt-loading technique
FER/1	20, ex.6	ethylenediamine	15	–	0.3x1	1	I[2]
FER/2	20, ex.19	pyrrolidine	15	–	–	1	I
FER/3	21, ex.8	piperidine	47	–	–	1	I
BETA/1	22	TEA[1]	30	28	0.2-0.3	1	I
BETA/2	23, ex.19	TEA	100	54	–	1	I
BETA/3	23, ex.19	TEA	20	12	0.5-1.0	1	I
BETA/4	23, ex.19	TEA	30	11	0.2-0.3	0.5	E
BETA/5	23, ex.19	TEA	30	11	0.2-0.3	0.5	I
USY/1	24	TEA	–	2.46	1	1	E
USY/2	24	TEA	–	2.46	1	0.5	E

1, tetraethylammonium; 2, I = impregnation of, and E = ion exchange with $Pt(NH_3)_4^{2+}$.

The zeolite powders were compressed and crushed in order to obtain pellets with a diameter of 0.3-0.5 mm. Activation of the catalyst was done in-situ in the reactor by oxydation in flowing oxygen at 673 K, followed by reduction in hydrogen at the same temperature. Isomerization and hydrocracking of n-decane was performed in a continuous flow tubular fixed bed microreactor with on-line analysis by capillary GLC. Product separation was done on a CptmSil5 capillary column of 50 m length. The reaction conditions are given in Table 2.

TABLE 2
Hydroconversion of n-decane - experimental conditions

Catalyst	P_{H_2} (MPa)	P_{n-C10} (kPa)	WHSV (h^{-1})
FER/1	0.1	1.3	0.3
FER/2	0.1	1.3	0.3
FER/3	0.1	1.3	0.3
BETA/1	1.0	1.3	0.4
BETA/2	0.1	1.1	0.7
BETA/3	0.4	1.1	0.8
BETA/4	0.7	0.8	0.6
BETA/5	0.7	0.8	0.6
USY/1	0.1	0.9	0.4
USY/2	0.7	0.8	0.6

RESULTS AND DISCUSSION

1. Requirements to obtain ideal bifunctional catalysts

For the application of the n-decane test, the porous structure should be transformed into an ideal bifunctional catalyst. Only for such materials are the two catalytic functions in balance and is the role of the metal phase restricted to catalyse exclusively hydrogenation-dehydrogenation reactions, while isomerization and cracking are catalysed by the acid sites (15). The reacting hydrocarbon molecules are then most sensitive to the geometrical constraints around the acid catalytic centers inside the zeolite pores (13). Moreover, on ideal bifunctional catalysts, the selectivity for isomerization of the feed n-paraffin is very high at low conversion levels. The formation of feed isomers is an essential condition for the application of four of the criteria of the n-decane test. The branching structure of the feed isomers reflects indeed very precisely the pore structure of the zeolite tested. On ideal bifunctional catalysts pure primary cracking of the feed isomers is observed, resulting in a symmetric molar distribution of the cracked products among their carbon numbers and no methane nor ethane are formed (15).

In order to transform acid zeolites into ideal bifunctional catalysts, loading with 0.5 % or 1 % by weight of platinum metal is often sufficient. This was shown for CaY (15), USY (24), ZSM-5 and ZSM-11 (25) as well as for zeolite BETA (22). A high selectivity for feed isomerization at low conversion and the virtual absence of methane and ethane in the cracked products at all levels of conversion are the prerequisites for the n-decane test to be applicable.

2. Determination of the nature of the feed isomers and hydrocracked products by shape-selective effects

In a previous report (13) it was shown that, depending on the zeolite crystal structure and on the free intracrystalline space available, the rate of formation of bulky feed isomers, like multibranched decanes and ethyloctanes is reduced. At 5 % feed conversion, 10-membered ring zeolites such as ferrierite, ZSM-5, ZSM-11 and clinoptilolite suppress completely the production of ethyloctanes and less than 10 % of the feed isomers are multibranched even when the yield of feed isomers is maximum. In similar conditions, the yield of ethyloctanes and multibranched decanes over USY is 12 and 35 %, respectively (13). The 12-membered ring zeolites and the cage-zeolites accessible through 8-membered rings show a behaviour intermediate between USY and the 10-membered ring zeolites. These differences are clearly confirmed in the present work (Figs. 1 and 2).

2.1. Ethyloctanes in the monobranched feed isomers

Fig. 1 shows the relative contribution of the ethyloctanes to the monobranched isomers at 5 % feed conversion for all samples tested. The different samples of zeolite BETA produce 5.5 to 7.5 % of ethyloctanes, which is markedly less than the 11.5 to 13 % obtained for zeolites USY/1 and USY/2. FER/3 produces no ethyloctanes, whereas FER/1 and /2 produce only traces of these compounds. According to this criterion, the three zeolite structures are easily distinguished and differences in Pt-loading, crystallite size, overall acidity and consequently of reaction temperature, influence these results only to a minor extent.

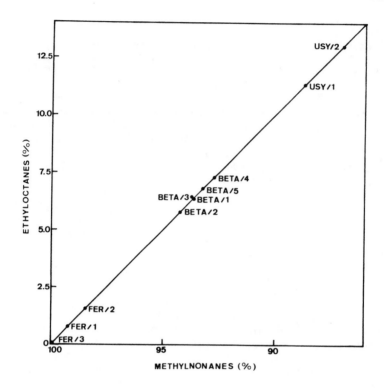

FIGURE 1
Contribution of the ethyloctanes and methylnonanes to the monobranched
isomers at 5 % conversion of n-decane (reaction conditions of Table 2).
Reaction temperatures (K) : FER/1, 423; FER/2, 448; FER/3, 430; BETA/1,
505; BETA/2, 429; BETA/3, 427; BETA/4, 446; BETA/5, 442; USY/1, 413 and
USY/2, 440.

2.2. Degree of branching of the feed isomers

        Fig. 2 shows the degree of branching of the feed isomers at
the highest conversion to feed isomers (Table 3). With the ferrierite
samples, the maximum yield of feed isomers was below 16 %, containing
less than 7 % multibranched ones. On the BETA and USY zeolites high
isomerization selectivity can be obtained. The latter fraction contains
then at least 30 % of multibranched decanes. The low yield of
multibranched isomers on the FER catalysts is probably not caused by
the low isomerization conversion levels. Indeed, FER/2 makes less
multibranched isomers at 16 % isomerization then FER/1 and FER/3 at
9 %.

TABLE 3
Conditions for highest isomerization conversion

Sample	reaction temp. (K)	yield of isomers (%)	conversion (%)
FER/1	434	9	71
FER/2	478	16	78
FER/3	441	9	63
BETA/1	546	48	67
BETA/2	465	38	59
BETA/3	465	56	82
BETA/4	475	15	59
BETA/5	473	25	50
USY/1	446	34	47
USY/2	474	30	57

This second criterion allows to separate the 10-membered ring zeolite from the two open pore materials. It is, however impossible to discriminate the different samples of each of the two open pore zeolites from each other using only this criterion.

2.3. Distribution of the methylnonanes

The amount of 2- and 5-methylnonane formed within the monobranched isomers is also extremely dependent on the zeolite framework used. It has been suggested that transition-state shape-selectivity is at the origin of the higher yields of 2-methylnonane in 10-membered ring zeolites (25), whereas the formation of this isomer is kinetically hindered in more "open" zeolites such as zeolite Y (15) and at low isomerization conversion should be significantly lower than its equilibrium value.

For all samples tested, the amount of 2- and 5-methylnonane found in the methylnonanes at 5 % conversion is given in Fig. 3. As expected for a 10-membered ring zeolite the FER samples produce more than 50 % of 2-methylnonane but less than 8 % of 5-methylnonane, which is the bulkier isomer. With BETA only between 23 and 32 % of 2-methyl-nonane and between 11 and 15 % of 5-methylnonane is obtained. The results with the two USY samples are very similar : almost 23 % of 2-methylnonane and 15 % of 5-methylnonane are obtained. It has been shown earlier that the hydroxyl groups at the crystal surface of some zeolites can distrub this distribution by secondary methyl-shifts (13). In case of zeolite BETA, however, the minor differences in crystal size do not seem to influence the product distribution. Indeed, samples BETA/1 and /3 which have different average crystallite sizes (Table 1), show the same 2-methylnonane yield. The reproducibility of the whole procedure is illustrated when BETA/4 and /5 are compared. These materials represent two different batches of the same sample prepared by the same procedure. The difference of the 2-methylnonane yield is lower than 4 %. Fig. 3 also shows that for most of the large pore zeolites, the 2-methylnonane content at 5 % isomerization is below thermodynamic equilibrium. BETA/2 and /5 for unknown reasons behave exceptionally.

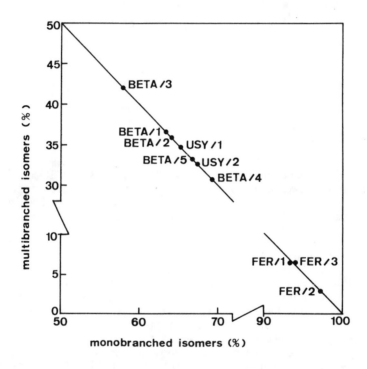

FIGURE 2
Mono- against multibranched isomers from n-decane at the highest conversion to feed isomers (reaction conditions of Table 2).

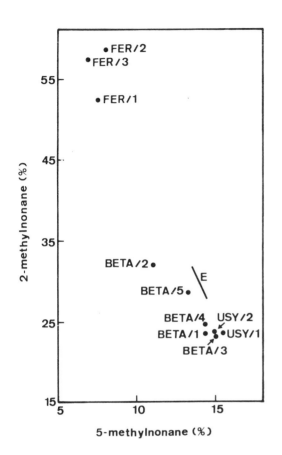

FIGURE 3
Yield of 2- and 5-methylnonane among the methylnonanes obtained at 5 % conversion of n-decane (reaction conditions of Table 2 - reaction temperatures see Fig. 1). E represents the thermodynamic equilibrium values in the temperature range of interest (26).

2.4. Isopentane yield in the hydrocracked products

The absolute yield of isopentane from decane is another parameter which is also very sensitive to structural effects and insensitive to secondary isomerization and cracking reactions. In large pores the hydrocarbon skeleton is highly branched before β-scission occurs. This results in a high probability for central β-scission and formation of much branched C5 fragments via hydrocracking of type A (27). Fig. 4 illustrates this for all FER, BETA and USY samples. The ferrierite catalysts produce only between 11 and 16 mol isopentane per 100 mol of feed hydrocracked. The samples of BETA on the other hand show a much higher selectivity for isopentane (± 55 mol), similar to that of the USY samples, except for BETA/1 which forms only 45 mol of isopentane. The ferrierite samples are again clearly distinguishable from BETA and USY, based on this criterion. BETA and USY again do not differ significantly from each other.

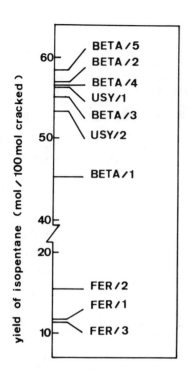

FIGURE 4
Yield of isopentane at 50 % cracking (reaction conditions of Table 2). Reaction temperatures (K) : FER/1, 434; FER/2, 478; FER/3, 441; BETA/1, 560; BETA/2, 471; BETA/3, 470; BETA/4, 475; BETA/5, 478; USY/1, 457 and USY/2, 485.

3. Evaluation of the criteria for zeolite characterization

All criteria but one separate nicely FER from the two other zeolites, but are unable to discriminate the latter two. This clearly indicates that zeolite BETA has large pores or cages. The formation of ethyloctanes is however significantly reduced in BETA compared to Y (Fig. 1), which indicates than its cavities or pores are smaller than the supercages of the faujasite structure. The catalytic similarity between both zeolites suggest a similar structure of the void space for both materials (cages accessible via windows). Since only the rate of formation of ethyloctanes is reduced, only the effective size of the cages can be smaller than in Y, their window aperture should be very similar.

The main differences between the FER samples are the following : i, they are obtained according to different synthesis methods (Table 1) and ii, the Si/Al_2 ratio of the synthesis gel of FER/3 was three times higher than for the other two. The results obtained with these three FER samples, tested in comparable conditions (Table 2), are very similar and the small differences in the results are not significant. From Figs. 1, 3 and 4, e.g., FER/3 appears to be the most shape-selective sample since it excludes totally the formation of ethyloctanes and produces the smallest amount of 5-methylnonane and isopentane. On the other hand, FER/3 produces more dibranched isomers than FER/2 (Fig. 2). It follows that for ferrierite-like materials, the synthesis method and the Si/Al_2 ratio is of minor influence on the product distribution.

When the same sample of zeolite BETA is either exchanged with the Pt complex (BETA/4) or impregnated (BETA/5), slightly different results are obtained in the same experimental conditions (Table 2). These differences however are considered not to be significant. From Figs. 1 and 3 it would appear that BETA/4 is more "open" than BETA/5, since it produces more ethyloctanes and less 2-methylnonane. The reverse could be deduced from Figs. 2 and 4, where it is seen that BETA/5 produces more multibranched isomers and isopentane.

The BETA/1 sample was tested at a temperature that was 50 to 90 K higher than the other BETA samples (Figs. 1 to 4). It shows a behaviour comparable to the others, except for its lower selectivity for isopentane (Fig. 4).

The zeolites BETA/2 and BETA/3 were synthesized following the same method, but the Si/Al_2 ratio of BETA/2 is 4.5 times higher than that of BETA/3 (Table 1). Both samples are loaded with the same amount of Pt by the same technique and tested at the same reaction temperatures (Figs. 1 to 4). To compensate for the higher acitivity of BETA/3 the partial pressure of H_2 was increased (Table 2). It should be noted that the reaction order in hydrogen equals - 1 (28). BETA/3 behaves as a slightly more open structure (Figs. 1 to 3), except for its lower selectivity for isopentane (Fig. 4).

The USY/1 and /2 samples were obtained according to the same procedure (Table 1). USY/2 was tested at a temperature which is 30 K higher than for USY/1. Only slightly different results are obtained with both catalysts. USY/1 produces some more multi-branched isomers (Fig. 2), 5-methylnonane (Fig. 3) and isopentane (Fig. 4), while USY/2 produces some more ethyloctanes (Fig. 1).

In conclusion, it appears that at least for the frameworks of ferrierite, zeolite BETA and USY, the nature of the synthesis method, restricted changes in chemical composition, the method of platinum loading, the amount of platinum and the reaction conditions do not obscure the results of the n-decane test reaction. When pronounced changes in chemical composition exist, however, as has been shown for zeolite Y (29), significant and systematic changes in the product distribution exist, reflecting the unit cell contraction which occurs upon dealumination. This work further shows that the criteria have to be used simultaneously in order to be able to discriminate between related void structures of the zeolite. It also results that only accurate information on the zeolite pore size and structure can be obtained from a relatively complex product distribution pattern.

ACKNOWLEDGMENTS

        J.A.M. and P.A.J. acknowledge N.F.W.O. for a research position as "Aspirant" and "Onderzoeksleider", respectively. J.P.P. is grateful to K.U.Leuven for a research grant. This work was performed in the frame of a concerted action on catalysis, sponsored by the Belgian Government.

REFERENCES

1. Weisz, P.B. and Frillette, V.J., J. Phys. Chem., 64, 1960, 382.
2. Weisz, P.B., Pure & Appl. Chem., 52, 1980, 2091.
3. Csicsery, S.M., in "Zeolite Chemistry and Catalysis", ACS Monograph No. 171, J.A. Rabo, Ed., Washington, 1976, 680.
4. Csicsery, S.M., Zeolites, 4, 1984, 202.
5. Derouane, E.G., in "Catalysis by Zeolites", B. Imelik et al., Eds., Elsevier, Amsterdam, 1980, Stud. Surf. Sci. Catal., 4, 1980, 5.
6. Derouane, E.G., in "Intercalation Chemistry", M.S. Whittingham and A.J. Jackson, Eds., Academic Press, New York, 1982, 101.
7. Derouane, E.G., in "Catalysis on the energy Scene", S. kaliaguine and A. Mahay, Eds., Elsevier, Amsterdam, 1984, Stud. Surf. Sci. Catal., 19, 1984, 1.
8. Derouane, E.G., in "Zeolites : Science and Technology", F. Ramôa Ribeiro et al., Eds., Nato ASI Series E, No. 80, Nyhoff, 1984, 347.
9. Frillette, V.J., Haag, W.O. and Lago, R.M., J. Catal., 67, 1981, 218.
10. Haag, W.O., Lago, R.M. and Weisz, P.B., Faraday Disc. Chem. Soc., 72, 1982, 317.
11. Dewing, J., J. Mol. Catal., 27, 1984, 25.
12. Gnep, N.S., Tejada, J. and Guisnet, M., Bull. Soc. Chim. France, 1982, 1-5.
13. Martens, J.A., Tielen, M., Jacobs, P.A. and Weitkamp, J., Zeolites, 4, 1984, 98.
14. Breck, D.W., Zeolite Molecular Sieves, Wiley, New York, London, Sydney, Toronto, 1974, 593.
15. Weitkamp, J., Erdöl, Kohle, Erdgas, Petrochemie und Brennstoff Chemie, 31, 1978, 13.
16. Vaughn, P.A., Acta Chrystallogr., 21, 1966, 983.
17. Kerr, I.S., Nature, 210, 1966, 294.
18. Wadlinger, R.L., Kerr, G.T., Rosinski, E.J., U.S.P. 3,308,069 (1967), assigned to Mobil Oil Corp.
19. Maugé, F., Auroux, A, Courcelle, J.C., Engelhard, Ph., Gallezot, P., and Grosmangin, J., in "Catalysis by Acids and Bases", B. Imelik et al., Eds., Elsevier, Amsterdam, 1985, Stud. Surf. Sci. Catal., 20, 1985, 91.
20. Plank, C.J., Rosinski, E.J. and Rubin, M.K., U.S.P. 4,016,245 (1977), assigned to Mobil Oil Corp.
21. Nanne, J.M., Post, M.F.M. and Stork, W.H.J., E.P. 12,473 (1980), assigned to Shell Int. Res.

22. Martens, J.A., Perez-Pariente, J. and Jacobs, P.A., Proceed. Int. Symp. Zeolite Catalysis, Siofok, Hungary, 1985, 487.
23. Valyocsik, E.W., E.P.A. 0,012,572 (1980), assigned to Mobil Oil Corp.
24. Steijns, M., Froment, G., Jacobs, P., Uytterhoeven, J. and Weitkamp, J., Ind. Eng. Chem. Prod. Res. Dev., 20, 1981, 654.
25. Jacobs, P.A., Martens, J.A., Weitkamp, J. and Beyer, H.K., Faraday Disc. chem. Soc., 72, 1982, 353.
26. Stull, R.D., Westrum, E.F. and Suike, G.C., "The Thermodynamics of Organic Compounds", Wiley, New York, 1969.
27. Martens, J.A., Jacobs, P.A. and Weitkamp, J., paper submitted to Applied Catalysis.
28. Steijns, M. and Froment, G., Ind. Eng. Chem. Prod. Res. Dev., 20, 1981, 660.
29. Jacobs, P.A., Martens, J.A., Beyer, H.K., in "Catalysis by Acids and Bases", B. Imelik et al., Eds., Elsevier, Amsterdam, 1985, Stud. surf. Sci. Catal., 20, 1985, 399.

COKE FORMATION ON PROTONIC ZEOLITES : RATE AND SELECTIVITY

M. GUISNET, P. MAGNOUX and C. CANAFF
UA CNRS 350, Catalyse en Chimie Organique, UER Sciences
40, avenue du Recteur Pineau
86022 Poitiers Cedex
France

ABSTRACT. The coking and the aging rates of four protonic zeolites
(HY, H mordenite, H offretite and HZSM-5) were determined during
n-heptane cracking at 450°C. On each of the zeolites, the coke was
characterized by conventional analysis (GC, HPLC, NMR, MS) of the
solvent extracts obtained after dissolution of the zeolite framework by
hydrofluoric acid. Coking, aging rates and coke composition are definitely dependent on the zeolite porous structure, which confirms that
coke formation is a shape-selective reaction. Thus the coking rate,
which is high when the available space near the active centers is
significant (e.g. larger cavities of HY) is slow due to steric constraints in the formation of reactional intermediates when this space is
reduced (e.g. ZSM-5). Deactivation is faster when molecule circulation
in the zeolite is unidimensional (activity highly dependent on pore
blockage) than when it is tridimensional. Coke is polyaromatic and very
heavy if the reactions which lead to its formation are but slightly
limited by steric constraints (e.g. HY, mordenite) ; it will be less
aromatic and lighter in the opposite case (gmelinite cages of offretite
and channels of ZSM-5 zeolite).

INTRODUCTION

   All the catalytic reactions of organic compounds are accompanied
by the formation of carbonaceous residues known as "coke". The rate of
this secondary reaction depends both on the operating conditions and on
the catalyst ; it can be much slower or close to the rate of the main
reactions : thus if the coke yield of a reforming unit is about 0.001 %
that of a cracking unit can attain 10 %. Rollmann and Walsh (1) have
shown that the greater the constraint index of a zeolite (which is a
characteristic of its porosity (2)) the smaller the coke yield : in the
same operating conditions it attains 1 % for large pore-size zeolites
(e.g. HY) and only about 0.05 % for intermediate pore-size zeolites
(e.g. ZSM-5) or for small pore-size zeolites (e.g. erionite).
   The activity of porous catalysts is decreased by coke deposits in
two different ways : site coverage and pore blockage (3-5) the deacti-

131

R. Setton (ed.), Chemical Reactions in Organic and Inorganic Constrained Systems, 131–140.
© 1986 by D. Reidel Publishing Company.

vation is faster in the second case, pore blockage being capable of
preventing access of the reactants to a significant number of active
centers. Zeolite deactivation by pore blockage will be obviously much
faster when molecule circulation is unidimensional than when access to
the active centers can occur by various paths (e.g. tridimensional
circulation).

If there are numerous works concerning coke formation rates and
catalyst deactivation (6,7) very few concerning the chemical nature of
coke are to be found. Coke characterization is limited most often to
the measurement of its atomic H/C ratio (7-10) . The value of this
ratio generally between 0.4 and 0.7 proves that coke is polyaromatic.
However in the case of zeolites, values higher than 1 have been found
(10). Various spectroscopic techniques : X ray analysis (8,9), IR
(11,12), high resolution ^{13}C NMR (13)... have been used to characterize
"in situ" the coke deposits. Again coke can be characterized by conventional analysis of the products either desorbed by a high temperature
treatment under gas flow (14) or extracted by solvents after dissolution of the mineral framework (14-16). It is this latter technique
which was employed to characterize the coke deposited on protonic
zeolites during n-heptane cracking. The zeolites chosen were 3 large
pore-size zeolites : HY, HOFF (H offretite), HM (H mordenite) and one
intermediate pore-size zeolite : HZSM-5 (table 1). The results obtained
confirm that coke formation is a shape-selective process : pore,
channel and cavity dimensions govern its rate and its selectivity.

EXPERIMENTAL

HY zeolite, H mordenite and H offretite were supplied respectively
by Union Carbide, by Norton and by Grace Davison ; HZSM-5 zeolite was
synthetized according to Mobil Patents (18). Table 1 gives the formulas
of the elementary unit cells of these zeolites.

TABLE 1. Characteristics of the zeolites

Zeolites	Size μm	Formulas	Channels of Na zeolites (17)
HY	4	$Na_{0.8}H_{47.7}Al_{48.5}Si_{143.5}O_{384}$	$\underline{12}$ 7.4***
HM	2	$Na_{0.5}H_{5.2}Al_{5.7}Si_{42.3}O_{96}$	$\underline{12}$ 6.7x7.0* ↔ $\underline{8}$ 2.9 x 5.7 *
HOFF	3	$K_{0.4}H_{3.7}Al_{4.1}Si_{13.9}O_{36}$	$\underline{12}$ 6.4* ↔ $\underline{8}$ 3.6x5.2**
HZSM-5	4	$Na_{0.0015}H_{2.1}Al_{2.1}Si_{93.9}O_{192}$	[$\underline{10}$ 5.4x5.6 ↔ 5.1x5.5]***

Coke formation was studied during the transformation of n-heptane
in a flow reactor at 450°C, p_{N_2}= 0.7 bar, p_{nC_7} = 0.3 bar, WWH = 2.6 hr^{-1}.

For the HZSM-5 zeolite more rigorous conditions had to be used :
$t = 450°C$, $p_{nC_7} = 1$ bar, $p_{N_2} = 0$, WWH = 3.4 hr^{-1} in order to obtain a
coke deposit sufficiently significant to be extracted.

H/C ratio was determined by combustion of coke in an oxygen flow
(10 cm^3/mn for 0.5 g of catalyst) at temperatures increasing by stages
from 200 to 500°C ; the water and the carbon dioxide liberated were
trapped respectively by anhydrone and by ascarite, the amounts being
determined by weighing. From the amount of water, was substracted the
amount liberated from a non-coked zeolite treated in the same conditions.

Coke extraction was carried out in two steps :
- treatment in a soxhlet apparatus for 24 hours in the presence of
chloroform at boiling point (60°C) ;
- extraction at room temperature, by methylene chloride (CH_2Cl_2) after
dissolution of the zeolite by a 40 % hydrofluoric acid solution. The
extracts were evaporated, dried and weighed before analysis.

Extracts were analyzed by NMR, HPLC, GC, MS. The proton NMR
spectrum was established with a Brucker 200 MHZ. The HPLC analysis was
carried out on a 4.6 by 250 mm column packed with Ultrasil NH_2 using
n-heptane as solvent. Elution of the aromatic hydrocarbon was monitored
at 254 nm with a visible UV absorption detector. Gas chromatography of
the extracts was performed with 20 m fused silica capillary column
coated with SE 52, by programming temperature from 120°C to 280°C at
5°C/mn. Mass spectra of the extracts were obtained with a Kratos MS 25
apparatus by direct introduction into the source (T = 350°C, $p = 10^{-5}$
torr, E ionisation = 10 eV).

RESULTS AND DISCUSSION

1. Coking and aging rates

From Fig. 1 we can compare the aging rates of the 4 zeolites. Practically HZSM-5 does not deactivate whereas on the contrary mordenite
loses rapidly nearly all its activity ; offretite deactivates rapidly
during the first hours of the reaction then maintains a practically
stable activity (20 % of its initial activity) ; HY zeolite deactivates
more slowly but uniformly. Fig. 2 shows that the aging rates of the
zeolites can be classified in the same order (HZSM-5 < HY < HOFF < HM)
if instead of the time on stream we consider the ratio of the coke
content to the maximum theoretical coke content of the zeolite.

The deactivation of HZSM-5 could be explained by the presence of a
coke deposit on the outer surface (19). It will be seen further on that
taking into account its nature the coke deposit is situated more likely
inside the pores. However, at least up to the maximum content obtained
here (1.5 %), this inner coke has practically no deactivating effect
and therefore does not prevent the access of n-heptane to the active
sites. If large pore-size zeolites are taken into consideration, the
slower deactivation of the HY zeolite can be related to the possibility of a tridimensional circulation through the pores ; deactivation
therefore occurs essentially by site poisoning. On the contrary, for

mordenite, deactivation is very fast due to blocking of the large channels because molecule circulation here is unidimensional. Offretite presents the most interesting situation : deactivation occurs both by pore blocking and by site poisoning. On the fresh catalyst it takes place quite rapidly by blocking of the large channels, the molecules of the reactants and of the products being able however to circulate through the narrow channels. Deactivation of narrow channel sites is very slow and offretite conserves after large channel blocking a signi-ficant activity almost constant with time (Fig. 1). These conclusions are confirmed by measurement of the adsorption characteristics of nitrogen and of various hydrocarbons for coked zeolites (20).

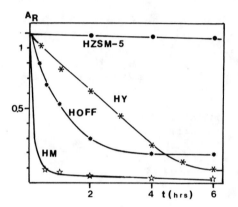 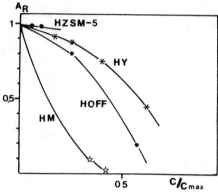

Figure 1. Change of A_R (activity at time t/activity after 5 mn) with reaction time t (hrs).

Figure 2. Change of A_R versus measured coke percentage/maximal coke percentage (C/C_{max}). C_{max} was estimated from the adsorption capa-city for n-hexane and by admitting the volumic mass of coke equal to 1 cm^3g^{-1}.

Fig. 3 shows that the coke formation rate decreases very rapidly with work-time in the case of large pore zeolites. For these zeolites the rate is initially (after 5 minutes reaction) equal to 2-3 x 10^{-3} moles of n-heptane transformed into coke per hour and per gram of cata-lyst, i.e. 5 to 10 times slower than the cracking rate. It decreases very rapidly and becomes equal to 0.5 x 10^{-3} on the HY zeolite, to 0.3 x 10^{-3} on the offretite and to less than 0.1 x 10^{-} on the morde-nite. On HZSM-5 it is much slower (0.01 x 10^{-3}) but decreases very little versus work-time. It can be noted moreover that coke formation versus work-time t obeys very correctly Voorhies' equation : % C = at^n (21) with n values of 0.35 for HY, of 0.21 for HOFF and of 0.15 for HM.

Fig. 4 shows that coke formation rate/cracking rate ratios decrease very rapidly versus work-time. Initially the ratio is 2.5 times greater for HOFF than for HY and HM and 420 times greater than for HZSM-5. This can be partly explained by the increase in steric

constraints during coke formation. Indeed the space available near the
active sites decreases from HY zeolite to HZSM-5 and the coke formation
reaction which requires very bulky intermediates (much more so than the
cracking reaction) is moreover more limited. The case of offretite is a
particular one and the very fast formation of coke could be related
either to the presence of stronger acid sites than on the other
zeolites (22) or of erionite faults in the large channels (23), which
could strongly favour coke formation.

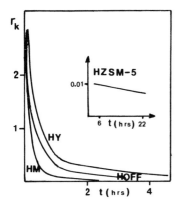
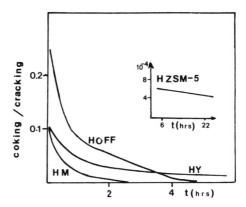

Figure 3. Change in coke formation rate r_k (10^{-3} mole of
n-heptane tansformed $hr^{-1}g^{-1}$)
against reaction time t (hrs).

Figure 4. Change in the ratio of rates
of n-heptane transformation into coke
and into hydrocarbons against reaction
time t (hrs).

2. Coke composition

The values of the H/C atomic ratio depend on the catalyst and on work-
time (table 2). On HY, HM and HOFF zeolites we have 0.4 < H/C < 0.7,
which is typical of polyaromatic compounds (H/C of coronene = 0.5). For
HZSM-5 coke, H/C is definitely higher, which proves that the heavy
compounds retained in this zeolite are not polyaromatic. As can generally be observed, H/C decreases with time on stream : at the beginning
of the reaction there is formation of coke precursors which adsorb on
the acid sites. After which, alkylation, cyclization, hydrogen transfer
cause, when the steric constraints do not prevent it, an increase in
the size and in the aromatic nature of these carbonaceous deposits.
    A very small part of the coke (less than 15 %) can be extracted
directly by treating the catalyst with chloroform (Table 2). It is a
"light" coke easily accessible to the solvents and which can therefore
be located on the outer surface of the zeolite or at pore apertures. A
more significant amount of coke can be extracted after dissolution of
the zeolite in hydrofluoric acid which proves that the coke deposit is
essentially internal. The case of HZSM-5 is quite clear since all the
coke is soluble in the solvents used and therefore probably located

inside the zeolite. Indeed, if the coke were formed on the outer acid sites it would be heavy and polyaromatic (therefore insoluble) for its formation would then not be limited by steric constraints.

The extraction yield decreases, as does the H/C ratio, with coking time (table 2) : thus on HY zeolite it passes from 65 % after 15 mn to 15 % after 6 hours. After 6 hours coking the extraction yield is very low on HY and on HM but remains remarkably high on HOFF. The two former cokes are essentially formed by very heavy compounds having probably over 8 aromatic rings since they are not soluble in methylene chloride and are found in the form of black particles after HF dissolution. These very heavy polyaromatic compounds have the space necessary to form in the large cavities of the HY zeolite and in the wide channels of HM and of HOFF. For this latter zeolite a significant amount of coke is extractible (about 50 %) ; this coke cannot then be located in the large channels and is probably found in the gmelinite cages of this zeolite.

TABLE 2.   Atomic ratio H/C of cokes. Percentage of coke extracted by solvent a) directly on the catalyst ; b) after dissolution of the mineral matrix in hydrofluoric acid.

		HY		HM		HOFF		HZSM-5	
Coking time (hrs)		0.25	6	0.25	6	0.25	6	6	23[*]
% Coke		4	16	3	4.5	4.5	10	0.5	4
H/C		0.6	0.4	0.7	0.6	0.65	0.5	2	1
% Extraction	a	15	0.7	6.7	2.2	2.2	1	–	14
	b	50	14	5	4.4	55.5	43	–	100
	a + b	65	14.7	11.7	6.6	57.7	44	–	114

*more severe conditions.

The composition of the extracts depends significantly on the catalyst and on the reaction time. To show this we have plotted the gas phase chromatograms of the extracts obtained for the 4 zeolites after 6 hours reaction (Fig. 5). The main information drawn from the GC, HPLC, MS and proton NMR analyses are given in Table 3.

TABLE 3.   Analysis of the solvent extracts obtained after dissolution of the mineral matrix in hydrofluoric acid.

	HY		HM		HOFF		HZSM-5
Coking time(hrs)	0.25	6	0.25	6	0.25	6	23
$\dfrac{\text{H aromatic}}{\text{H alkyl}}$ (NMR)	1.5	1.8	ε	ε	1.8	1.8	0.25-0.35
n_{carbones} (MS)	19-28	16-32	16-23	14-22	14-25	14-25	17-21
$n_{\text{aromatic rings}}$ (HPLC)	3-7	3-7	0	0	3>4	4>3	1-2

The HY zeolite and the offretite extracts are very polyaromatic, those of the HY zeolite presenting 3 to 7 aromatic cycles and those of the offretite 3 to 4. Associated with these polyaromatics are alkyl groups and/or saturated rings as we were able to show by mass spectrometric identification of the main products (24) ; the proportion of alkyl groups is more significant in the case of HOFF. These polyaromatic compounds with 3 or 4 rings are probably not formed in the large offretite channels which can receive heavier compounds but rather in the gmelinite cages of the narrow channels. These cages are effectively large enough to contain these polyaromatic compounds with 3 or 4 rings (ex. phenanthrene or chrysene).

The HZSM-5 extract has not a very marked aromatic character. The products present 1 to 2 aromatic rings with alkyl and napthenic groups. This confirms that the porous structure of HZSM-5 does not allow the formation of polyaromatic molecules. These compounds very slightly polar are probably only weakly adsorbed on the acid sites and will be therefore easily displaced by the reactant molecules. The low "toxicity" of the coke deposited on HZSM-5, found in n-heptane cracking (Fig. 2) could thus be explained.

It is curious to note that the mordenite extract contains only very few aromatic compounds but essentially alkyl and partially unsaturated cyclic hydrocarbons. These compounds result most likely from the transformation of compounds emprisoned between two plugs of coke which, through lack of hydrogen acceptor partners, were not able to evolve towards polyaromatic compounds.

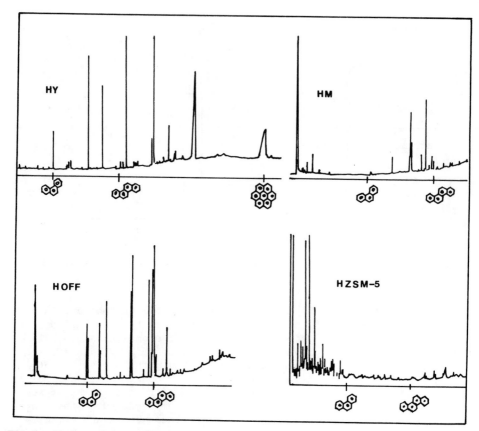

Figure 5. Gas phase chromatograms of CH_2Cl_2 extracts after dissolution of the zeolites by HF treatment. For reference, the retention of phenanthrene and chrysene are indicated.

CONCLUSION

The main conclusions of our study are summarized in Table 4 :
    i) Coke formation rate is fast when the space available near the active centers is significant and therefore the steric constraints are weak (e.g. large cavities in HY zeolite and large channels of HM and HOFF). It will be slow in the opposite case (ZSM-5 and small channels in OFF).
    ii) Coke composition depends also on the available space near the active sites : coke will be polyaromatic and very heavy if the reactions leading to its formation are not limited by steric constraints (e.g. HY and HM) whereas it will be less aromatic and lighter in the opposite case (e.g. ZSM-5 and gmelinite cages of HOFF).

iii) Deactivation rate is fast when molecule circulation is unidimensional (e.g. mordenite), coke deposit being capable of blocking the access of reactants to a large number of active sites ; on the contrary the deactivation rate will be slow when molecule circulation is tridimensional.

TABLE 4.

Zeolite	Porous structure	Coking rate	Aging rate	Selectivity (a)
HY	***, cavities	high	medium	A (> 8) P-A (4-7)
HM	*, tubular	high	high	A (> 8) P-N-O
HOFF	***, large and small channels	high→low	medium→low	A (> 8) P-A (3-4)
HZSM-5	***, tubular	low	low	P-N-A (1-2)

*Unidimensional ; ***tridimensional ; A aromatic ; P paraffinic ;
N naphtenic ; O olefinic ; (a) number of aromatic rings.

Coke formation is hence a shape-selective process since the porous structure of the zeolite determines for a large part the coking and the aging rates as well as the coke composition. We must not underestimate however the influence of other factors in particular the strength and the density of the acid sites.

REFERENCES

1. L.D. Rollmann and D.E. Walsh, *J. Catal.* 1979, **56**, 139.
2. V.J. Frilette, W.O. Haag, R.M. Lago, *J. Catal.* 1981, **61**, 218.
3. J.W. Beekman and G.F. Froment, *Ind. Eng. Chem. Fund.* 1979, **18**, 245.
4. J.W. Beekman and G.F. Froment, *Chem. Eng. Sci.* 1980, **35**, 805.
5. E.G. Derouane, *Catalysis by Acids and Bases* , Studies in Surface Science and Catalysis **20**, B. Imelik et al Eds, Elsevier Amsterdam, Oxford, New-York, 1985, p. 221.
6. J.B. Butt, *Adv. Chem. Ser.* 1972, **109**, 259.
7. E.E. Wolf and F. Alfani, *Catal. Rev. - Sci. Eng.* 1982, **24**(3), 329.
8. W.G. Appleby, J.W. Gibson and G.M. Good, *Ind. Eng. Chem., Process Des. Dev.* 1962, **1**, 102.
9. R.C. Haldeman and M.C. Botty, *J. Phys. Chem.* 1959, **63**, 489.

10. L.D. Rollmann and D.E. Walsh, *Progress in Catalyst Deactivation* ,
    J.L. Figueireds Ed., NATO Asi Series E, **54**, Martinus Nijhoff
    Publishers, The Hague, Boston, London, 1982, p. 81.
11. D. Eisenbach and E. Gallei, *J. Catal.* 1979, **56**, 377.
12. D.G. Blackmond, J.G. Goodwin, Jr. and J.E. Lester, *J. Catal.* 1982,
    78, 34.
13. E.G. Derouane, J.P. Gilson and J.B. Nagy, *Zeolites* 1982, **2**, 42.
14. H.S. Bierenbaum, R.D. Partridge and A.H. Weiss, *Molecular Sieves* ,
    W.M. Meier et al Eds, *Adv. Chem. Ser.* 1973, **121**, 605.
15. P.B. Venuto and L.A. Hamilton, *Ind. Eng. Chem. Prod. Res. Dev.*
    1967, **6**, 190.
16. P.B. Venuto, L.A. Hamilton and P.S. Landis, *J. Catal.* 1966, **5**,
    484.
17. W.M. Meier and D.H. Olson, *Atlas of Zeolite Structure Type* ,
    Structure Commission of the International Zeolite Association,
    Juris Druck + Verlag, Zurich, 1978.
18. R.J. Argauer and R.G. Landolt, U.S. Patent, 3.702.886/1972.
    C.D. Chang, W.H. Lang and A.J. Silvestri, U.S. Patent
    3.894.106/1975.
    S.A. Butter, W.W. Kaeding and A.T. Jurewicz, U.S. Patent
    3.894.107/1975.
    C.D. Chang, A.J. Silvestri and R.L. Smith, U.S. Patent
    3.928.483/1976.
19. P. Dejaifve, A. Auroux, P.C. Gravelle, J.C. Vedrine, Z. Gabelica
    and E.G. Derouane, *J. Catal.* 1981, **70**, 123.
20. P. Cartraud, A. Cointot, M. Dufour, P. Magnoux and M. Guisnet, to
    be published.
21. A. Voorhies, *Ind. Eng. Chem.* 1945, **37**, 318.
22. J. Tejada, G. Bourdillon, N.S Gnep, C. Gueguen and M. Guisnet,
    Proceedings of the 9th Iberoamerican Symposium on Catalysis II ,
    Lisboa, 1984, 1408.
23. N.Y. Chen, J.L. Schlenker, W.E. Garwood and G.T. Kokotailo,
    J. Catal. 1984, **86**, 24.
24. P. Magnoux, C. Canaff and M. Guisnet, to be published.

ACID-CATALYZED REACTIONS FOR THE CHARACTERIZATION OF THE POROUS
STRUCTURE OF ZEOLITES

F.R. Ribeiro[*], F. Lemos[*], G. Perot[**], M. Guisnet[**]
* Grupo de Estudos de Catalise Heterogenea, Instituto Superior
  Tecnico, Av. Rovisco Pais, 1096 Lisboa Codex, Portugal
** UA CNRS 350, Catalyse en Chimie Organique, UER Sciences,
   40,avenue du Recteur Pineau, 86022 Poitiers Cedex, France

ABSTRACT. The selectivities of reactions on acid zeolites depend in
particular on the diffusional path of molecules (channel sizes and
tortuosities...) and on the space available for the formation of reactional intermediates. These selectivities can therefore be used to
characterize the zeolite porous structure. In two examples (the constraint index and the selectivity of m-xylene transformation) we show
that numerous other factors can influence these selectivities which
makes the characterization complicated. This is especially the case
with the strength and the density of the acid sites, but also with the
presence on the zeolite of active sites other than acid sites or with
the presence of impurities in the reactional mixture or again with the
rapid formation of a coke deposit. The conditions for a precise characterization of the zeolite porous structure will be specified.

INTRODUCTION

Zeolites are very interesting as catalysts for numerous reactions of
acid catalysis : indeed these perfectly crystallised silicoaluminates
are much more active than the amorphous silica alumina (up to 10.000
times more (1)) ; moreover their porous structure gives them a particular selectivity (shape selectivity). As a great number of zeolites
which differ by their porous structure can be synthetized and as their
acidity as well as their porosity can be modified by a great number of
treatments, (exchange, dealumination...) the chemist can hope to obtain
by a just choice of the zeolite and of its treatments a catalyst perfectly selective for the transformation of a given organic compound.
        The shape selectivity of zeolites, first reported by Weisz and
Frilette (2) is due to the fact that the acid sites are internal and
that their channels and cavities are more or less of the same size as
organic molecules. Reviews on shape selective catalysis by zeolites
have been published by Csicsery (3,4), Weisz (5), Derouane (6,7).
        Three main types of shape selectivity can be defined :
        1. Reactant selectivity in the case of zeolite pores too small to

141

R. Setton (ed.), Chemical Reactions in Organic and Inorganic Constrained Systems, 141–150.
© 1986 by D. Reidel Publishing Company.

allow some reactants to reach the active centers.

    2. Product selectivity when the diffusion of certain products from the active sites to the external surface of the crystallite is limited.

    3. Restricted transition state selectivity when steric constraints can limit the formation of certain too bulky intermediates or transition states.

    Reaction selectivity will depend therefore on the respective sizes (and shapes) :

    i) of the channels and of the reactant and product molecules.

    ii) of the space available near the active sites (cavities or channels intersections) and of the reactional intermediates.
Conversely it is in principle possible, if the sizes and the shapes of the reactants, of the products and of the reactional intermediates are known, to characterize the zeolite porous structure with the help of reaction selectivity (8-12). It will be noticed however that the selectivity of zeolites is determined not only by their porous structure but quite obviously by the characteristics of their active sites : in acid catalysis the nature, the strength and the density of the acid sites. It will be therefore most important on the one hand to know how the characteristics vary as a function of the mode of preparation of the zeolite and on the other to discriminate between the influences on the selectivity of the acidity and of the porosity. We shall show with two examples the difficulties met during the characterization of the porous structure of zeolites by means of model reactions.

ACIDITY OF ZEOLITES

The type of zeolite as well as the undergone treatments determine its acidic characteristics : nature (Brönsted or Lewis), strength and density of the acid sites. Most zeolites are synthetized in sodic form and as such, are insufficiently acid for the catalysis of most organic reactions. The necessary acidity is generated by exchange of sodium for protons or for bi or trivalent cations (13,14). In the first case exchange can occur directly through an acid treatment if the porous network is sufficiently resistant or in two stages : exchange for ammonium ions followed by calcination :

$$e.g. \quad NaZ + NH_4^+ \longrightarrow Na^+ + NH_4Z$$

$$NH_4Z \longrightarrow NH_3 + HZ$$

The temperature for complete elimination of ammonium ions depends on the zeolite : the stronger the acid sites formed, the higher the temperature. At high calcination temperatures zeolite dehydroxylation occurs with formation of Lewis acid and basic sites. If the nature of the acid sites depends therefore essentially on the calcination conditions, their strength increases generally with the protonic exchange rate. This is distinctly shown by the change of the adsorption heat of ammonia on zeolites against their protonic exchange rate : thus for a mordenite, it decreases from 140 to 110 kjoule per mole for the strongest acid sites when the protonic exchange rate decreases from 92 to 37 % (15). One single exception was found : namely that of NaHZSM-5 zeolites for which the strength of the acid sites does not apparently depend on the protonic exchange rate since their activity for the iso- merization of xylenes and for the conversion of dimethylether into hydrocarbons is proportional to the protonic exchange rate (16). We shall notice that the strength of the acid sites depends not only on the zeolite exchange rate but also on their Si/Al ratio : the higher the Si/Al ratio (and therefore the lower the density of the acid sites) the stronger the acid sites.

The exchange of alkaline ions for bi or trivalent cations also leads to acid zeolites. The appearance of Brönsted sites (verified in particular by IR) can be explained by the dissociation of coordinati- vely bound water molecules under the action of the electrostatic field associated to the bi or trivalent cation.

$$M^{2+} + H_2O \longrightarrow M^{2+}\text{---}O\underset{H}{\overset{H}{\diagup}} \longrightarrow \left[M\text{-}O^{\diagup H}\right]^{+} + H^{+}$$
$$\text{to lattice } O^{2-}$$

The acid characteristics of these zeolites are essentially different from those of the protonic zeolites. Thus by IR analysis of pyridine adsorption, it can be observed that a HY zeolite treated under vacuum at 450°C presents mainly very strong Lewis sites (retaining pyridine above 450°C) whereas a CeHY zeolite presents mainly very strong Brönsted sites (17).

ACTIVE SITES FOR HYDROCARBON TRANSFORMATIONS

Active sites in hydrocarbon reactions can be classified into two categories :
      - acidobasic sites which catalyze facile reactions such as cis trans isomerization of olefins and double-bond shift and,
      - Brönsted acid sites responsible for more difficult transforma- tions such as skeletal isomerization of olefins, isomerization and cracking of alkanes, isomerization and disproportionation of aromatics (18,19). The Lewis acid sites alone do not seem to be active but can increase the strength and the activity of the Brönsted acid sites (20,21).
      The stronger the Brönsted acid site the greater its activity : thus a site capable of retaining adsorbed pyridine up to 300°C cannot transform isooctane at 350°C whereas a site capable of retaining it up to 400°C can transform 300 to 400 molecules of isooctane per hour (21).
      Moreover the acid strength required for the catalysis of a reac- tion depends a great deal on its difficulty. Thus the skeletal isome- rization of 3,3-dimethyl-1 butene which is very fast at 250°C requires sites capable of retaining pyridine up to 290°C while isobutane dispro- portionation requires sites capable of retaining it up to 550°C (21).
      Acid site density is for certain reactions (generally bimolecular)

a determining factor. This would particularly be the case for dipropor-
tionation of propane (23), of butanes (24) or of aromatics (25-28),
reactions which could require for their catalysis two adjacent acid
sites.

MODEL REACTIONS FOR THE CHARACTERIZATION OF ZEOLITE POROUS STRUCTURE

1. Constraint index (CI)

Frilette et al proposed to characterize zeolite porous structure by the
ratio of the cracking rate constants of n-hexane and 3-methylpentane
(C.I.). The large pore-size zeolites such as Y, mordenite ... have,
like amorphous silica alumina, CI values between 0.4 and 0.6. This
corresponds to the intrinsic selectivity of acid sites : 3-methyl-
pentane capable of leading by adsorption to a tertiary carbenium ion
cracks faster than n-hexane capable of leading only to secondary
carbenium ions. CI depends probably on the strength of the acid sites
of the catalyst, since n-hexane cracking requires stronger acid sites
than 3-methylpentane (22).
     Erionite, a small pore-size zeolite (8 ring window) has a very
high CI value : 3-methylpentane being too bulky to attain the acid
sites (reactant selectivity) can therefore react only on the outer acid
sites. In this case, CI will have an infinite value if the outer active
centers are eliminated, for example by poisoning with a base too bulky
to penetrate the zeolite pores.
     Intermediate pore-size zeolites (10 ring window) such as ZSM-5 and
ZSM-11 have an intermediate CI value (for example about 8 for ZSM-5 and
ZSM-11 à 316°C). In this case the difference of reactivities of
n-hexane and 3-methylpentane is not due to configurational limitations.
Indeed Frilette et al (8) and Haag et al (29) show that CI does not
depend on the size of the zeolite crystallites. Steric constraints
limiting the formation of carbenium ion (occurring by a bimolecular
step : hydride transfer from 3-methylpentane to a preadsorbed carbenium
ion) would explain the low reactivity of 3-methylpentane.
     Other factors can also influence the CI value. With ZSM-5, CI
decreases when the temperature increases becoming equal to 1.5 at
540°C. According to Haag and Dessau (30) this decrease could be related
to a change from the conventional cracking mechanism (with bimolecular
formation of the carbenium ion) to a monomolecular mechanism. As the
relative participations of the two mechanisms to cracking depend not
only on the temperature but also on the density of the acid sites (31)
CI must depend on this latter parameter.
     The presence of dehydrogenating impurities on the zeolite can also
influence the CI value. Thus CI (measured in the absence of H_2 so as to
avoid bifunctional catalysis) increases from 4 to 15 when 1 % of
platinum is introduced into ZSM-5 (32). This increase is not related to
that of steric constraints since the activity increases significantly
instead of decreasing. This increase in the activity can be explained
by a promoting influence of olefins formed by dehydrogenation of the
reactants on platinum sites. Moreover, in order to give an explanation

of the CI increase the authors suggest that the preadsorbed carbenium ions to which hydrides are transferred from alkanes are C_6 cations on PtHZSM-5 instead of C_3 cations on HZSM-5. This would create more steric constraints during the formation of the transition state of this bimolecular step.

The coke deposit on zeolite can also modify the CI value (33). This is particularly the case with offretite, a zeolite which presents large channels connected by smaller ones. For a protonic offretite (22) the CI value increases from 6 to over 50 after a one-hour reaction (Figure 1). This result can be related to the rapid deposit in the large channels of carbonaceous compounds which prevent 3-methylpentane from attaining the active sites. n-Hexane capable of circulating in the small channels continues to crack but without producing branched compounds.

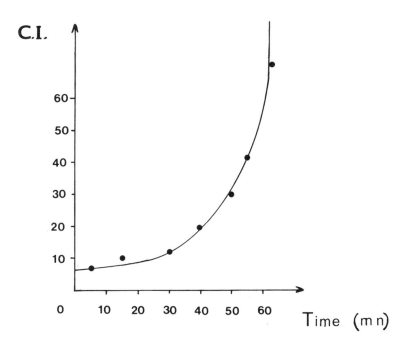

Figure 1. Change of the constraint index CI of HOFF with the time on stream (22).

The CI value measured in the presence of pyridine is very great because the only sites not poisoned are those inside the gmelinite cages which are accessible to n-hexane but not to 3-methylpentane. It must be noted that for this zeolite the CI value must increase when the protonic exchange increases since the acid sites of the gmelinite cages are formed only at high values of the protonic exchange rate (34).

To summarize, the constraint index of a zeolite can depend on its acidity (strength and density of sites, relative number of inner and outer active sites), on the channel sizes and on the space available near the active centers ; other factors such as dehydrogenating impurities or coke deposits can modify its value. It is therefore necessary to complete the characterization of the zeolite by constraint index by using either other model reactions or physicochemical techniques (adsorption...).

2. Selectivity of m-xylene transformation

On the acid catalysts m-xylene undergoes two main reactions : isomerization into ortho and paraxylenes and disproportionation into toluene and trimethylbenzenes. Three types of selectivities therefore can be defined :

i) isomerization selectivity measured by the formation rate ratio para/ortho,

ii) disproportionation selectivity given by the trimethylbenzene distribution and

iii) disproportionation/isomerization selectivity.

We can notice first that m-xylene cannot penetrate inside the small pore-size zeolites and therefore cannot be used to characterize them.

Isomerization selectivity is interesting for characterizing the diffusional path for contrarily to the constraint index, it depends in principle neither on the strength of the acid sites nor on the space available near the active centers : indeed the isomerizations of m-xylene into ortho and paraxylenes occur practically at the same rate on the non-shape selective catalysts whatever be their acid strengths ; the intermediates (benzenium ions) of these reactions are little bulky and practically of the same size. This selectivity should not be influenced by the presence of secondary products since both reactions occur through the same mechanism. Nevertheless this selectivity will depend as did the CI on the relative activities of the inner and outer acid sites. The value of the selectivity characteristic of the porous structure of the zeolite will hence be the one determined after a selective elimination of the activity of the outer sites. Obviously the coke deposit can modify the value of this selectivity : by increasing the configurational limitations it will provoke an increase in the selectivity for p-xylene, the smaller isomer. To summarize, the isomerization selectivity is a convenient means of characterization of the diffusional path in a zeolite (9). This path depends obviously not only on the porosity of the zeolite (width and tortuosity of the channels) but also on the size of its crystallites.

m-Xylene, contrarily to o- and p-xylenes, can lead by disproportionation directly to the three trimethylbenzene isomers (Fig. 2).

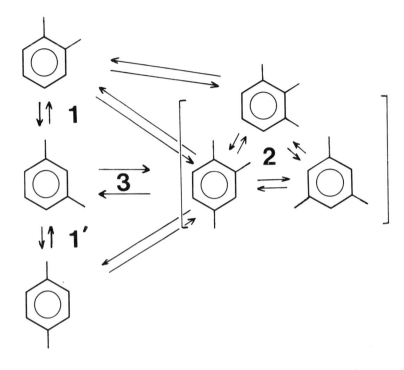

Figure 2. Scheme of m-xylene transformation on acid catalysts.

As these three reactions occur through the same mechanism, the
disproportionation selectivity does not depend on the strength of the
acid sites. It must be noted that the exact composition of the
trimethylbenzene mixture which results from the m-xylene disproportionation becomes very difficult to estimate if isomerization (Figure 2,
reactions 1, 2) is faster than disproportionation (reaction 3). Given
these isomerization reactions, the representative value of this composition will be obtained by extrapolation at zero conversion of the
trimethylbenzene formation curves. As the trimethylbenzene sizes are
different (1,2,4 TMB < 1,2,3 TMB < 1,3,5 TMB) the disproportionation
selectivity will depend, like isomerization selectivity, on the
diffusional path. It will depend moreover on the space available near
the active centers since the three bimolecular intermediates (I_1, I2,
I_3) which give respectively the TMB isomers (1,2,4 TMB ; 1,2,3 TMB ;
1,3,5 TMB) are of different sizes (35,9) : $I_1 < I_2 < I_3$. The influence
of the steric constraints, like that of the configurational limitations, will be to favor the formation of 1,2,4 TMB at the expense of
1,2,3 and above all 1,3,5 TMB. It will be observed that the study of
trimethylbenzene isomerization allows to specify which of these two
parameters determines the disproportionation selectivity.
      The disproportionation/isomerization selectivity (D/I) depends
both on the acidity and on the porosity of the zeolite. It is quite

normal for D/I to depend on the acidity since the two reactions involve
different intermediates : benzylic carbocations and benzenium ions with
diphenylmethane skeleton for disproportionation and monomolecular
benzenium ions for isomerization (36). The determining parameter does
not seem to be the strength of the acid sites but rather their density
(11). D/I will depend largely on the space available near the acid
sites since disproportionation involves bulky bimolecular intermediates
and is very sensitive to steric constraints, contrarily to intramolecular isomerization. It is without doubt mostly for this reason that
m-xylene isomerization on ZSM-5 is not accompanied by disproportionation. However configurational limitations to TMB desorption can also
explain this lack of disproportionation since ZSM-5 is inactive in
1,2,4 TMB isomerization. Other factors can also influence the D/I
value. Thus by adding isoalkanes to the reactant, the disproportionation rate decreases but not the isomerization (37) with consequently a
decrease of D/I. This inhibiting effect can be explained by a rapid
reaction between isoalkanes and the benzylic carbocations intermediates
of the disproportionation with reduction of their concentration :

$$\text{[ring]}-CH_2^+ \; + \; R_1-\underset{\underset{R_3}{|}}{CH}-R_2 \; \longrightarrow \; \text{[ring]} \; + \; R_1-\underset{\underset{R_3}{|}}{\overset{+}{C}}-R_2$$

Hydrogen preactivated on metallic sites has the same effect for the
same reason (38).

CONCLUSION

The two examples which we have developed show that the selectivity of
the reactions on acid zeolites can depend not only on the three well-
known factors :
    a) characteristics of acid sites particularly strength and density
    b) diffusional path of the molecules (sizes and tortuosities of
    channels...)
    c) space available near the active centers
but also on certain other factors such as :
    d) impurities in the reactant, sites of the catalysts other than
    the acid ones...
The porous structure of the zeolites could be characterized by means of
model reactions provided the determining factor is known unambiguously.
For this characterization the influence of factors a and d must be
known exactly and the information obtained must apply only to factor b
or to factor c. In order to characterize the diffusional path, it would
be preferable to use an intramolecular reaction so as to limit the
steric constraint effects (factor c) whereas to characterize the space
available near the active sites, the sizes of the reactants and of the
products as well as the reaction conditions must be chosen so that the
diffusional limitations can be neglected (factor b). Competitive transformation of one reactant is to be preferred to that of several reac-

tants for the selectivity thus obtained (reactivity ratio) depends on the characteristics of the acid sites.

REFERENCES

1. J.C. Miale, N.Y. Chen and P.B. Weisz, *J. Catal.* 1966, **6**, 278.
2. P.B. Weisz, V.J. Frilette, *J. Phys. Chem.* 1960, **64**, 382.
3. S.M. Csicsery in *Zeolite Chemistry and Catalysis* , J.A. Rabo Ed., ACS Monograph 171, American Chemical Society, Washington 197, p. 680.
4. S.M. Csicsery, *Zeolites* , 1984, **4**, 202.
5. P.B. Weisz, *Proc. 7th Intern. Congress Catal.* , T. Seiyama, K. Tanabe Eds, Kodansha, Tokyo, Elsevier Amsterdam, Oxford, New-York, 1981, p. 3.
6. E.G. Derouane, *Catalysis by Zeolites* , Studies in Surface Science and Catalysis **5**, B. Imelik et al Eds, Elsevier Amsterdam, Oxford, New-York, 1980, p. 5.
7. E.G. Derouane, *Zeolites Science and Technology* ,F. Ribeiro et al Eds, NATO Asi Series E, **80**, Martinus Nijhoff Publishers, The Hague, Boston, Lancaster, 1984, p. 347.
8. W.J. Frilette, W.O. Haag, R.M. Lago, *J. Catal.* 1981, **67**, 218.
9. N.S. Gnep, J. Tejada, M. Guisnet, *Bull. Soc. Chim.* 1982, p. 5.
10. M. Guisnet and G. Perot, Symposium on Shape Selective Catalysis, Division of Fuel Chemistry, ACS Meeting Seattle, 1983.
11. M. Guisnet, *Chemia Stosawana* 1984, 369.
12. J.A. Martens, M. Tielen, P.A. Jacobs and J. Weitkamp, *Zeolites* , 1984, **4**, 98.
13. M.L. Poutsma, in *Zeolite Chemistry and Catalysis* , J.A. Rabo Ed., ACS Monograph 171, American Chemical Society, Washington, 1976, p. 680.
14. D. Barthomeuf, in *Zeolites : Science and Technology* , F. Ribeiro et al Eds, NATO Asi Series E, **80**, Martinus Nijhoff Publishers, The Hague, Boston, Lancaster, 1984, p. 317.
15. P. Cartraud, A. Cointot, M. Dufour, N.S. Gnep, M. Guisnet, G. Joly and J. Tejada, to be published.
16. M. Guisnet, F. Cormerais, Y. Chen, G. Perot, E. Freund, *Zeolites* 1984, **4**, 108.
17. F. Lemos, F.R. Ribeiro, M. Kern, G. Giannetto and M. Guisnet, to be published.
18. H.A. Benesi and B.H.C. Winquist, *Advances in Catalysis* , D.D. Eley, H. Pines, P.B. Weisz Eds, Academic Press, New-York, San Francisco, London, 1978, **27**, 97.
19. M. Guisnet, in *Catalysis by acids and bases* , B. Imelik et al Eds, Studies in Surface Science and Catalysis **20**, Elsevier, Amsterdam, Oxford, New-York, Tokyo, 1985, 283.
20. C. Mirodatos and D. Barthomeuf, *J. Chem. Soc., Chem. Comm.* 1981, 39.
21. P.D. Hopkins, *J. Catal.* 1968, **12**, 325.
22. G. Bourdillon, Thèse Poitiers 1985.
23. F. Avendano, Thèse Poitiers 1984.

24. M. Guisnet, F. Avendano, C. Bearez and F. Chevalier, *J. Chem. Soc.*, *Chem. Comm.* 1985, 336.
25. R.P. Absil, J.B. Butt and J.S. Dranoff, *J. Catal.* 1984, **85**, 415.
26. G.W. Pukanic and F.E. Massoth, *J. Catal.* 1973, **28**, 304.
27. N.S. Gnep and M. Guisnet, *Appl. Catal.* 1981, **1**, 329.
28. M.L. Martin de Armando, N.S. Gnep and M. Guisnet, *J. Chem. Res.* 1981 (S) 8 ; (M) 243.
29. W.O. Haag, R.M. Lago and P.B. Weisz, *Faraday Discuss. Chem. Soc.* 1981, **72**, 317.
30. W.O. Haag and R.M. Dessau, *Proc. 8th Intern. Congress Catal.* 1984, **III**, 305.
31. G. Giannetto and M. Guisnet, to be published.
32. M. Guisnet, G. Giannetto, P. Hilaireau and G. Perot, *J. Chem. Soc.*, *Chem. Comm.* 1983, 1411.
33. T. Mahtout, Thèse Institut Français du Pétrole, 1985.
34. C. Mirodatos and D. Barthomeuf, *J. Catal.* 1979, **57**, 136.
35. S.M. Cscicsery, *J. Catal.* 1971, **23**, 124.
36. M. Guisnet and N.S. Gnep, *Zeolites Science and Technology* , F. Ribeiro et al Eds, NATO Asi Series, Martinus Nijhoff Publishers, The Hague, Boston, Lancaster, 1984, p. 571.
37. N.S. Gnep and M. Guisnet, *React. Kinet. Cat. Lett.* 1983, **22**, 237.
38. N.S. Gnep, M.L. Martin de Armando and M. Guisnet, *Spillover of Adsorbed Species* , G.M. Pajonk, S.J. Teichner, J.E. Germain Eds, Studies in Surface Science and Catalysis **17**, Elsevier, Amsterdam 1983, 309.

PILLARED CLAYS: SYNTHESIS AND STRUCTURAL FEATURES.

Thomas J. Pinnavaia
Department of Chemistry
Michigan State University
East Lansing, Michigan 48824 (U.S.A.)

ABSTRACT: This paper summarizes some of the recent advances that have been made in the synthesis and structural characterization of pillared clays. The usual synthetic approaches to clays pillared by polyoxocations depend on the formation of suitable cationic oligomers through hydrolysis reactions. More versatile routes less dependent on hydrolysis chemistry are currently being developed which make use of the in situ oxidation of metal cluster cations within the clay galleries. Pillared clays that are well ordered in the direction normal to the silicate layers exhibit several orders of 00 ℓ x-ray reflection and zeolitic adsorption properties. However, pillared clays can also contain delaminated fractions in which some of the layers are believed to aggregate in a card-house fashion. In the limit of such aggregation the delaminated clays are essentially x-ray amorphous. Information concerning the nature of the pillaring aggregates and their interaction with the silicate layers is beginning to come to light through magic angle spinning (MAS) NMR studies. MAS NMR investigations of aluminum pillared clays indicate that depending on the nature of the host clay, the alumina aggregates may chemically react with the silicate layers in forming a cross-linked structure. In other types of pillared clays the silicate layers do not become involved in chemical bond formation through reaction with the pillaring aggregate.

1. HISTORICAL

In 1955, Barrer and McLeod (1) first demonstrated that permanent porosity could be induced in smectite clay (eg., montmorillonite) by replacing the gallery Na^+ ions with alkylammonium ions. The alkylammonium ions functioned as molecular props or pillars between the silicate layers and exposed the intracrystal surfaces for adsorption and possible catalysis. Subsequent experiments, more recently reviewed by Barrer (2), showed that by varying the size and shape of the alkylammonium ions a range of zeolitic pore sizes could be achieved.

Bicyclic amine cations (3,4) and tris metal chelates (4,5) also have been used successfully as pillaring agents. However, the thermal stability of these pillaring cations, as with the alkylammonium ions, is limited by the

151

R. Setton (ed.), Chemical Reactions in Organic and Inorganic Constrained Systems, 151–164.

presence of carbon-carbon bonds. For example, intercalated o-phenanthrolein complexes of the type $M(phen)_3{}^{2+}$ decompose at about 350°C. Even lower decomposition temperatures are observed for clays pillared by alkylammonium ions.

The thermal instability problem exhibited by clays pillared with organic or organometallic cations was solved by using robust polyoxycations as pillaring reagents. Three different groups, those of Brindley (7), Vaughan (8), and Lahav and Shabtai (9), independently reported the intercalation of aluminum polyoxycations. These first examples of "alumina" pillared clays were structurally stable to 500°. Other polyoxycations, such as those of chromium (10) and zirconium (11), have also been used as pillaring reagents and efforts are still underway to develop new synthetic routes to such intercalates. In fact, the term "pillared clay" is being used more frequently in reference to clays interlayered with thermally stable polyoxy aggregates.

2. SYNTHESIS

Most of the polyoxycation pillared clays reported to date have been prepared by the reaction of the Na^+ form of the smectite clay with a solution of the polyoxy cation. In general the reaction proceeds by direct ion exchange of the surface Na^+ ions in the native clay by the polyoxycation. There is an advantage to using the Na^+ form of the clay over an exchange form containing more highly charged cations. Typically, Na^+-smectites in aqueous suspension are highly dispersed with the tactoid thickness ranging from one to a few silicate layers. Thus, most of the Na^+ exchange cations are readily accessible and the exchange reaction is facile.

As an example of the direct ion exchange method for the preparation of pillared clays one may cite the pillaring of montmorillonite by a polyoxycation of aluminum such as $Al_{13}O_4(OH)_{24}{}^{7+}$ (12):

$$7Na^+\text{-Mont.} + Al_{13}O_4(OH)_{24}{}^{7+} \longrightarrow Al_{13}O_4(OH)_{24}{}^{7+}\text{-Mont.} + 7Na^+ \qquad (1)$$

where a typical unit cell formula for montmorillonite might be $[Al_{3.3}Mg_{0.7}]Si_{8.0}O_{20}(OH)_4$. It should be noted, however, that the polyoxocations may undergo hydrolysis within the clay galleries to give species of reduced net charge, eg., $[Al_{13}O_4(OH)_{24+x}]^{(7-x)+}$. Such hydrolysis reactions on the intracrystal surfaces of the clay can be important in determining the stoichiometry of the final product.

Another approach to the pillaring of clays, due in part to the efforts of Brindley (13,14), involves the in situ hydrolysis of metal ions within the clay galleries:

$$Ni^{2+}\text{-Mont.} \xrightarrow{\ Ni^{2+},\ OH^-\ } [Ni_x(OH)_{2x-y}]^{y+}\text{-Mont.} \qquad (2)$$

The idea here is to form polyoxyoligomers in the clay galleries which would not be formed in aqueous solution because of the more favorable precipitation of the metal hydroxide at the OH^- concentration utilized in the reaction. Partitioning the metal ion between the acidic clay galleries and the aqueous phases avoids the formation of a metal hydroxide precipitate. Although Brindley was able to achieve modest gallery heights (∿5 A) by using the in situ hydrolysis method, the products tended to be chloritic and exhibited relatively low surface areas.

The above methods of pillared clay synthesis are limited to metal ions with hydrolysis chemistries that lead to the formation of polyoxy oligomers. An alternative and potentially more versatile route involves the use of metal complex cations as pillar precursors. In this latter approach the intercalated metal complex precursor is converted in situ to a molecular-size oxide aggregate. The process of converting metal complex precursors to oxide aggregates should be applicable to a variety of metal ion systems. The viability of the approach was demonstrated by Endo et al. (15,16) and Manos et al. (17) in the synthesis of a silica-pillared montmorillonite by the in situ hydrolysis of tris(acetylacetonato)silicon cations in the clay galleries. The silica-pillared clay possessed excellent thermal stability (> 500°C), but the interlayer free spacing was relatively low ($\sim$ 3.0 Å). The extent of pillaring was most likely limited by the amount of silicon that could be intercalated (1.0 Si per equivalent of clay).

More recently, we have examined the intercalation of niobium and tantalum clusters of the type $M_6Cl_{12}^{n+}$ (n = 2,3) in montmorillonite and their subsequent in situ oxidation to oxide aggregates at elevated temperatures (18). In these systems there are 2.0 or 3.0 metal atoms per exchange equivalent. Consequently, the interlayer free spacings are larger than those found for silica pillared clays. The following paragraphs illustrate the use of cluster cations as precursors to metal oxide pillaring agents.

Figure 1 illustrates the binding of $Nb_6Cl_{12}^{2+}$ and $Nb_6Cl_{12}^{3+}$ in 1:1 ratio, henceforth abbreviated $Nb_6Cl_{12}^{2+,3+}$, to Na^+-montmorillonite. The relationship between the Na^+ released to solution and the amount of $Nb_6Cl_{12}^{2+,3+}$ bound is shown in Figure 2. These two sets of data indicate that the binding of cluster cations occur by the displacement of interlayer Na^+ ions. The inital ion-exchange reaction may be written as

$$M_6Cl_{12}^{n+}(aq) + \overline{nNa^+} \longrightarrow \overline{M_6Cl_{12}^{n+}} + nNa^+(aq) \quad (3)$$

where the horizontal lines represent the silicate layers. As the extent of exchange increases the average charge per cluster decreases. The decrease in cluster charge results in a binding limit which exceeds the value expected on the basis of the clay ion exchange capacity and formal cluster charge. This reduction in cluster charge has been attributed to the hydrolysis of the hydrated ion, $Nb_6Cl_{12}(H_2O)_6^{n+}$, where the water molecule occupying terminal positions on the central Nb_6 core. Thus, the position of the hydrolytic equilibrium shown in eq. 4

$$\overline{Nb_6Cl_{12}(H_2O)_6^{n+}} + \underset{\rightleftharpoons}{\overset{H_2O}{}} \overline{Nb_6Cl_{12}(OH)_x(H_2O)_{6-x}^{(n-x)+}} + H_3O^+ \quad (4)$$

shifts to the right as the cluster loading increases, and the resulting hydrogen ions are displaced by more cluster cations until a complete monolayer of hydrated cations occupy the interlayer surfaces. Analogous results were observed for $Ta_6Cl_{12}^{2+}$ in montmorillonite.

The $Nb_6Cl_{12}^{n+}$ cations themselves function as pillars, as evidenced by the d_{001} spacings of 18.4 ± 0.2 Å and BET N_2 surface area of 58 m^2/g

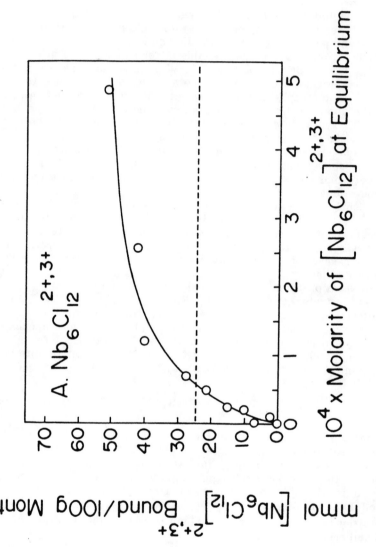

Figure 1.   Binding of $Nb_6Cl_{12}{}^{2+,3+}$ to Na^+-montmorillonite in aqueous suspension at 25°C. The dashed line is the expected loading based on the cation exchange capacity of the clay and the formal charges on the cluster cations.

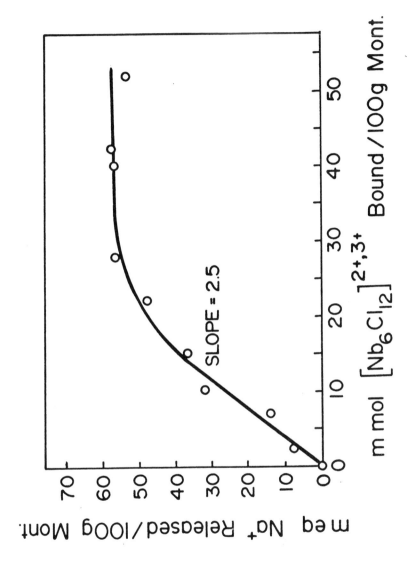

Figure 2.   Relationship between the Na$^+$ released and the Nb$_6$Cl$_{12}$$^{2+,3+}$ bound by Na$^+$–mont–morillonite.

at a cluster loading of 33 mmol/100 g. However, the expected surface area
is 180 m^2/g, suggesting that the clusters may tend to concentrate near the
edge of certain interlayers and restrict access to those interlayers. Regardless
of the distribution of $Nb_6Cl_{12}{}^{n+}$ on the interlamellar surfaces, the clusters
can be oxidized to oxide aggregates. Oxidation occurs upon heating the
intercalated cluster cations under vacuum at 240°C. For clusters with a
net charge of +1 the proposed reaction involves water functioning as the
oxidizing agent:

$$Nb_6Cl_{12}(OH)(H_2O)_5 + 9H_2O \longrightarrow 3Nb_2O_5 + H^+ + 12HCl + 8H_2 \qquad (5)$$

The oxide-intercalated product exhibits a lattice expansion of 9.0 Å and
a surface area of 64 m^2/g, appreciably larger than the 10-15 m^2/g surface
area of the parent Na^+-montmorillonite.

3.   STRUCTURE

Pillared clays prepared by direct ion exchange presumably contain
intercalated polyoxycations of the same nuclearity as the related cations
in aqueous solution. However, at elevated temperatures where the ions
undergo dehydration and dehydroxylation, the resulting oxide aggregates
may have a nuclearity different from that of the parent polyoxycation. The
best-studied type of pillared clays are those containing polyoxycations of
aluminum. ^{27}Al NMR spectroscopy has been particularly helpful in examining
the nature of the polyoxyaluminum ions in both the solution and the solid
state.

Two types of aluminum pillaring reagents have been used in the synthesis
of alumina pillared clays. One solution, known as aluminum chlorohydrate
(ACH) is a commercial product prepared by the reaction of aqueous $AlCl_3$
with Al metal. The second type of solution is a base-hydrolyzed $AlCl_3$ solution
prepared at OH^-/Al^{3+} ratios in the range 1.0-2.5. Pinnavaia et al. (19) showed
that the ^{27}Al NMR spectra of both types of pillaring reagents exhibited
a sharp resonance near 63 ppm which was consistent with the presence of
the tetrahedral Al site in the $Al_{13}O_4(OH)_{24}{}^{7+}$ ion (cf., Figure 3). In addition,
the ACH solutions exhibited broad tetrahedral and octahedral resonances
indicative of higher polymers. The ACH and based-hydrolyzed $AlCl_3$ solutions
also showed resonances near 0 ppm indicative of some $Al(H_2O)_6{}^{3+}$ and
$(Al_2(OH)_2(H_2O)_8)^{4+}$ dimer. Pinnavaia et al. (19) also demonstrated that
these two very different types of pillaring reagents afforded very similar
pillared products in terms of d-spacings, composition, and surface areas.
On the basis of these similarities it was suggested that the $Al_{13}O_4(OH)_{24}{}^{7+}$
species was preferentially bound to the intracrystal surfaces and that it
was this species which acted as the initial pillaring agent, in agreement
with the original suggestion of Vaughan (9).

Recently, Plee et al. (20) reported a ^{27}Al magic angle spinning (MAS)
NMR investigation of pillared hectorite and laponite prepared from
base-hydrolyzed $AlCl_3$. For hectorite and laponite there is no structural
aluminum present in the layers. These workers observed a sharp resonance
near 60 ppm which they attributed to the tetrahedral aluminum in the Al_{13}

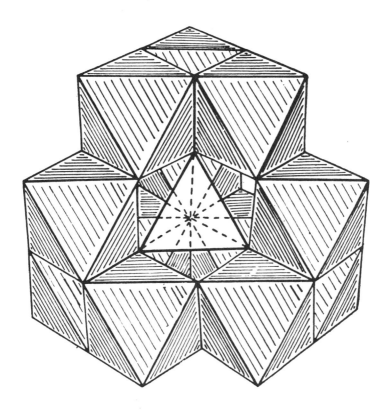

Figure 3. Schematic structure of $Al_{13}O_4(OH)_{24}(H_2O)_{12}{}^{7+}$ according to Johansson (12). The central Al is a tetrahedral environment of 4 oxygen atoms. The remaining 12 Al occupy octahedral positions defined by bridging OH groups and terminal H_2O molecules.

polymer. Also, they observed a resonance near 7 ppm which they attributed to the octahedral aluminum in the Al_{13} oligomer. In solution, the octahedral resonances are too broad to observe, as expected for this quadrupolar nucleus in a strong electric filed gradient. In pillared clays this octahedral resonance is intense, relative to the tetrahedral Al signal but not as intense as the 1:12 ratio expected for the structure shown in Figure 3. Apparently, the electric field gradients for a fraction of the octahedral aluminum sites in the intercalated Al_{13} oligomers allow these sites to be NMR observable. Dehydration of the pillared clays at 350° leads to a decrease in the relative intensity of the octahedral resonance. Nevertheless, the tetrahedra resonance is maintained indicating a structural relationship between the pillar at room temperature and the oxide aggregate formed at 350°. The structural relationships indicated by NMR are consistent with the fact that the d_{001} spacings of these pillared clays are ~ 18.5 Å at room temperature and 18-17.5 Å after calcination.

Plee et al. (20) also investigated the ^{27}Al MAS NMR spectra of air-dried and calcined pillared beidellite. In beidellite, there is some substitution of silicon by aluminum in the tetrahedral sheet, as well as aluminum in the octahedral sheet. As shown in Figure 4, the tetrahedral aluminum in the layers gives rise to a resonance at 69.3 ppm, whereas the octahedral aluminum in the layers occurs near 3 ppm. Upon pillaring with base-hydrolyzed $AlCl_3$, the resonance due to the tetrahedral aluminum in the Al_{13} oligomer is apparent at 62.3 ppm (cf., Figure 4). Upon calcining the pillared beidellite, the ^{27}Al MAS-NMR spectrum is greatly modified (cf., Figure 4). The most dramatic effects of calcining are (1) there is no defined peak at the chemical shift expected for the tetrahedral aluminum of the pillar, (2) the intensity of the resonance for the tetrahedral aluminum in the layer is decreased, and (3) a new resonance appears as a shoulder near 56.5 ppm. The new resonance has also been observed by Diddams et al. (21).

Plee et al. (20) have interpreted the above ^{27}Al MAS NMR results for calcined pillared beidellite in terms of a reaction between the Al_{13} pillars and the tetrahedrally substituted layers of beidellite. Their model for the reaction involves the inversion of an aluminum tetrahedron in the tetrahedra sheet and the linking of this inverted tetrahedron to the pillaring aggregate by formation of Al-O-Al bonds. A schematic representation of the Al_{13} pillar crosslinking to the layers is given in Figure 5. Statistically, the bonding between pillar and layer would involve three inverted tetrahedra per pillar. The first two oxygen (or OH or H_2O) layers in the cross-linked pillars form six aluminum octahedra sharing edges. The second and third oxygen layers define four edge-sharing octahedral aluminums and the third and fourth oxygen layers define three edge-sharing aluminum octahedra. Thus the cross-linking Al_{13} aggregate no longer contains a tetrahedral Al site. Finally, it should be noted that the above model was supported by ^{29}Si MAS NMR data. The most important ^{29}Si MAS NMR result was that an upfield shift occurred for the resonance for the silicon adjacent to the inverted Al tetrahedra, but the intensity of the resonance was similar to that found for the native beidellite. This latter result showed that the silicon tetrahedra were not inverted nor directly linked to the pillaring aggregate.

The nature of the layer aggregation in pillared clays has become the subject of some discussion. It has generally been presumed that the layers are aggregated in a stacked, face-to-face fashion. Evidence for such

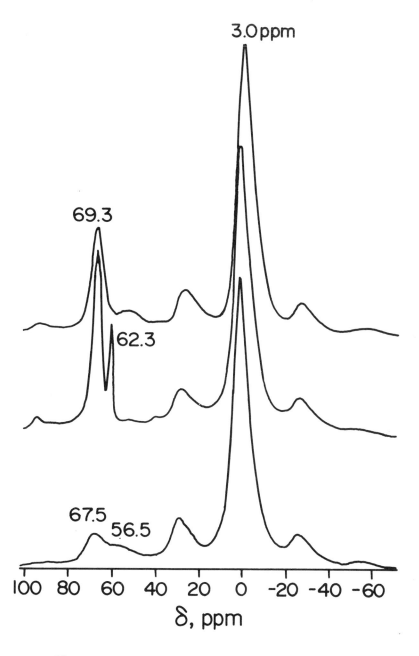

Figure 4. ^{27}Al MAS NMR spectra (130.2 MHz) for beidellite (top), pillared beidellite (middle), and pillared beidellite calcined at 350°C (bottom). Spectra are from Plee et al. (20).

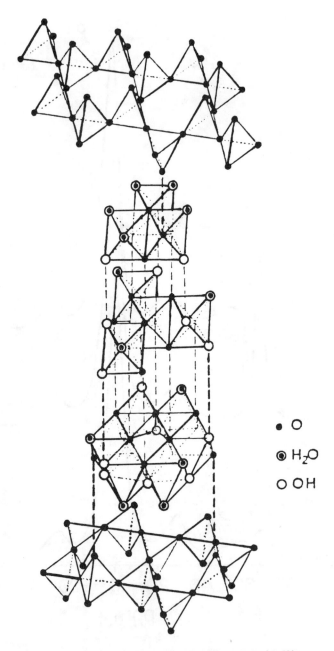

Figure 5.   Schematic structure of calcined pillared beidellite as proposed by Plee et al. (20). Exploded view of the thermally modified Al_{13} pillars sharing oxygen atoms with three inverted Al tetrahedra of two tetrahedral clay layers.

aggregation is provided directly by x-ray diffraction data. Typically, pillared clays exhibit a few orders of 00ℓ reflection. However, Pinnavaia et al. (19) have demonstrated that the sorption properties of clays pillared by polyoxycations of aluminum, whether derived from ACH or base-hydrolyzed $AlCl_3$, depend dramatically on the method of drying the reaction products. Products dried in air at room temperature were more regularly stacked (as judged from the number of $00\,\ell$ reflections) and exhibited a smaller pore size than products which were freeze-dried. Table I compares the adsorption data for air-dried and freeze-dried pillared montmorillonite containing 2.87 pillaring Al^{3+} ions per unit cell. Note that the freeze-dried products adsorbs a significant amount of $(F_9C_4)_3N$ with a kinetic diameter of 10.4 Å whereas the air-dried product does not adsorb 1,3,5-triethylbenzene with a kinetic diameter of 9.2 Å.

The above observations led to the proposal (19,22) that air-drying tends to optimize face-to-face aggregation, whereas freeze-drying tends to preserve some of the card-house structure (ie., edge-face and edge-edge aggregation) present in the flocculated clay. It was further proposed that the larger molecules with kinetic diameters > 6.2 Å were adsorbed on the surfaces of the delaminated fraction. Evidence for delamination was provided by studies of the reaction of ACH with a laponite, a synthetic hectorite with very small, lath-shape, morphology. The x-ray powder patterns indicated the freeze-dried products to be substantially amorphous, as expected for a layer aggregation model with extensive edge-face interactions.

Recently, Van Damme and Fripiat (23) analyzed the adsorption data of Pinnavaia et al. (19) for alumina pillared clays in terms of a fractal model. By applying the concepts of fractal dimensionality of solid surfaces to pillared clays, they derived the relationship

$$\ln[(N_{mo}/t) - (N_s/\Delta)] = \ln(n_p\sigma_r^{-D/2}t^{-1}) + (D/2)\ln\sigma_p \qquad (6)$$

where N_{mo} is the number of molecules needed to cover a given fraction of surface in absence of pillars, N_s is the number of molecules needed to occupy the volume between two surfaces separated by an average distance Δ, where Δ is the pillar height, n_p is the number of pillars, t is the statistical thickness of the adsorbate of cross sectional area σ_p, and D is the fractal dimension of the surface. Plots of $\ln N_s$ vs. $\ln R$, where R is the adsorbate radius, shown in Figure 6, gave D values of 2.0 for both air-dried and freeze-dried pillared montmorillonite. This result means that the surfaces of these clays are molecularly smooth with regard to the size of the adsorbates over the range (4.0 - 10.4 Å). It also means that the pillar distribution is non-random, otherwise there would be regions of high pillar density and local molecular sieving which would tend to increase the value of D. Thus, the differences between freeze-dried and air-dried pillared clays cannot be due to two types of surfaces, pillared and delaminated. Of course, the results do not mean that delamination does not occur in freeze-dried samples. They do suggest, however, that the pillar separation and, perhaps, the pillar constitution is influenced by the method of drying.

Table I.   Adsorption Data for Air-Dried and Freeze-Dried Pillared Montmorillonite[a]

Drying Method	Amount Adsorbed[b], mmol/g			
	C_6H_6 (5.8 Å)	$C(CH_3)_4$ (6.2 Å)	$1,3,5-Et_3C_6H_3$ (9.2 Å)	$(F_9C_4)_3 N$ (10.4 Å)
Air-Dry	1.72	1.12	0.00	0.00
Freeze-Dry	1.61	1.61	0.71	0.43

[a] Pillared products containing 2.87 intercalated Al per $O_{20}(OH)_4$ unit were prepared using ACH as the pillaring reagent. Samples were activated in vacuum at 350°C for 3 h prior to adsorption at 25°C.

[b] Equilibrium pressures (torr) were as follows: C_6H_6, 70; $C(CH_3)_4$, 728; $1,3,5-Et_3C_6H_3$, 0.18; $(F_9C_4)_3N$, 0.22.

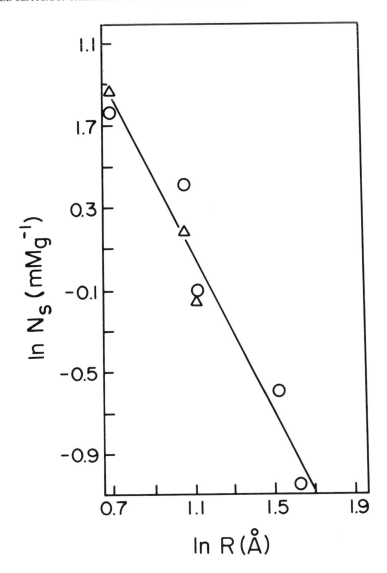

Figure 6.   Plot of the amount of adsorbate bound to Al$_{13}$-montmorillonites at saturation vs. kinetic radius of the adsorbate (22). The pillared clay was prepared from base-hydrolyzed AlCl$_3$ according to the method described in ref. 19.   Triangles refer to air-dried samples, circles to freeze-dried samples.

4. ACKNOWLEDGMENT

The partial support of this research by the National Science Foundation through grant CHE-8306583 is gratefully acknowledged.

5. REFERENCES

1. Barrer, R.M.; McLeod, D.M. Trans. Faraday Soc. **1955**, 51, 1290.
2. Barrer, R.M., "Zeolites and Clay Minerals as Sorbents and Molecular Sieves", Academic Press, New York, 1978, 407–486.
3. Mortland, M.M.; Berkheiser, V.E. Clays Clay Miner. **1976**, 24, 60.
4. Shabtai, J.; Frydman, N.; Lazar, R. Proc. 6th Intern. Congr. Catal. **1976**, London, The Chemical Society, London **1977**, p. 660.
5. Knudson, M.I.; McAtee, J.L. Clays Clay Miner. **1973**, 21, 19.
6. Traynor, M.F.; Mortalnd, M.M.; Pinnavaia, T.J. Clays Clay Miner. **1978**, 26, 319.
7. Brindley, G.W.; Sempels, R.E. Clay Miner. **1977**, 12, 229.
8. Lahav, N.; Shani, U.; Shabtai, J. Clays Clay Miner. **1978**, 26, 107.
9. Vaughan, D.E.W.; Lussier, R.J. Preprints 5th Intern. Conf. on Zeolites, Naples, Italy, June 2-6 (1980).
10. Brindley, G.W.; Yamanaka, S. Amer. Mineral **1979**, 64, 830.
11. Yamanaka, S.; Brindley, G.W. Clays Clay Miner. **1979**, 27, 119.
12. Johansson, G. Acta Chem. Scand. **1960**, 14, 771.
13. Yamanaka, S.; Brindley, G.W. Clays Clay Miner. **1978**, 26, 21.
14. Brindley, G.W.; Kao, C.C. Clays Clay Miner. **1980**, 28, 435.
15. Endo, T.; Mortland, M.M.; Pinnavaia, T.J. Clays Clay Miner. **1980**, 28, 105.
16. Endo, T.; Mortland, M.M.; Pinnavaia, T.J. clays Clay Miner. **1981**, 29, 153.
17. Manos, C.G.; Mortland, M.M.; Pinnavaia, T.J. Clays Clay Miner. **1984**, 32, 93.
18. Cristiano, S.P.; Wang, Jialiang; Pinnavaia, T.J. Inorg. Chem. **1985**, 24, 1222.
19. Pinnavaia, T.J.; Tzou, M.S.; Landau, S.D.; Raythatha, R.H. J. Molec. Catal. **1984**, 27, 143.
20. Plee, D.; Borg, F; Gatineau, L.; Fripiat, J.J. J. Amer. Chem. Soc. **1985**, 107, 2362.
21. Diddams, P.A.; Thomas, J.A.; Jones, W.; Ballantine, J.A.; Purnell, J.H. J. Chem. Soc. Chem. Commun. **1984**, 1340.
22. Pinnavaia, T.J. in "Heterogeneous Catalysis", B.L. Shapiro, Ed., Texas A&M University Press, College Station, TX, **1984**, pp. 142-164.
23. Van Damme, H.; Fripiat, J.J. J. Chem. Phys. **1985**, 82, 2785.

PILLARED MONTMORILLONITE AND BEIDELLITE. ACIDITY AND CATALYTIC PROPERTIES

G. PONCELET and A. SCHUTZ
Groupe de Physico-Chimie Minérale et de Catalyse
Place Croix du Sud 1
B-1348   Louvain-la-Neuve
Belgium

ABSTRACT. The acidic and catalytic properties of montmorillonite and beidellite pillared with hydroxy-aluminum polymers are compared. Both pillared clays exhibit thermally resistant 18 Å spacings and the amounts of Al fixed (probably as Al_{13} oligomers) are similar ($\sim$ 2 mmoles/g). However, the specific surface area and pore volume of pillared beidellite (PB) are higher than for pillared montmorillonite. Infrared data show that these clays also differ by the nature, amount, strength and thermal stability of the Brönsted acid sites. In PB, the acidity is associated with SiOH groups that are created by the proton attack of Si-O-Al linkages of the tetrahedral layer, whereas in PM, H_3O^+ is the main source of acidity. Upon dehydration of PM, the protons migrate towards the octahedral layer and are no longer available as catalytic acid sites. Microcatalytic tests show that pillared beidellites are definitely more active catalysts than pillared montmorillonites.

INTRODUCTION

Pillared interlayered smectites are considered with a growing interest since the last ten years. As far as we are aware, hydroxy-Al interlayered clays with basal spacings of 17.4 Å after calcination at 450 °C were first reported by Turner et al. (1).

Brindley et al. (2) prepared hydroxy-aluminum beidellites with basal spacings of 17 Å that did not collapse upon thermal treatment by adding the clay to hydroxy-aluminum solutions with OH/Al molar ratios between 2.0 and 3.0.

Lahav et al. (3) obtained cross-linked montmorillonites with d(001) values of 18 Å using a pillaring solution with an OH/Al molar ratio of 2.3 that was aged for at least 6 days before addition to a dilute clay suspension.

Nearly 19 Å spacings were reported by Rengasamy et al. (4) upon adsorption of Al-polycations in the interlamellar space of bentonite.

Zr-OH pillars confering to montmorillonite similar thermally stable spacings were intercalated by Yamanaka et al. (5).

R. Setton (ed.), Chemical Reactions in Organic and Inorganic Constrained Systems, 165–178.
© 1986 by D. Reidel Publishing Company.

Thermal decomposition of intercalated acetato-hydroxy iron complexes was used by Yamanaka et al. (6) to prepare Fe-pillared montmorillonite with somewhat smaller spacings (16.7 Å).

Vaughan et al. (7) prefered a commercial solution of Al-chlorhydrol to produce pillared clays with 18 Å spacings.

Recently, Shabtai et al. (8) reported similar d values for Al-pillared hectorite and fluor-hectorite.

Whatever the preparational procedure adopted by these authors, polyhydroxy-aluminum interlayered clays develop surface areas between 200 and 350 m^2/g and are characterized by a microporous system and high sorption capacities (7, 9-11).

Both Brönsted and Lewis-type acidities have been evidenced by IR spectroscopy, using pyridine as a probe molecule (12-14).

Pillared interlayered clays have soon shown interesting catalytic potentialities, especially as catalysts for the conversion of hydrocarbons (15-23) as well as molecular sieves (24).

Shabtai et al. (25) found the pillared clays superior to conventional Y-zeolite for the cracking of bulky molecules, due to the larger pore opening of the former ones.

Pillared montmorillonites used as supports for metals in composite catalysts containing zeolites exhibit excellent hydrocracking activity and selectivity to light furnace oil, as observed by Occelli et al. (26).

These solids show interesting activities in several other catalytic reactions, namely in the synthesis of amines, esters and ethers (27). However, as compared to commercial catalysts containing 15 % Y-zeolite, pillared montmorillonites deactivate more rapidly due to the collapse occurring above 540 °C and the resulting drop in their surface area.

Pinnavaia et al. (10) pointed out that freeze-drying had a beneficial effect on the catalyst longevity in the cracking of isopropylnaphtalene, as well as on the sorption capacities (11).

In a recent paper, Occelli et al. (28) showed that pillared bentonite has a high initial activity in the conversion of methanol to olefins, but a rapid deactivation occurs due to coke formation. In the alkylation of toluene with ethylene, the pillared clays are moderately active as compared to RE-Y-zeolites, but both catalysts deactivate evenly. In the cracking of gas oil, the same authors observed that pillared bentonite shows a good selectivity for light cycle gas oil, accounted for by the open microporous system and the acidity generated, according to these authors, by the pillars.

In a previous paper, Plee et al. (13) showed some evidence that the acidity of pillared beidellite was stronger as compared to that of pillared montmorillonites.

Recently, Plee et al. (29) proposed from high-resolution MAS-NMR of ^{27}Al and ^{29}Si data that the structural modifications occurring upon calcining pillared beidellite could be interpreted as the growth of a tridimensional network grafted on the bidimensional network of the clay.

The present paper aims to compare some physico-chemical and catalytic data obtained on pillared beidellite and montmorillonite.

EXPERIMENTAL

. Materials : preparation and characterization

Beidellite was synthesized in hydrothermal conditions from an amorphous gel having the following chemical composition (after calcination at 500 °C): $Na_{0.7}Al_{0.7}Si_{7.3}O_{22}$.
    200 g of gel and 640 ml of 0.1 M NaOH solution were introduced in a 1 l autoclave which was heated up to 320 °C (130 bar) within 24 h and maintained at that temperature for 5 days. The coarse unreacted fraction was separated from the clay fraction (yield: 90 % wt) by sedimentation in distilled water.
    The crystallinity of the beidellite was controlled by XRD. The CEC of the NH_4^+-exchanged form was 94 meq/100 g, i.e. near the value (96 meq/100 g) expected for beidellite with the following structural formula :
$$[Si_{7.33}Al_{0.67}]^{IV} [Al_4]^{VI} O_{20} \cdot (OH)_4 \cdot Na_{0.67} nH_2O$$
    IR spectra of NH_4^+-exchanged beidellite showed   the absorption band at 3030 cm^{-1} which, according to Chourabi et al. (30), is typical for NH_4^+ ions in exchange position in smectites where the isomorphic substitutions are located within the tetrahedral layers.
    The structural formula of Wyoming montmorillonite is :
$$[Si_{7.95}Al_{0.05}]^{IV} [Al_{3.0}Fe_{0.35}Mg_{0.65}]^{VI} O_{20}(OH)_4 Na_{0.7} nH_2O$$
The CEC value was 80 meq/100 g.

. Pillaring of montmorillonite and beidellite

Adequate volumes of 0.5 M NaOH were added to 0.2 M $AlCl_3.6H_2O$ solutions under stirring in order to prepare hydroxy-aluminum solutions with different OH/Al molar ratios between 0.6 and 2.4. Distilled water was then supplied in order that the final solutions were 0.1 M in Al.
    These solutions were then added to clay suspensions in volumes such as to introduce between 5 and 30 meq Al/g of clay. After allowing to stand for 30 min, the suspensions were dialyzed against distilled water untill free of excess ions. The pillared clays were freeze-dried.

. Characterization of the pillared clays

The d(001) spacings of the pillared clays were determined from the XRD patterns, using a Philips equipment (Cu anticathode, Ni filters).
    The IR spectra were recorded on self-supporting wafers (10-20 mg) with a Perkin-Elmer 180 instrument. The samples were outgassed and heated in an IR cell. Pyridine was used as a probe molecule for the characterization of the acidity of the materials.
    The BET surface areas and pore volumes were obtained from the N_2 adsorption-desorption isotherms (at liquid N_2 temperature),measured in a conventional glass volumetric apparatus provided with Bell & Howell pressure gauges.

. Catalytic evaluation

The catalytic tests were performed in a continuous flow micro-reactor
operated at atmospheric pressure, by flowing the carrier gas (He, H_2)
through a saturator kept at constant temperature. The catalyst (200 mg)
was sandwiched between two layers of pure fine-grained quartz (0.2 mm).
Activation was done in situ. The analysis of the effluent was
achieved by means of a monitored 6-way sampling valve coupled with a
HP 5880 gas chromatograph; the products were separated using high resolution capillary columns.

RESULTS

. XRD characterization of the pillared smectites

Influence of the OH/Al molar ratio of the pillaring solutions
on the spacings

Typical values of the d spacings obtained for the pillared clays are
given in Table 1.
As can be seen, stable pillars are formed in the interlamellar
space of the two smectites for a relatively broad range of OH/Al ratios
in the pillaring solutions, provided the washing step is carefully
carried out. A more detailed study on the preparation of pillared
clays will be published elsewhere (31).

TABLE 1. Basal spacings (in Å) of pillared montmorillonites (PM) and
beidellites (PB) prepared with different Al-OH solutions

PILLARED MONTMORILLONITES				
[Al] meq/g	15		30	
OH/Al	120°C	300°C	120°C	300°C
0.6	17.7	16.7	17.7	17.0
	14.2	14.0	14.7	15.0
0.8	18.4	16.7	18.4	17.7
1.0	18.6	17.7	18.6	18.0
1.2	18.5	17.2	18.6	18.2
1.4	18.6	17.7	18.8	18.2
1.6	18.8	17.7	18.8	18.2
1.8	18.6	17.7	18.8	18.0
2.0	18.6	17.3	18.6	17.8
2.2	18.4	17.7	18.6	17.7
PILLARED BEIDELLITES				
1.2	18.8	17.0		
1.6	18.4	17.4		
2.0	18.8	17.4		
2.4	18.8	17.4		

Note that the values observed after heating PM at 300 °C are
slightly higher when the Al concentration amounts to 30 meq/g.

Influence of the concentration of Al in the final suspension

Beidellite and montmorillonite were pillared with Al-OH solutions
(OH/Al = 2.0) in order to introduce 5, 10, 15, 20 and 30 meq Al/g of
clay.
    The basal spacings observed after heating at 110 and 400 °C are
shown in Table 2.

TABLE 2.   Basal spacings (Å) of pillared montmorillonites and beidelli-
           tes: influence of the Al concentration in meq Al/g of clay

meq Al/g	PILLARED MONTMORILLONITES d(001) in Å		PILLARED BEIDELLITES d(001) in Å	
	110°C	400°C	110°C	400°C
5	18.8	17.6	18.4	17.4
10	18.8	17.6	18.4	17.6
15	19.0	17.6	18.8	17.6
20	19.0	17.6	18.8	17.6
30	18.8	17.3	18.8	17.6

    As can be seen, high and heat resistant spacings are obtained
even when supplying only 5 meq Al/g. As previously, the pillared clays
were washed using the dialysis method.
    The Al content that is intercalated as hydroxy-Al species with-
in the interlamellar space of the two smectites amounts to approximately
2 mmoles/g, and the residual CEC, after NH_4-exchange is about 0.1 meq/g
of pillared clay.

. Specific surface area and pore volume

The specific surface areas (BET method) and pore volumes of the 18 Å
clays are compared in Table 3. The N_2 isotherms were established on samples calcined at 400 °C under vacuum.

TABLE 3. BET specific surface area (So, m^2/g) and pore volumes (Vp,cm^3/g)
         of pillared montmorillonite (PM) and pillared beidellite (PB)

Sample	So, m^2/g	Vp_{tot},cm^3/g	Vp_{micro},cm^3/g
PM	250	0.21	0.10
PB	320	0.25	0.12

    The specific surface area of PM is significantly lower than
that of PB, although as seen earlier, both materials exhibit similar
d spacings and intercalate identical amounts of Al as hydroxy-polymers.
    The pore volumes of PB are also slightly but significantly
higher than those of PM. These differences may be indicative of a more
homogeneous distribution of the pillars in PB than in PM.

. Infrared investigation of the surface acidity

Fig. 1 compares the IR spectra in the OH stretching region recorded on thin films of PB and PM that have been treated under vacuum at increasing temperatures.

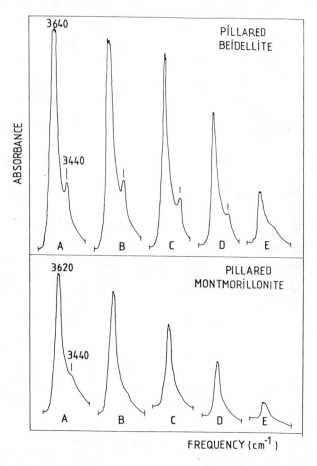

Fig. 1. IR spectra in the OH stretching region of pillared beidellite
and montmorillonite as a function of the outgassing temperature:
A, 200°C; B, 300°C; C, 400°C; D, 500°C and E, 600°C.

Two distinct bands are observed for PB, at 3640 and 3440 cm^{-1},
whereas for PM, beside the band at 3620 cm^{-1}, the absorption at 3440 cm^{-1}
shows up as a small shoulder.

The stretching OH band at 3440 cm^{-1}, also observed for H-beidellite (a Na-beidellite treated with 0.05 M HCl) and in decationed NH$_4$-beidellite, is attributed to Si-OH...Al entities that are produced by the proton attack of tetrahedral Si-O-Al linkages (13), as it occurs similarly in decationed Y-zeolites (32). In PM, (as well as in heat treated NH$_4$-montmorillonite), the shoulder at 3440 cm^{-1} indicates the existence of a small amount of Si by Al substitutions in the tetrahedral layers.

Fig. 2 shows the evolution of the integrated intensity of the OH stretching band at 3620-3640 cm^{-1} as a function of the calcination temperature under vacuum for PM (upper figure) and PB (lower figure). For comparison, the evolutions observed for the NH$_4$-exchanged clays treated similarly are also shown.

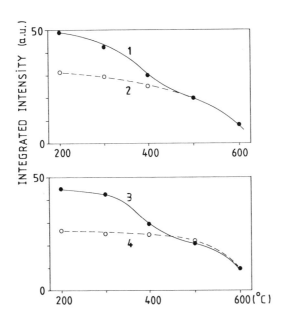

Fig. 2. Evolution of the integrated intensity (a.u.) of the band at 3640-3620 cm^{-1} of pillared montmorillonite (1), NH$_4$-montmorillonite (2), pillared beidellite (3), and NH$_4$-beidellite (4), as a function of the outgassing temperature.

The differences in the intensities observed between the pillared samples and the NH$_4$-exchanged forms constitute the contribution of the OH groups belonging to the pillars. As can be seen, the evolutions are similar and the main loss in intensity occurring between 300 and 400 °C is due to the dehydroxylation of the pillars. This has been confirmed by DTA and TGA measurements (not reported here).

Similarly, the evolution of the OH stretching band at 3440 cm^{-1} has been followed for pillared beidellite. As shown in Fig. 3, a maximum in the intensity is reached when the clay is outgassed at 300°C. At higher temperatures, dehydroxylation of the Si-OH groups is beginning. However, even a calcination at 600 °C does not suppress this band.

The Brönsted acid nature of these tetrahedral Si-OH groups is clearly evidenced with pyridine. Indeed, upon adsorption of the base on PB (previously outgassed at 300 °C), the absorption at 3440 cm^{-1} completely vanishes whereas the band characterizing the pyridinium ion (proton acidity) at 1540 cm^{-1} appears, as shown in Fig. 4. These spectra were recorded after thermal treatment under vacuum at increasing temperatures.

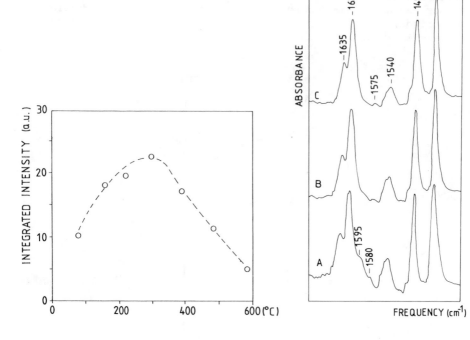

Fig. 3. (Left) Pillared beidellite: evolution of the integrated intensi-
        ty (a.u.) of the band at 3440 cm^{-1} as a function of the outgas-
        sing temperature.

Fig. 4. (Right) IR spectra of pyridine adsorbed on pillared beidellite
        (previously outgassed at 300°C) after heating under vacuum at
        150°C (A), 250°C (B) and 310°C (C).

Noteworthy is the intense absorption at 1454 cm^{-1} which characterizes pyridine in interaction with Lewis acid sites. This band does
not show up in the spectra of the proton exchanged clays and, therefore,
is associated to Lewis sites present, even at low temperature, on the
pillars.

The amount of protons in pillared clays which reacts with pyridine is affected by the calcination temperature.

The integrated intensities of the band at 1540 cm^{-1} are plotted
in Fig. 5, against the outgassing temperature. As can be seen, at 200°C,
the intensities are comparable for both pillared clays. Increasing the
calcination temperature prior to pyridine adsorption results in a steep
drop in the proton content in the case of PM, while PB keeps its acidity.
The slight increase observed between 350 and 500°C, if significant, might
be attributed to the contribution of the protons liberated during the
dehydroxylation of the pillars, which after Vaughan et al. (7), may be
represented as follows :

$$[Al_{13}O_4(OH)_{24}(H_2O)_{12}]^{7+} \rightarrow 7H^+ + 6.5\ Al_2O_3 + 20.5\ H_2O$$

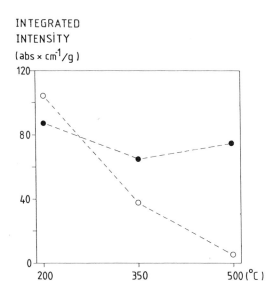

Fig. 5. Integrated intensities of the band at 1540 cm^{-1} for pillared
        beidellite (●) and pillared montmorillonite (O) at different
        temperatures of precalcination.

assuming that the pillars correspond to the Al_{13} species described by
Johansson et al. (33).

        The dramatic disappearance of the Brönsted sites observed for
PM would be due to the fact that, upon thermal activation, the protons
migrate into the octahedral layer of the clay where they induce a prema-
turated dehydroxylation (30). These observations are in general agreement with the results of Occelli et al. (12, 34).

        It is thus clear, from these results, that the nature,strength
and amount of the proton content of PB is largely different than for PM.

. Catalytic properties of pillared montmorillonite and beidellite

        Cracking of cumene

This reaction is commonly used as an evaluation test of the acid properties of a catalyst, mainly because cumene is decomposed, under normal
conditions, into benzene and propylene, via the breaking of the alkyl
group of the intermediate carbocation formed on the protonated site.

        50 mg of pillared beidellite (PB) and montmorillonite (PM) were
calcined at 350°C and reacted with cumene at 350°C. The weight hourly
space velocity (WHSV) was 2.5 g of cumene per gram of catalyst and per
hour. The conversion was followed as a function of time.

        Fig.6 compares the results obtained for PB and for a series of

PM. The difference between the two clays is obvious. Not only the pillared montmorillonites are less active than pillared beidellite, but they
also deactivate more rapidly than PB. These results could be expected
from the IR investigation data.

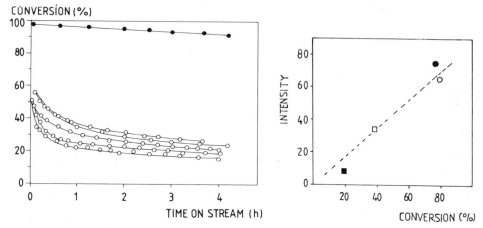

Fig. 6. (Left). Cracking of cumene: evolution of the conversion vs time
         at 350°C for pillared beidellite (●) and different pillared
         montmorillonites (O) precalcined at 350°C.
Fig. 7. (Right). Relation between the intensity of the 1540 cm^{-1} band
         (pyridinium) and the conversion of cumene (in %) for pillared
         beidellite O, ● and pillared montmorillonite □ , ■ . Open sym-
         bols : simples calcined at 350°C; full symbols : samples calci-
         ned at 500°C.

        The difference between the two pillared clays is still more
pronounced when comparing the conversions achieved at 350°C (in temperature programmed conditions, not reported here) on samples that have been
calcined at 500°C. Nearly 75% of cumene is transformed on PB while only
20% on PM. The integrated intensities of the pyridinium band at 1540 cm^{-1}
are plotted, in Fig. 7, against the conversions observed at 350°C for
the two clays calcined at 350 and 500°C. This figure again shows the
superiority of PB as well in activity as in stability.

        Isomerization of m-xylene

On acid catalysts, dimethylbenzene may undergo isomerization and/or dis-
mutation into toluene and trimethylbenzenes.
        A few comparative tests have been carried out on the pillared
clays and on an ultrastable Y-zeolite (USY) prepared by a steaming treatment of NH$_4$-Y zeolite at 750°C.
        The catalysts were previously calcined at 400°C under air for
2 h. The WHSV was 1.37 g of m-xylene per gram of catalyst and per hour.
The detailed results of these tests will be published at large elsewhere
(35). The main conclusions may be summarized as follows :
        . The activity sequence observed is USY > PB > PM.

. All the catalysts deactivate with time on stream in the reverse sequence as the one above.

. Substantially more disproportionation products are obtained
for USY, whereas higher selectivities in isomers (% isomers/% isomers +
% disproportionation products) for the pillared clays.

. The para/ortho ratio is generally > 1 for the pillared clays
and < 1 for USY.

. Hydroconversion of n-heptane

Hydroisomerization-hydrocracking of n-paraffins on H-zeolites impregnated with a group VIII metal (principally Pt) has been largely investigated (36-40) because this reaction yields a good evaluation of the
strength of the acid sites, and according to the chain length of the
paraffin, the distribution and nature of the isomers, it provides informations on the pore size and shape of the catalyst.

In the case of $n-C_7$, the reaction products are easily separated
and identified using GLC fitted with a capillary column.

The freeze-dried pillared clays were impregnated with the Pt II
tetrammine complex in order to have a loading of 1 % (by wt) of Pt. After
being dried at 110°C, the catalysts were calcined in air at 400°C for
2 h, followed by a hydrogen reduction at the same temperature. Three
catalysts were taken as references: an ultrastable Y-zeolite (USY),
H-ZSM-5 and a commercial silica-alumina containing 25 % Al_2O_3.

The reaction was carried out in temperature-programmed conditions at a heating rate of 2°C/min using 0.2 g of catalyst. The WHSV was
0.9 g $n-C_7$ per gram of catalyst and per hour, with a molar ratio $H_2/n-C_7$
of 16.

Fig. 8 shows the evolution of the conversion vs reaction tempe-
rature. As observed for the other test-reactions, PB is more active than
PM, and the former one is almost as active as the reference zeolites.
Silica-alumina is the least active.

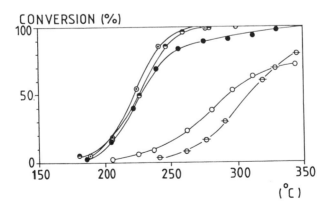

Fig. 8. Temperature-programmed reaction of $n-C_7$: evolution of the conversion for ⊙: USY; ●:H-ZSM-5; ●: PB; O: PM; ⊖: silica-alumina.

    If now we consider the selectivities in isomerization products,
pillared beidellite performs much better than the zeolites. An illustration of it is shown in Fig. 9, which gives the yields (in %) of isomers
versus reaction temperature for the different catalysts.

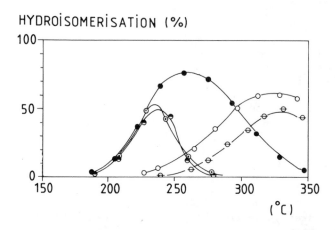

Fig. 9. Percentages of C_7 isomers in the reaction products as a function
        of reaction temperature: ⊙ : USY; ⊖ : H-ZSM-5; ● : PB; O : PM;
        ⊖ : silica-alumina

    The higher selectivities in isomers obtained for PB indicates
that the acid strength is lower than that of the zeolites for which more
cracked products are also formed.
    As far as pore shape and size are concerned, n-C_7 and its iso-
mers are not bulky enough molecules to yield informations. There are
no geometrical hindrances and the products can diffuse freely in the
pore volume of the interlayered materials. Data obtained from n-C_{10}
hydroconversion which yields multibranched and also bulkier isomers will
be reported elsewhere (41).

    In conclusion, the results of the microactivity tests definite-
ly show that the pillared beidellites are much more active and keep
their catalytic activities at higher temperatures than the pillared
montmorillonites. They also confirm the differences in acid properties
(nature, strength, number of active sites) evidenced by infrared spectroscopy.

ACKNOWLEDGEMENTS

    The authors gratefully acknowledge the Compagnie Française de
Raffinage for financial support of this research. They also thank Dr. P.
A. Jacobs (K.U.-Leuven) for helpful discussions.

REFERENCES

1. R.C. Turner and J.E. Brydon, Soil Sci. 100, 176-181 (1965).
2. G.W. Brindley and R.E. Sempels, Clay Miner. 12, 229-237 (1977).
3. N. Lahav, U. Shani and J. Shabtai, Clays and Clay Miner.26,107-115 (1978).
4. P. Rengasamy and J.M. Oades, Austr. J. Soil Res.16, 53-66 (1978).
5. S. Yamanaka and G.W. Brindley, Clays and Clay Miner. 27,119-124 (1979).
6. S. Yamanaka, T. Doi, S. Sako and M. Hattori, Mat. Res. Bull. 19, 161-168 (1984).
7. D.E.W. Vaughan and R.J. Lussier, in Proc. Vth Int. Conf. on Zeolites, Ed. L.V. Rees, Heyden & Sons, 94-101 (1980)
8. J. Shabtai, M. Rosell and M. Tokarz, Clays and Clay Miner. 32, 99-107 (1984).
9. M.L. Occelli, F. Hwu and J.W. Hightower, Div. Petrol. Chem. Amer. Chem. Soc., 672-688 (1981).
10. T.J. Pinnavaia, M.S. Tzou, S.D. Landau and R.H. Raythatha, J. Mol. Catal. 27, 195-212 (1984).
11. T.J. Pinnavaia, in Heterogeneous Catalysis, IInd Symp. Ind.-Univ. Coop. Chem. Progr., Publ. IUCCP, Texas A & M Univ. Press, 142-164 (1984)
12. M.L. Occelli and R.M. Tindwa, Clays and Clay Miner. 31,22-28 (1983).
13. D. Plee, A. Schutz, G. Poncelet and J.J. Fripiat, in Catalysis by Acids and Bases, Elsevier Sci. Publ. Co., Eds B. Imelik, C. Naccache, G. Coudurier, Y. Ben Taarit and J.C. Vedrine,343-350 (1985).
14. D. Tichit, F. Fajula, F. Figueras, J. Bousquet and C. Gueguen, in Catalysis by Acids and Bases. Elsevier Sc. Publ. Co., Eds B. Imelik et al., 351-360 (1985)
15. D.E.W. Vaughan, R.J. Lussier and J.S. Magee, U.S. Pat. 4.176.090 (1979).
16. D.E.W. Vaughan, R.J. Lussier and J.S. Magee, U.S. Pat. 4.271.043 (1981).
17. D.E.W. Vaughan, R.J. Lussier and J.S. Magee, U.S. Pat. 4.248.739 (1981).
18. M.G. Reed and J. Jaffe, U.S. Pat. 4.060.480 (1977).
19. M.G. Reed and J. Jaffe, U.S. Pat. 3.796.177 (1977).
20. M.Occelli and J.V. Kennedy, Brit. Pat. 2.116.202 (1983).
21. M. Occelli and J.V. Kennedy, Brit. Pat. 2.116.062 (1983).
22. P. Jacobs, G. Poncelet and A. Schutz, Br. Fr. 81.16387 (1981).
23. D. Plee, A. Schutz, F. Borg, G. Poncelet, P. Jacobs, L. Gatineau and J.J. Fripiat, Br. Fr. 84.06482 (1984).
24. J. Shabtai and N. Lahari, U.S. Pat. 4.216.188 (1980).
25. J. Shabtai, R. Lazar and A.G. Oblad, VIth Int. Cong. Catalysis Japan, paper B8 (1980).
26. M.L. Occelli and R.J. Rennard, Am. Chem. Soc. Div. Fuel Chem. 30-39 (1984).
27. P.A. Diddams, J.M. Thomas, W. Jones, J.A. Ballantine and J.H. Purnell, J. Chem. Soc. Chem.Comm., 1340-1342 (1984).

28. M.L. Occelli, R.A. Innes, F.S.S. Hwu and J.W. Hightower, Appl.
        Catal. 14, 69-82 (1985).
29. D. Plee, F. Borg, L. Gatineau and J.J. Fripiat, J. Am. Chem. Soc.
        107, 2362-2369 (1985).
30. B. Chourabi and J.J. Fripiat, Clays and Clay Miner. 29,260-268
        (1981).
31. Article in preparation (to Clays and Clay Miner.)
32. J.B. Uytterhoeven, L.G. Christner and W.K. Hall, J. Phys. Chem.
        29, 2117-2125 (1965).
33. G. Johansson, G. Lundgren, L. Sillen and R. Söderquist, Act. Chem.
        Scand. 14, 769-773 (1960).
34. M.L. Occelli and J.E. Lester, Ind. Eng. Chem. Prod. Res. Div. 24,
        27-32 (1985).
35. Article in preparation.
36. P.A. Jacobs,in Carboniogenic Activity od Zeolites, Elsevier (1977).
37. P.A. Jacobs, J.A. Martens, J. Weitkamp and H.K. Beyer, Farad. Disc.
        Chem. Soc. 72, 353-384 (1982).
38. M. Steijns, G. Froment, P. Jacobs, J. Uytterhoeven and J. Weitkamp,
        Ind. Eng. Chem. Prod. Res. Div. 20, 654-660 (1981).
39. J. Weitkamp, P.A. Jacobs and J.A. Martens, Appl. Catal. 8, 123-141
        (1983).
40. P.A. Jacobs and J.A. Martens, This NATO Workshop.
41. A. Schutz, D. Plee, F. Borg, P. Jacobs, G. Poncelet and J.J.Fripiat,
        Paper to be presented at the Eighth Int. Clay Conf. Denver,
        Co, U.S.A. (1985).

INTRACRYSTALLINE COMPLEXATION BY CROWN ETHERS AND CRYPTANDS IN CLAY MINERALS

E. Ruiz-Hitzky and B. Casal
Consejo Superior de Investigaciones Científicas
Instituto de Físico-Quimica Mineral
c/ Serrano 115 bis
28006 Madrid
Spain

ABSTRACT. The ability of macrocyclic compounds such as crown ethers and cryptands to intercalate in layered materials as smectites, open way towards a particular chemistry of these compounds in intracrystalline environments. They form very stable polydentate coordination compounds in the interlayer space, replacing water molecules which belong to the hydratation sphere of the exchangeable cations. Stoichiometry, geometry, packing arrangement, acidic-base reactions, etc., are determined by the nature of the interlayer cations, as well as the mineral charge and steric limitations imposed by the host lattice.

INTRODUCTION

Low charged 2:1 phyllosilicates, can act as host lattice for neutral organic guests, so that intercalation compounds are formed, with the intracrystalline cations as adsorption centers for direct coordination with water, alcohols, ketones, etc. (1). The clay minerals montmorillonite, hectorite and vermiculite, are natural examples among these silicates. They are built up by the staking of negatively charged silicate layers which are compensated by exchangeable cations located in the interlayer space.
    A great number of macrocyclic polyethers such as crown ethers and cryptands, have been synthetized in the course of the two last decades. These compounds are of a remarkable interest because of theirs excellent properties as complexing agents (2,3).Various cations can be introduced into the macrocyclic cavity of these ligands, the stability of the resulting complexes depending upon the nature (charge and size) of the cations and of certain characteristics of the macrocyclic molecule, such as its geometry, type of heteroatoms of the ring, etc. (4,5).
    In this contribution, we try to give an overview and some recent results of the coordination chemistry for the interaction between different macrocyclic compounds and cations, both of them located within the interlayer space of 2:1 phyllosilicates.

R. Setton (ed.), Chemical Reactions in Organic and Inorganic Constrained Systems, 179–189.
© *1986 by D. Reidel Publishing Company.*

INTERLAYER COMPLEXATION

Crown ethers and cryptands penetrate into the interlayer space of smectites (montmorillonite and hectorite) or vermiculite to form complexes
with their interlayer cations. This reaction occurs as a spontaneous process under mild experimental conditions, the adsorption taking place from
diluted solutions of macrocycles in anhydrous organic solvents (6).
    Complexation of the intracrystalline cations by macrocyclic com-
pounds causes a substitution of the water molecules which originally belong to the coordination shell of the cations, and produces an increase
of the interlayer distance as represented in the Scheme 1.

Scheme 1

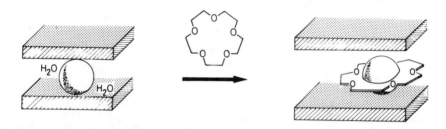

    The change of the hydration sphere of the interlayer cations by
macrocyclic ligands is clearly evidenced by IR spectroscopy (7). Fig.1
shows the IR difference spectrum obtained from a Na-montmorillonite film
and from the same sample after treatment with a methanolic solution of
the 18-crown-6 (18C6). Thus, in the "negative absorbance" region appear
ν_{OH} and δ_{HOH} vibrations of water molecules in the natural silicate. In the
"positive absorbance" region, the IR bands correspond to the characteristic vibrations of the intercalated 18C6. Quantitative measurements indicate that the interlayer water content decreases linearly with the amount
of the adsorbed macrocyclic compounds as determined by conventional car-

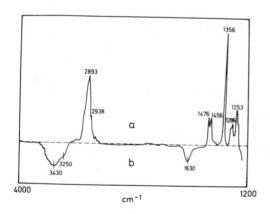

Figure 1. IR difference spectrum
obtained from the IR spectra of
Na-mont./18C6 complex and of the
starting Na-montmorillonite.

a)positive region:adsorbed 18C6
b)negative region:desorbed H_2O

(After Casal and Ruiz–Hitzky, in
reference 7).

bon microanalysis. So, in Fig.2,the decrease of the water content in Na-montmorillonite is plotted with respect to the amount of 18C6 intercalated. A certain quantity of water remains in the silicate even when the maximum of the 18C6 has been intercalated. This indicates that a fraction of the cations can not be complexed, due to the steric hindrance imposed by the organic ligands in the interlayer space.

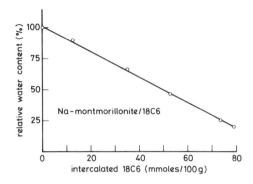

The substitution of water molecules during intercalation of macrocyclic polyethers is an irreversible process, in contrast to most other clay organic complexes obtained from adsorption of neutral organic species, particularly if the exchangeable cation are of the alkaline group. This is in agreement with the unusual stability which characterizes the complexes between macrocyclic polyethers and alkaline-cations formed in homogeneous media (8).

Figure 2. Decrease of H_2O content in Na-montmorillonite as a function of 18C6 intercalated.

The coordination between the macrocyclic ligands and the interlayers cations has been supported by IR and UV spectroscopies. The perturbation of the characteristic absorption bands of the macrocyclic compounds can be correlated with the cation-ligand interactions (7). These interactions can also be observed when the interlayer cation (i.e., NH_4^+ ions) presents absorption bands in the mid-infrared region. The spectral modifications in the 2800-3500 cm^{-1} ν_{NH} stretching region, have been explained as a result of the lowering of the T_d symmetry, characteristics of free ammonium cations, to C_{3v} or C_{2v} symmetries (9). This symmetry changes have been interpreted as a consequence of hydrogen bonding interactions between ammonium ions and oxygen atoms of the intercalated macrocyclic compounds.

The most satisfactory evidence of the ligand-cation coordination in the intracrystalline space of clays has been recently provided by laser microprobe mass analysis (LAMMA). This is a "soft ionization" technique, mainly developed in the last decade which permits to analyze solid state samples directly by mass spectrometry. Using this technique we have clearly recognized the peaks corresponding to the intact cation-ligand complexes and some fragments issued from their decomposition, along with the fragmentation ions of the silicate host lattice (10). Fig.3 shows the LAMMA spectrum of Na-montmorillonite/15C5 complex,where the intense signal at m/z 243 is assigned to (15C5-Na)$^+$ species. Similar results were also obtained for alkali-cryptand intercalations.

Interesting observations for the complexation of cryptands have been made when transition metal cations are the interlayer cations in the starting clays. In this case, protonation of the adsorbed cryptand is induced by a proton transfer reaction which involves dissociated water molecules in the interlayer space, as represented by the Scheme 2.

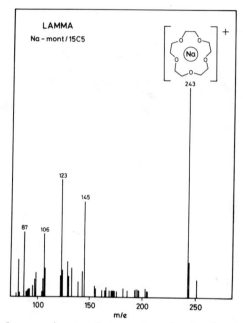

Figure 3. Laser microprobe mass analysis (LAMMA) in the positive ion detection mode of the Na–montmorillonite/15C5 complex.

Scheme 2

$$M^{2+}(H_2O)_n \rightleftharpoons [M(OH)]^+(H_2O)_{n-1} + H^+$$

$$[\,\text{cryptand}\,] \overset{\cdot}{\vdots} N \;+\; H^+ \rightleftharpoons [\,\text{cryptand}\,] \overset{\cdot}{\vdots} N^+ - H$$

Protonation of the cryptand in the interlayer space could be demostrate by IR spectroscopy (7). Quantitative chemical analysis shows that the initial content in interlayer cations does not change after adsorption of the macrocyclic. Thus, one may suppose that $M(OH)^+$ species coexist with monoprotonated cryptands, in order to balance the electrical charge in the system.

The source of protons can also be ammonium or alkylammonium ions if they are the saturating interlayer cations (7). In these cases, the free bases, i.e. ammonia or alkylamines, are displaced from the interlayer space to the solution when the cryptand (L) is intercalated, according to the following equation:

$$[\,\text{silicate}\,] - RNH_3^+ \;+\; L \rightleftharpoons [\,\text{silicate}\,] - LH^+ \;+\; RNH_2$$

Similar proton transfer reactions have already been reported for the interlayer adsorption of organic bases (pyridine, methylamine, etc.) in NH_4^+ saturated clays (11,12).

STOICHIOMETRIC CONSIDERATIONS

Adsorption isotherms at 298 K, obtained in diluted methanolic solutions, are of the H type according to the Giles <u>et al</u> classification (13), showing the high affinity of the host lattice for the guest species. The "plateau" of the isotherms may serve to calculate the experimental monolayer capacities (x_m) for each homoionic phyllosilicate. As an example, Fig.4 shows the adsorption isotherms of different macrocyclic compounds in Ba-montmorillonite, with x_m values (expressed in mmole/g) of 0.9 for 15C5 and 0.5 for 18C6 and cryptand C(222).

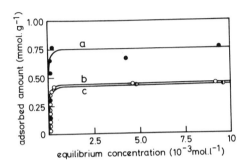

Figure 4. Adsorption isotherms (methanol, 298 K) for the intercalation of 15C5 (o), 18C6 (o) and cryptand C-222 (o) into Ba-montmorillonite.

Since the number of adsorption centers, given by the cation exchange capacity (CEC), is 0.45 mmole/g for a homoionic M^{2+}-montmorillonite, the observed x_m indicates that either 1:2 or 1:1 cation/ligand complexes are formed in the interlayer space, in agreement with results from X-ray diffraction analyses (14). We define here the <u>interlayer coordination ratio</u> (ICR) as the ratio between the amount of adsorbed macrocyclic compounds and the quantity of interlayer cations in the silicate, i.e. ICR= $z.x_m$/CEC. The ICR values obtained for intercalation of several macrocyclic species into homoionic montmorillonites are shown in Table I. It is important to remark that the ICR concept differs from the stoichiometry of the cation/ligand interlayer complexes. In general, the ICR values don't agree with the stoichiometry found in complexes obtained in homogeneous medium. The main cause of this fact may be sterical restrictions imposed by a constrained system such as the intracrystalline environment of phyllosilicates. Thus, the charge of the silicate, and consequently the available area per cation, determines the extent of interlayer complexation as well as the arrangement of the macrocyclic ligands in the interlayer space.

In fact, when homoionic montmorillonites saturated with <u>monovalents</u> <u>cations</u> (z=1) are used, the ICR is in general, less than unity because the area occupied by the macrocyclic molecule exceeds the available area per cation, and a fraction of these remains uncomplexed (15). Exceptions are the complexes formed by adsorption of the smallest crown ethers (12C4 and 15C5). In this cases, coordination with cations larger than the cavity size of the ligand, gives rise to formation of 1:2 cation/ligand complexes, which have also been found in homogeneous medium (4).

TABLE I

Macrocyclic ligands		Interlayer cations								
		Na$^+$	K$^+$	NH$_4^+$	Mg^{2+}	Sr^{2+}	Ba^{2+}	Co^{2+}	Ni^{2+}	Cu^{2+}
	n = 1 : 12C4	2.0	0.8	1.0	•	–	1.8	•	•	•
	n = 2 : 15C5	0.8	0.7	0.7	•	2.3	2.0	•	•	•
	n = 3 : 18C6	0.9	0.8	0.8	•	1.1	1.2	•	■	•
	n = 1 : DB18C6	0.8	0.6	0.7	•	–	0.7	•	•	•
	n = 2 : DB24C8	0.6	0.5	0.7	•	–	0.8	•	■	■
	n = 0 : C(2 2 1)	0.8	0.7	0.7	0.8	1.0	1.1	–	1.3	0.9
	n = 1 : C(2 2 2)	0.7	0.7	0.7	int.	1.3	1.1	1.0	1.0	–

Interlayer coordination ratio (ICR) values obtained for intercalation of different macrocyclic compounds into homoionic montmorillonites.

Nevertheless, complexation involving K$^+$ or NH$_4^+$ cations showed a differents behaviour depending on whether it occurred on the interlayer space or in a homogeneous medium. In the second case, the size relation between the cations ($r_i \sim 1.4$ Å) and the cavity of the macrocyclic ligands ($r_c = 0.7$ Å and 1.0 Å for 12C4 and 15C5 respectively) favours formation of the 1:2 cation/crown complexes, in agreement with the ion–radius–cavity concept, which has been used to explain the stoichiometry of the complexes obtained in homogeneous media (4). So, a "sandwich" type complex has been reported when complexation between potassium and a five oxygen member ring such as benzo–15–crown–5 takes place (16). For the interlayer complexation, on the contrary, ICR values as well as X–ray diffraction analysis, indicate that 1:1 complexes are formed by intercalation of crown ethers in K$^+$ or NH$_4^+$ saturated montmorillonites. This behaviour may be explained by the strong tendency of these cations to become coordinated to the "hexagonal holes" of the silicate layers. In this way, K$^+$ or NH$_4^+$ forms "sandwich" type complexes between the crown ether from one side and the "hexagonal holes" of the silicate from the other.

Intercalations of voluminous crown ethers as dibenzo–24–crown–8 (DB24C8), leads to a decrease of the ICR, as would be predicted from steric arguments with those ligands. 2:1 ion–paired bimetallic complexes are obtained for complexation of Na$^+$ and K$^+$ by DB24C8 in homogeneous media (17), but a similar coordination does not appear probable in the interlayer space of a montmorillonite, because the average distance between neighbouring cations (>10 Å) is larger than the cavity diameter of the ligand (~ 4 Å).

In homoionic clays saturated by <u>divalent cations</u>, the available area per cation is twice as large as for monovalent cations and consequently the ICR values increase for both crown ethers and cryptands intercalations as shown in Table I. Although Mg^{2+}/crown ether complexes have recently been described (4), numerous attemps to intercalate these macrocyclic ligands in Mg^{2+} saturated clays were so far unsuccessful, even under drastic experimental conditions. Similar results are obtained for clays saturated with transition metal cations as Co^{2+}, Ni^{2+} and Cu^{2+}. This result can be explained by the strong polarizing effect of these cations with prevents a substitution of their hydration shell by the crown ether.

Macrobicyclic compounds (cryptands C-222 and C-221) have spheroidal three-dimensional cavities, able to encapsulate selectively a large variety of metal cations, in 1:1 complexes (cryptates), even when high charge density cations as Ca^{2+}, Mg^{2+}, Co^{2+} or Cu^{2+} are involved (18). The so called "cryptate effect" is invoked to explain the high stability of these complexes compared with the binding properties of the crown ethers (4,5,18). Intracrystalline complexation by cryptands gives also 1:1 coordination complexes, the extent of the intercalation being limited by steric hindrance (Table I).

STRUCTURAL FEATURES

Intercalation of macrocyclic compounds causes an increase of the interlayer spacing (Δd_{001}) which can be observed by X-ray diffraction. Crown ether intercalation (Table II) is characterized by values of Δd_{001} between 4 and 9 Å. These values can be explained by arrangement and stoichiometry of the interlayer complexes, which are ruled by:

      a) charge and size of the interlayer cations
      b) cavity radius of the crown ethers
      c) steric hindrance of the ligands

TABLE II

MACROCYCLIC LIGANDS (cavity radius, Å)	$\Delta d_{001} (Å)^a$				
	Na^+ (0.95)^b	K^+ (1.33)	NH_4^+ (1.36)	Sr^{2+} (1.13)	Ba^{2+} (1.35)
12 - crown - 4 (0.7)	8.2	4.5	4.6	6.8	6.7
15 - crown - 5 (1.0)	4.1	4.5	4.6	7.9	8.1
18 - crown - 6 (1.4)	6.4	4.5	4.3	3.9	4.0
DB 12 - crown - 6 (1.4)	9.0	8.0	8.5	-	7.8
DB 24 - crown - 8 (2.0)	8.3	9.1	8.4	-	7.2

a) $\Delta d_{001} = \bar{d}_{001} - 9.5$
b) in parentheses : ionic radius (Å)

Increase in the basal spacing (Δd_{001}) of homoionic montmorillonites produced by intercalation of crown ethers.

The ratio between cavity radius (r_c) and ionic radius (r_i) is a determining factor in the structure and stoichiometry of the crown ether interlayer complexes. If the r_c/r_i ratio is greater than 1, then the cation can be enclosed in the cavity of the macrocyclic compound and 1:1 complexes are formed. In this case, Δd_{001} is close to 4 Å, which correspond to the thickness of a crown ether molecule, supposing a planar arrangement of the ligand with respect to the plane defined by the silica-

te layers (Fig. 5 a). A monodimensional Fourier synthesis for a Na-montmorillonite /15C5 complex also supports this disposition, since Na$^+$
is found at the same height as the cavity of the crown ether which
lies parallel to the layer surface (14). Moreover, a dichroic effect for
the CH$_2$ rocking bands in the IR spectra of the intercalates confirms
this planar arrangement of the crown ethers in the 4 Å phases (7).

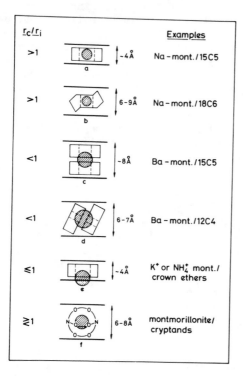

Figure 5. Schematic interlayer arrangement proposed for homoionic smec-
tites/crown ethers (or cryptands) complexes.

     Other orientations of the ligands with respect to the silicate
layers are also found in 1:1 complexes, e.g. in the Na-montmorillonite/
18C6 complex. In this case the observed Δd_{001} (6.4 Å) can be explained
by a distorsion of the cavity of the crown in order to suit the size of
the Na$^+$ cation (Fig. 5b), in analogy to what is known from similar complexes in homogeneous media (19). For DB18C6 and DB24C8 crown ether intercalations, the high values of Δd_{001} obtained can be due to: i)"fold-
ing" of the ligand around the cation in order to obtain maximum coverage
and/or ii) steric hindrance of the macrocyclic ligand in the interlayer
space which causes overlaping of adjacent molecules. At the present, it
is not possible to decide which is the main cause of this Δd_{001} observed.

     When the cation size exceeds the cavity size of the crown ether,
i.e. $r_c/r_i < 1$, 1:2 interlayer complexes are formed, at it has been de-

duced from X-ray diffraction and chemical analysis. As already said, exceptions are the K^+ and NH_4^+-crown ether-montmorillonite complexes, in wich 1:1 intracrystalline complexes are found (Fig. 5 e).

The Δd_{001} values observed for the 1:2 complexes are generally compatible with two ligands "sandwiching" the cation either on a coplanar ($\Delta d_{001} \sim 8$ Å) or on a tilted ($\Delta d_{001} \sim 6$-7 Å) configuration with respect to the silicate layers (Figures 5 c and 5 d). Both configurations have been supported by the observation of dichroic effects on the IR spectra (7).

Intercalations of <u>macrobicyclic cryptands</u> C(222) and C(221) into homoionic smectites (montmorillonite and hectorite) produces 1:1 complexes (Fig. 5 f) with an increase of the interlayer spacing comprised between 6 and 8 Å (Table III).

TABLE III

INTERLAYER CATION	IONIC RADIUS (Å)	Δd_{ool} (Å)[a]	
		C(221) $r_c = 1.16$Å	C(222) $r_c = 1.4$Å
Na^+	0.95	7.6	7.9
K^+	1.33	(*)	8.0
NH_4^+	1.36	(*)	7.3
Mg^{2+}	0.65	5.9	Int.
Ca^{2+}	0.99	6.8	6.6
Sr^{2+}	1.13	7.0	6.9
Ba^{2+}	1.35	6.8	7.2
Cu^{2+}	0.72	6.3	6.9
Ni^{2+}	0.68	6.3	7.0
Co^{2+}	0.79	5.9	7.0

a) $\Delta d_{ool} = \bar{d}_{ool} - 9.5$
r_c = cavity radius
(*) : Segregation of two crystalline phases ($\Delta d_{ool} = 7$-9Å)
Int : Interstratified material

Increase in the basal spacing (Δd_{001}) of homoionic montmorillonites produced by intercalation of criptands.

The observed differences in the Δd_{001} values may be correlated with the electrical charge of the interlayer cations. So, for monovalent cations the increase of the interlayer distances was found to be greater than for divalent cations. Conformation charges of the cryptate molecule might be induced by the nature of the interlayer cation and the considered constrained system but predictions appear to be difficult at present, and further investigations need to be made.

CONCLUSIONS

Results are given concerning the ability of macrocyclic poliethers to bind, in heterogeneous conditions, cations located in a intracrystalline environment such as the interlayer space of charged 2:1 phyllosilicates. The counter-ion is in this case a macroanion constituted by the

layers of the silicate with a diffuse negative charge wich determines
the number of cations in the interlayer space. In such a system, complex
ation is mainly ruled by the silicate charge density, the nature of the
interlayer cations as well as the steric hindrance of the macrocyclic
ligands. This characteristics together with the two-dimensional organization of the solid, impose restrictions on the stoichiometry. packing
arrangement and structural features of the intracrystalline complexes.

The most remarkable features of this family of intercalation compounds are their high stability and the ion-exhange properties without
an appreciable desorption of ligands (14). These properties are very
suitable for studing these materials as membrane models, and they open
the way to a variety of applications such as selective ion-exchangers
and phase transfer catalysts.

ACKNOWLEDGEMENTS

Helpfull discussions with Professors J.M. Serratosa and J.A. Rausell are
gratefully acknowledged. We wish to thank Dr. M. Gregorkiewitz for cri-
tically reviewing the manuscript. This work was partially supported by
the Comision Asesora de Investigación Científica y Técnica (CAICYT),
Spain.

REFERENCES

1. THENG, B.K.G. The Chemistry of clay-organic reactions, Adam Hilger,
      London, (1974).
2. PEDERSEN, C.J. 'Cyclic polyethers and their complexes with metal
      salts', J. Am. Chem. Soc., 89, 7017 (1967).
3. DIETRICH, B., LEHN, J.M. and SAUVAGE, J.P., 'Diaza-polioxa-macrocycles
      et macrobicycles', Tetrahedron Lett., (34), 2885 (1969).
4. POONIA, N.S. 'Multidentate macromolecules: Principles of complexion
      with alkali an alkaline earth cations' in Progress in Macro-
      cyclic Chemistry, Vol. 1, (R.H. Izatt and J.J. Christensen,
      Eds.), John Wiley and sons, N.Y. (1981).
5. LEHN, J.M. 'Cryptates: The Chemistry of macropolycyclic inclusion
      complexes', Acc. Chem. Res, 11, 49 (1978).
6. RUIZ-HITZKY, E. and CASAL, B. 'Crown ether intercalations with phy-
      llosilicates',Nature, 276, 596 (1978).
7. CASAL B. and RUIZ HITZKY, E. 'Estudio por espectroscopía IR y UV de
      la adsorción interlaminar de poliéteres macrocíclicos en filo-
      silicatos', Op. Pur. Apl., 18, 49 (1985).
8. LAMB, J.D., IZATT, R.M. and CHRISTENSEN, J.J. 'Stability constants
      of cation macrocyclic complexes and their effects on facili-
      tated membrane transport rates' in Progress in Macrocyclic
      Chemistry, Vol. 2, (R.M. Izatt and J.J. Christensen, Eds.)
      John Wiley and sons, N.Y. (1981).
9. CASAL, B., RUIZ HITZKY, E., SERRATOSA, J.M. and FRIPIAT, J.J. 'Vibra-
      tional spectra of ammonium ions in crown-ether-NH_4^+-montmori-
      llonite complexes', J. Chem. Soc., Faraday Trans. I, 80, 2225
      (1984).
10. RUIZ HITZKY, E., CASAL, B., VAN VAECK, L. and ADAMS, F., Unpublished

results.
11. RUSSELL, J.D., CRUZ, M.I. and WHITE, L. 'Adsorption of 3-amino-tri-
    azole by montmorillonite', J. Agr. Food Chem., 16, 21 (1968).
12. RAMAN, K.V. and MORTLAND, M.M. 'Proton transfer reactions at clay
    mineral surfaces', Soil Sci. Soc. Am. Proc., 33, 313 (1969).
13. GILES, C.H., Mac EWAN, T.H., NAKHMA, S.N. and SMITH, D. 'Studies in
    adsorption. Part. XI. A system of classification of solution
    adsorption isotherms, and its use in diagnosis of adsorption
    mechanism and in measurement of specific surface areas of so-
    lids', J. Chem. Soc., 3973 (1960).
14. CASAL, B. 'Estudio de la interaccion de compuestos macrocíclicos
    (éteres-corona y criptandos) con filosilicatos', Ph. D. Thesis,
    Univ. Complutense, Madrid 1983.
15. CASAL, B., and RUIZ HITZKY, E. 'Interlayer adsorption of macrocyclic
    compounds (crown-ethers and cryptands) in 2:1 phyllosilicates
    I. Isotherms and kinetics', Clay Min. (in press).
16. PEDERSEN, C.J. and FRENSDORFF, H.K. 'Macrocyclic poliehters and their
    complexes', Angew. Chem., Internat. Edit., 11, 16 (1972).
17. FENTON, D.E., MERCER, M. POONIA, N.S. and TRUTER, M.R. 'Preparation
    and crystal structure of a binuclear complex of potassium with
    one molecule of cyclic polyether: bis(potassium thiocyanate)
    dibenzo-24-crown-8', J.C.S. Chem. Comm., 66 (1972).
18. DIETRICH, B. 'Cryptate complex' in Inclusion compounds, Vol. 2,
    (Atwood, J.L., Davies, J.E.D. and Mac Nicol, D.D., Eds.) Aca-
    mic Press (1984).
19. DOBLER, M., DUNITZ, S.D. and SEILER, P. 'Hydrated sodium thiocyanate
    complex of 1,4,7,10,13,16-hexaoxacyclooctadecane'. Acta Crys-
    tallogr., 30 B, 2741 (1974).

SELECTIVITY INDUCED BY HETEROGENEOUS CONDITIONS IN THE ALKYLATION OF
ANIONIC REAGENT

G. BRAM
Laboratoire des Réactions Sélectives sur Supports
UA CNRS 478, Bâtiment 410, Université de Paris-Sud
F 91405 ORSAY (France)

ABSTRACT. It is possible, starting from anionic species, to perform
efficiently, and in absence of solvent, organic synthesis in different
heterogeneous conditions. These " dry media " reactions not only exhibit
interesting reactivity, in very mild experimental conditions, but also
give rise to peculiar selectivity induced by the heterogeneous media.
Some selective anionic alkylations performed in different heterogeneous
conditions will be presented and discussed : reactions on solid inorganic supports, solid-liquid " without solvent " phase transfer catalysis,
and uncatalysed solid-liquid heterogeneous reactions.

It is well known that organic synthesis is usually performed in homogeneous media : reactions are effected after the reactants have been
dissolved in a suitable solvent . One of the major development in
organic synthesis during the last ten years has been the use of heterogeneous condition of reactions : the more and more widespread
utilization of phase transfer catalysis (PTC)[1-4] and the use of solid
organic[5] and inorganic[6] supports are illustrative . We have recently
developped three different types of heterogeneous conditions of reactions in the field of anionic activation and particularly anionic
alkylation :
        -alkylation on solid inorganic supports
        - solid-liquid phase transfer catalysis (PTC) "without added
solvent".
        -uncatalysed solid-liquid reactions.
    These heterogeneous processes exhibit two common features : reac-
tions are performed in the absence of any solvent in the so-called
"dry media"[7] conditions and at temperature kept below or equal to 85°C.
By this way it is possible not only to achieve a good reactivity in
very mild and economical conditions with a very simple work-up, but also
to induce selectivities of reaction impossible to obtain in homogeneous
media.

R. Setton (ed.), Chemical Reactions in Organic and Inorganic Constrained Systems, 191–196.

Alkylation of ambident anions

The alkylation of acetoacetate anion 1 (figure 1)presents two types of selectivities : ambident reactivity, C-versus O-alkylation, depending on orbital control versus charge control for the reaction, and mono-C versus di-C selectivity depending on the shift of an equilibrium involving acetoacetate and mono-C alkylated acetoacetate anions .

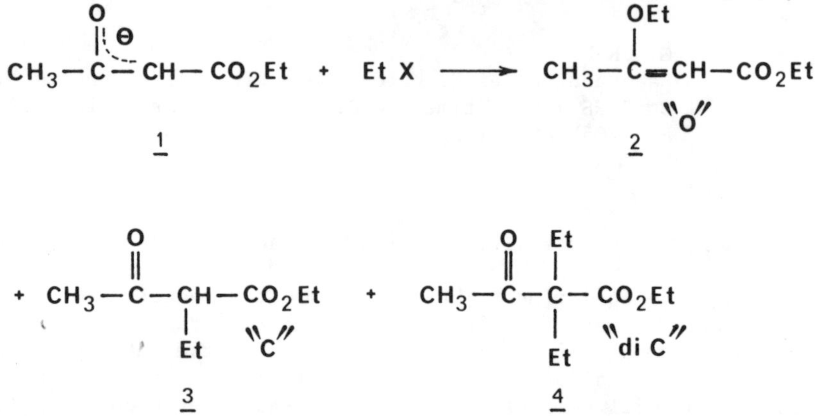

$$CH_3-C=CH-CO_2Et \quad + \quad Et\,X \quad \longrightarrow \quad CH_3-C=CH-CO_2Et$$

1
"O"
2

$$+ \quad CH_3-C-CH-CO_2Et \quad + \quad CH_3-C-C-CO_2Et$$

Et "C"
Et "di C"
3
4

Figure 1. Alkylation of acetoacetate anion

We found that when the reaction [9,10] is performed on alumina in "dry media " conditions at 25°C mono C-alkylated product 3 is formed essentially selectively, even with Et_2SO_4 only traces of di-C alkylated product 4 and O-alkylated compound 2 are observed. At 85°C a more reactive but less selective system is obtained; appreciable quantities of di-C-alkylated compound 4 are produced, but still only a trace of enol ether is formed (Table 1)

Table 1 : Alkylation of acetoacetic ester anion 1

alkyl. agent	temp.	time (days)	total yield %	2"O"%	3"C"%	4"diC"%
Et_2SO_4	85°C	1	80	2	63	35
	25°C	5	76	2	96	2
EtBr	"	"	53	1	97	2
EtI	"	"	52		97	3

The anion of β - naphtol, like the other phenoxide ions, is an ambident anion [8,11] . By alkylation for instance with benzylbromide, four products can be formed (figure 2) and usually all are formed competitively.

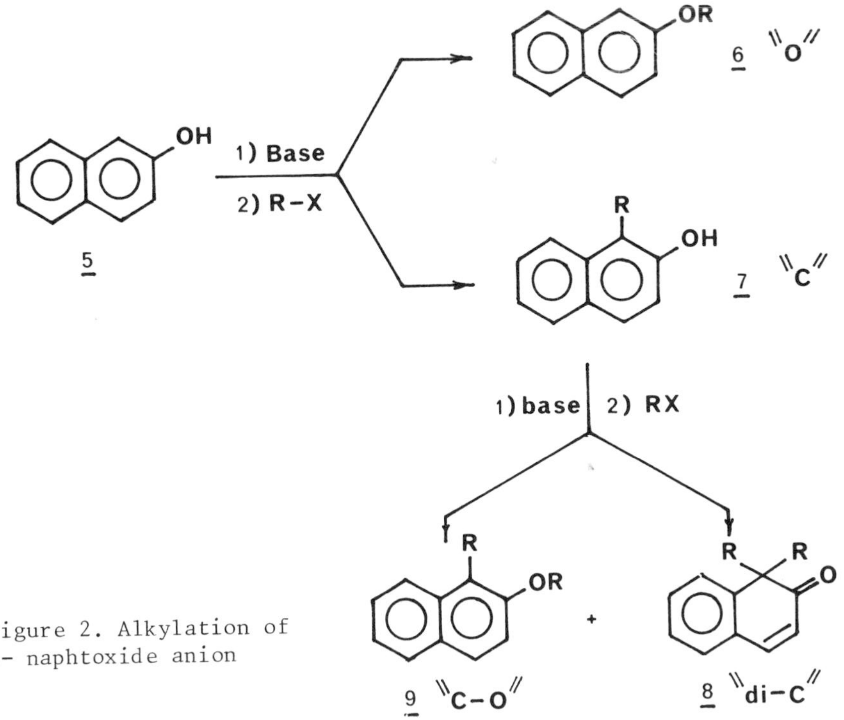

Figure 2. Alkylation of β- naphtoxide anion

It is well known[8] that C-alkylation of phenoxides anions by SN_2 process
is difficult : contrary to enolates, O-alkylation of phenoxides is
highly favoured because aromaticity is lost in the transition state
leading to the mono C-alkylated product 7 and when the di C-alkylated
compound 8 is formed. Actually, if many syntheses of O-benzylated ether
6 are described, there is, to our knowledge, no preparative method
giving selectively each of the four different alkylation products .
We were able to carry out[12] such selective syntheses by using heterogeneous solid-liquid reactions performed in the absence of any organic
solvent. The observed selectivities depend on the nature of the solid
base used (Table II)

Table II Benzylation of β - naphthoxide anion

a	Base	t(h)	Temp.	6"O"%	7"C"%	8"di-C"%	9"C O'%
1:1:1	KOH-Aliquat(40%)	3	60°C	85[b] (75)[c]			
1:2:2	LiOH	2	85°C		96[b] (90)[c]		
1:3:3	LiOtBu	2	85°C			88[B] (80)[c]	10[b]
1:1.5.1	2 +KOH-Aliquat (40%)	2	85°C				85[b] (74)[c]

a: ratio naphthol: base : $PhCH_2$. b:Yield byVPC c:Yield in isol.
product

- The O-alkylation to give 6 is due to solid-liquid PTC alkylation without added solvent[13], solid KOH being the anion-generating base and Aliquat 336,essentially (nC$_8$H$_{17}$)$_3$ N$^+$CH$_3$Cl$^-$ the phase transfer catalyst.

- Mono C-alkylated naphtol 7 is obtained in the absence of catalyst with LiOH as base

- Di C-alkylated compound 8 is directly and selectively obtained when the stronger base LiOtBu is used, in the absence of catalyst .

- C- and O-dialkylated product 9 is easily obtained when 7 is O-alkylated in the presence of solid KOH. In every case good yields are obtained under mild and economical conditions and with a very easy work-up .

Selective N-propargylation of heterocycles

N-propargyl derivatives 11 of heterocycles 10 are of interest because of their pharmacodynamic properties, but their usual synthesis (figure 3) in basic media gives rise to low yields of 11, together with allenes 12 and ynamines 13 ; compounds 12 and 13 being resonance stabilized[13].

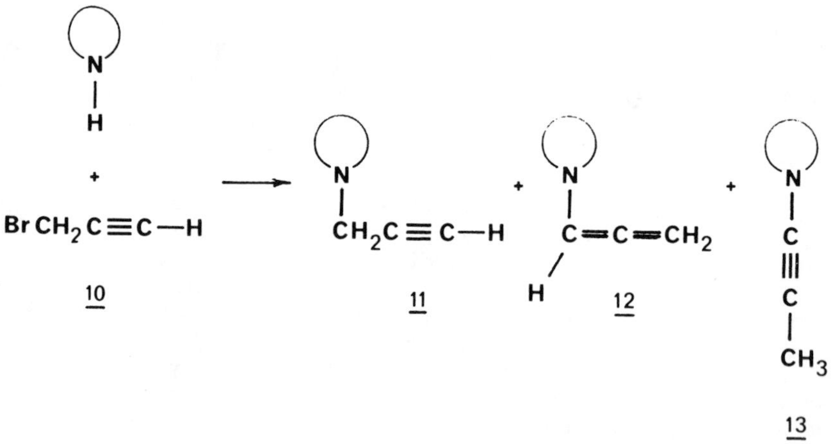

Figure 3. Alkylation of N- heterocycles by propargyl bromide .

We have found[14] that N-propargyl dérivatives 11 are selectively obtained when the heterocycles 10 are alkylated with propargyl bromide in solid-liquid PTC without solvent or on alumina in " dry media " conditions. In some cases use of solid-liquid PTC without solvent in the presence of mineral oxides works efficiently (Tables III).

Table III   Alkylation of heterocycles 10 by propargyl bromide

Exp	10	Method[a]	Temp°C	Yields % 11	12
1	pyrrole	A	20	20	46
2	pyrrole	B	20	55	0
3	indole	A	20	83	2
4	imidazole	A	20	35	45
5	imidazole	A	55	0	81
6	imidazole	B	20	85	12
7	imidazole	C	20	74	1
8	imidazole	D	20	0	0
9	imidazole	F	20	76	0
10	benzimidazole	A	20	81	0
11	benzimidazole	E	20	55	0
12	2-methyl benzimidazole	A	20	45	34
13	2-methyl benzimidazole	B	20	72	0
14	acridanone	A	20	95	0
15	acridanone	F	20	42	0

a) A:Solid-liquid PTC; B:solid-liquid PTC + alumina;
   C:Solid-liquid PTC +titanium dioxide; D:solid-liquid + zinc
   oxide; E:solid-liquid PTC + silica; F:reaction on alumina

In the alkylation of acridanone and benzimidazole, solid-liquid PTC without any solvent exclusively leads to propargyl derivatives 11, but variable amounts of 1-allenyl isomers 12 are formed in the case of pyrrole ,
imidazole, and methyl-2 benzimidazole (Expts : 1,4,12). The formation
of ynamines 13 is never observed. On the other hand, alkylation of heterocycles on alumina in " dry media" (Expts : 9,15) is always selective
but appears to be less efficient (Expt: 15). Furthermore, addition of
small amounts of mineral oxides into solid-liquid without solvent PTC
reaction médium is noteworthy : silica reduces the yields (Expt :11)
as it was previously reported in the case of indole alkylation , alumina
or titanium dioxide enhances the selectivity of the reaction , the yield
being only slightly reduced (Expts : 2, 6, 13). Zinc oxide completely
inhibits the reaction, propargyl bromide being converted to allenylbromi-
de. This property of zinc oxide to catalyse allene formation is well
documented [15]

References
1- Weber W.P., Gokel G.W., " Phase Transfer Catalysis in Organic Chemistry". Springer- Verlag.New York 1977
2- Starks, C.M. Liotta C." Phase Transfer Catalysis". Academic Press.
New York 1978.
3- Dehmlov, E.V.; Dehmlov, S.S. " Phase Transfer Catalysis" 2nd edition
Verlag Chemie . Weinhem 1983
4- Caubère, P; " Le tranfert de phase et son utilisation en chimie

organique. Masson Paris 1982

5- Hodge,P; Sherrington, D.C., " Polymer Supported Reactions in Organic synthesis "   J. Wiley and Sons. New York 1980.

6- McKillop, A; Young, V.W. Synthesis, 1979, 401, 481.

7- Keinan, E; Mazur,Y. J. Amer. Chem. Soc. 1977,99, 1861.

8- Reutov,O.A; Beletskaya, I.P; Kurts, A.L. " Ambident Anions " Consultants Bureau New York 1983 .

9- Bram, G; Fillebeen-Kahn, T; Geraghty, N. Synthetic Comm.1980, 10,274.

10- Bram, G; Geraghty,N; Nee, G; Seyden-Penne,J. J.C.S. Chem. Comm. 1980, 325.

11- Kornblum, N; Smiley, R.A.; Blackwood, Iffland, D.C. J. Amer. Chem Soc 1955, 77, 6269.

12-, G; Loupy,A; Sansoulet, J; Vaziri- Zand,F. Tetrahedron Lett. 1984,25, 5035.

13- Theron, F; Verny, M; Vessière, R. in S. Patai " The Chemistry of carbon- carbon triple bond" J. Wiley New York 1978. Part I p. 389 .

14- Galons, H; Bergerat,I; Combet-Farnoux C; Miocque, M; Decodts, G; Bram, G. submitted for publication .

15- Saussey, J; Lavalley, J.C. J. Chem. Phys. 19 78, 75, 505.

THE REACTIONS IN CLAYS AND PILLARED CLAYS

James A. Ballantine
Chemistry Department
University College of Swansea
Singleton Park
Swansea SA2 8PP
United Kingdom

ABSTRACT. This article reviews the extraordinary variety of organic chemical processes which can be catalysed heterogeneously by clay materials.

INTRODUCTION

Acid-treated and cation-exchanged montmorillonites have been known for the last 50 years to possess the ability to catalyse organic reactions. In the period between the 30's and 60's acid-treated montmorillonites were major catalysts used in petroleum processing, but were later superseded by the more thermostable zeolites.
    In the past seven years, however, there has been a resurgence of interest in the use of modified montmorillonites as heterogeneous catalysts for a wide variety of organic reactions.
    The object of this presentation is to survey the present position of montmorillonites as heterogeneous catalysts for organic reactions.

EARLY INDUSTRIAL USES OF CLAYS.

The role of clays as heterogeneous catalysts in industry is really quite a modern one, although clay materials have been used for centuries for a variety of important industrial purposes such as the manufacture of pottery and house bricks and have found recent use as solid lubricants in the drilling muds for oil exploration.
    One of the earliest uses of montmorillonites in the chemical industry was as "bleaching agents" in clean-up processes (eg. Fuller's earth). The clays were used to absorb polar, coloured materials selectively to give products their desired clean appearance. This was obviously making use of the selective intercalation properties of the clays.

R. Setton (ed.), Chemical Reactions in Organic and Inorganic Constrained Systems, 197–212.

CLAYS AS CRACKING AND ISOMERISATION CATALYSTS.

Amongst the earliest catalytic roles for clay minerals was their use
as cracking catalysts for petroleum processing.   As early as 1930 a
Bavarian montmorillonite was used by Eugene Houdry as the initial
catalyst for the first 10 tonne catalytic cracker in the U.S.A. [1].
Since that time the catalytic cracking of gas oil was developed using
montmorillonites and modified clay catalysts in efforts to increase
the proportion of useful gasoline fractions.

The catalytic activity of natural montmorillonites was improved
considerably by treatment with hot acid followed by drying to give
material which was highly active as a cracking catalyst.   This hot
acid treatment was considered to leech out almost half of the
octahedral aluminium atoms in the lattice and to deposit them, on
heating, as Al_2O_3 on the catalyst surface [2,3].   These hot acid
treated montmorillonites were more stable at the high temperatures
required for cracking and retained their acidity better than
untreated or cold acid treated montmorillonites.

Hundreds of patents have been taken out describing the
production and properties of cracking catalysts by the modification
of natural clays.   Amongst the most active companies in this field
were Standard Oil Co. and the Houdry Process Co. in the 1940's, and
the Great Lakes Carbon Co. in the 1950's.   Catalyst development
proceeded in the 1960's and 70's with the introduction of synthetic
alumina-silicates made by hydrothermal methods and incorporating a
host of additional metals.   These catalysts were also efficient for
the hydrocracking and hydroforming of gas oil in the presence of
hydrogen.   One of the most active companies in this phase of
development was Chevron Research Co. who published numerous patents
in the 1970's.

The major problem of the clay cracking catalysts was their
hydrothermal instability.   Coke tended to build up in the pores of
the catalyst after some time on-stream and the usual way to remove
this coke was to treat the catalyst "in situ" with steam at high
temperatures.   This treatment was necessary but was detrimental to
the life and activity of the catalyst and clays were completely
replaced by zeolites in the 1970's, because of the much improved
stability of zeolites to this steaming process.

However, with the recent development of pillared clays in which
the clay sheets are crosslinked, or separated by metal oxide clusters
to prevent swelling [4,5,6,7], the petrol companies are having
another look at the efficiency and economics of using pillared clays
as catalysts for petroleum processing.

The catalytic isomerisation of straight chain saturated alkanes
over montmorillonites has been demonstrated by Chevron [8,9] and
Shell [10].   For example a C_{4-7} straight chain alkane mixture was
converted in 60% yield to iso-products over an acidified
montmorillonite having Ni and Co substitution at 250° in the presence
of hydrogen [10].

The organic chemistry of the cracking and reforming processes is
obviously very complex but it has been pointed out that there is some

relationship between the acidity of the clay and its cracking ability
and it is considered that carbocations are involved in many of the
processes [2,11,12].

METHANOL AND SYN-GAS CONVERSION.

Methanol can be transformed into a mixture of C_{2-4} alkenes over
cation exchanged montmorillonites [13] and over tetramethyl ammonium
montmorillonites into which silica has been deposited [14].
Methanol can also be converted into methyl methanoate over a
synthetic flortetrasilicic mica [15].

$$MeOH \quad \xrightarrow{\text{Al(III)-Mont } 350^{o}} \quad C_{2-4} \text{ Alkenes}$$

$$MeOH \quad \xrightarrow{\text{Cu-TSM } 300^{o}} \quad HCO_2Me \quad 68\%$$

Methane can be obtained from a mixture of CO and hydrogen
(Syn-gas) over a Ni and Ru catalyst [16,17]. Syn-gas can also be
converted into a hydrocarbon mixture rich in the C_{5-12} region by use
of a transition metal intercalated montmorillonite or Co substituted
montmorillonite [18,19].

$$CO \quad + \quad H_2 \quad \xrightarrow{\text{Co Mont } 175^{o}} \quad \begin{array}{ccc} C_{1-4} & C_{5-11} & C_{12+} \\ 22\% & 54\% & 24\% \end{array}$$

mainly iso-alkanes and alkenes

CLAYS AS CATALYSTS FOR OLIGOMERISATION REACTIONS

Alkenes undergo facile oligomerisation in the presence of acid
treated or ion-exchanged montmorillonites.   The process involves
protonation to give a stabilised carbocation which then reacts with
further alkene molecules to give dimers and oligomers after
deprotonation.

The process works best with heavily substituted alkenes and Dow Chemical Co. devised a process for removing tertiary alkenes from an alkene product stream by the selective dimerisation of the tertiary alkenes [20], to provide a stream free of tertiary alkenes after fractionation.

Gulf Research describe an oligomerisation process whereby a mixture of C_{2-5} alkenes in a stream from a catalytic cracker is treated with a calcined montmorillonite [21] to provide useful C_{5-24} products.

Styrene has been polymerised using a variety of ion-exchanged montmorillonites [22,23] and α-substituted styrenes have been dimerised to both linear dimers by a Monsanto process [24] and to indane derivatives using a variety of cation-exchanged montmorillonites [25,26,27].

$$H^+\text{-Mont}\quad 100^\circ$$

90%

One of the earliest reported clay catalysed dimerisation reactions involved trans-stilbene which was intercalated in Cu(II)-exchanged montmorillonite and heated to 200° to give a dimer. The dimer was examined by mass spectrometry and was shown to have an identical spectrum to 1,2,3,4-tetraphenylcyclobutane but the compound was not completely characterised [28].

Dimerisation of triphenylamine within montmorillonite is followed, on heating, by a rearrangement of the benzidine type to yield N,N,N',N'-tetraphenylbenzidine as the sole product [29].

$$Ph_3N \xrightarrow{\text{Na-Mont}\quad 100^\circ} Ph_2N-C_6H_4-C_6H_4-NPh_2$$

Aromatic compounds undergo oligomerisation on heating, after treatment with a variety of cation exchanged montmorillonites; eg. Cu(II), Fe(III) or Ru(III).   The reactions are thought to proceed through the production of an aromatic cation radical species [30]. Oligomerisation of benzene [31], toluene [28], phenols [30,32,33,34] and aniline [35] have been studied.

Dienes can also undergo dimerisation in a self Diels-Alder type reaction and in this way Shell used a montmorillonite with lattice incorporation of Cu(I) and Cs(I) to cyclise butadiene to vinylcyclohexene [36].

   Acid treated montmorillonite has been used also to catalyse the
Diels-Alder reactions of activated dienophiles [37]. Recently a
number of Diels-Alder reactions have been performed using Fe(III)
doped montmorillonite ("clayfen").   The addition of 10% of
4-t-butylphenol to the catalyst allowed low temperature dimerisations
of such unactivated dienes such as   cyclohexadiene and
2,4-dimethylpenta-1,3-diene to proceed in high yield [38].
Improvements in reaction rate and selectivity of Diels-Alder
reactions were reported in low temperature reactions using Fe(III)
doped montmorillonite by modification of the organic solvent [39],
and successful low temperature reactions with furan derivatives have
been reported with this catalyst [40].   Acid treated
montmorillonites have also been successful in ene-reactions with
diethyl oxomalonate to produce γ-lactones [41].   These reactions are
to be discussed elsewhere at this meeting.

CLAYS AS CATALYSTS FOR DEHYDRATION REACTIONS

Both Al(III)- and proton-exchanged montmorillonites are efficient
catalysts for the intramolecular dehydration of secondary and
tertiary alcohols to the corresponding alkenes (eg. cyclohexanol to
cyclohexene in 88% yield) [1,42]. Al(III)-exchanged montmorillonite
is also effective for the elimination of ethanoic acid from
cyclohexyl ethanoate to cyclohexene in 98% yield [27].
   Acid treated montmorillonite was used by Gulf Oil Co. to
dehydrate propan-2-ol to di-(prop-2-yl) ether and propene in a flow
system [43].   A study of the elimination of water from primary,
secondary and tertiary alcohols over Al(III)-exchanged
montmorillonite has shown that primary alcohols gave mainly
di-(alk-1-yl) ethers with little alkene production, whereas secondary
and tertiary alcohols, other than propan-2-ol, gave the corresponding
alkene almost exclusively [42].

$$R\text{-}CH_2\text{-}CH_2\text{-}OH \xrightarrow[200^{\circ}]{Al(III)\text{-}Mont} R\text{-}CH_2\text{-}CH_2\text{-}O\text{-}CH_2\text{-}CH_2\text{-}R + R\text{-}CH=CH_2$$

                                          63%   3%

                                          4%   88%

   The fact that primary alcohols gave the di-alk-1-yl ether almost
exclusively must preclude the intermediacy of carbocation
intermediates in the production of the ether, as rearrangement to the
more stable alk-2-yl carbocation would certainly take place resulting
in a different product mix [42].

The diol, ethylene glycol, can be polymerised with Al(III)-exchanged montmorillonite at 200° to give polyethylene glycol mixtures and diethylene glycol can be cyclised to give a high yield of 1,4-dioxan in addition to oligomerisation [42].

$$HO-CH_2-CH_2-OH \xrightarrow{\text{Al(III)-Mont } 200^{\circ}} H-(-O-CH_2-CH_2-)_n-OH$$
$$60\%$$

$$HO-CH_2-CH_2-O-CH_2-CH_2-OH \xrightarrow{\text{Al(III)-Mont } 200^{\circ}}$$ $$60\%$$

Benzyl alcohol undergoes polymerisation with Al(III)-exchanged montmorillonite to yield poly(phenylenemethylene) by elimination of water via an electrophylic aromatic substitution route [42].

Primary amines and thiols also undergo similar elimination reactions with Al(III)-exchanged montmorillonite. Primary amines are converted to di-alk-l-ylamines by loss of ammonia [44] and thiols gave di-alk-l-yl thioethers by elimination of hydrogen sulphide [45].

$$2 \ R-CH_2-XH \xrightarrow[\text{Al(III)-Mont}]{-H_2X} R-CH_2-X-CH_2-R$$

Where X = O, NH or S.

Cycloalkyl amines gave high yields of the corresponding dicycloalkyl amines under these conditions, in strict contrast to the situation with cycloalkyl alcohols where intramolecular elimination to the cycloalkene is almost exclusively preferred [42,44].

The cyclic secondary amine, pyrrolidine, gave unexpected products on treatment with Al(III)-exchanged montmorillonite and a protonic mechanism has been proposed for this transformation which has no analogy in the solution chemistry of amines [44].

41%   14%

CLAYS AS CONDENSATION CATALYSTS.

The intermolecular elimination of water between two reactants is a very common acid catalysed organic process and can be successfully catalysed by montmorillonites.

Esters can be prepared in good yield from the reaction of alcohols with carboxylic acids in the vapour phase over acid treated montmorillonites [46].

Amides can be formed from amines and carboxylic acids. Glycine, activated by intercalation with Cu(II)-exchanged montmorillonite, can be coupled with benzoic acid to give low yields of hippuric acid [47].

It has been suggested that the polypeptides necessary for life could have been formed by condensation reactions in which amino-acids, activated in the form of aminoacyl-adenylate residues, coupled together to form peptide bonds in the interlamellar area of a montmorillonite.[48,49]. Simple model experiments have shown that amino-acids can be polymerised by this process in the presence of a montmorillonite [50]. The presence of the clay is essential for the polymerisation and Paecht-Horowitz has suggested a mechanism whereby the aminoacyl-adenylate binds to an outside edge site of the montmorillonite so that the ammonium terminal residue resides in the highly charged interlamellar zone and is activated towards a coupling reaction [51].

$$(aa)_n-AMP \; + \; (aa)_m-AMP \; \xrightarrow{\text{Mont}} \; (aa)_{n+m}-AMP \; + \; AMP$$

aminoacyl-adenylates   polypeptide-adenylate

Acetals are easily prepared by the use of acid treated or cation-exchanged montmorillonites. Cyclic acetals are obtained from carbonyl compounds and 1,2-diols [52], or from enol ethers and 1,2-diols [53], and from carbonyl compounds and ethylene oxide [54].

70%

An extremely rapid reaction occurs between carbonyl compounds and trimethyl orthoformate at room temperature in the presence of acid treated or Al(III)-exchanged montmorillonite to give the dimethyl acetal in quantitative yield in a few minutes [55].

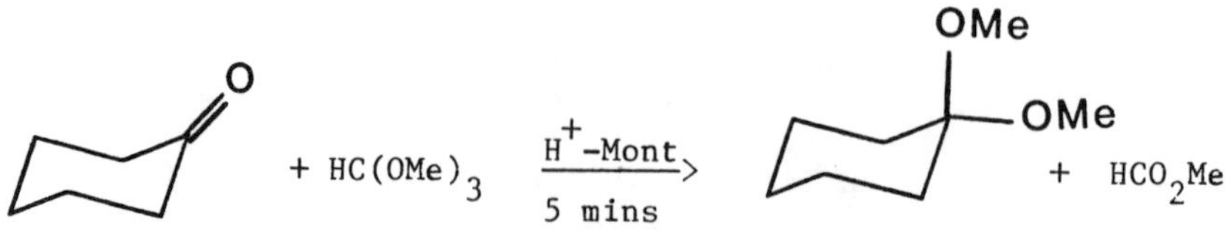

100%

Acetals of formaldehyde can be prepared conveniently from the
alcohol, dichloromethane and aqueous sodium hydroxide in the presence
of a quaternary ammonium exchanged montmorillonite as a phase
transfer catalyst [56].

Diacylate acetals of carbonyl compounds can be prepared in high
yields from the carbonyl compound and acetic anhydride using acid
treated montmorillonites [57] and a very useful acetal derivative for
the protection of alcohol groups during synthesis can be prepared by
the reaction with formaldehyde diethyl acetal in the presence of acid
treated montmorillonite by alcohol exchange [58].

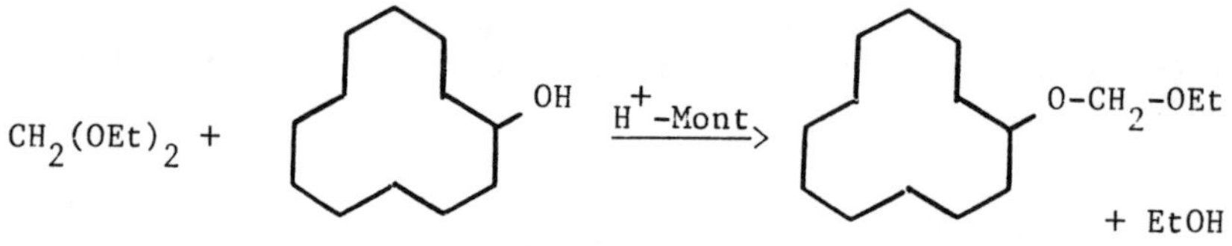

80%

Enamines can be efficiently prepared by the condensation of
carbonyl compounds with secondary amines, such as morpholine, in the
presence of acid treated montmorillonites in refluxing benzene [59].

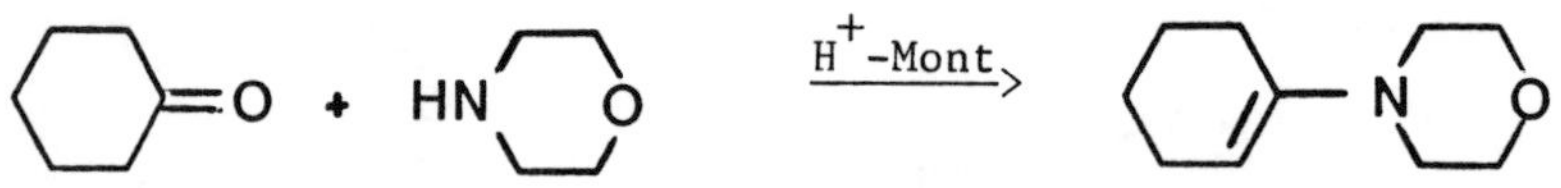

80%

CLAYS AS CATALYSTS FOR ADDITION REACTIONS

Acid treated and cation-exchanged montmorillonites are effective
catalysts for a whole range of addition reactions.   In the case of
alkenes the first step is the addition of either a proton , or Lewis
acid site.   The intermediate carbocation can then rearrange to give
more stable forms before reaction with a nucleophile.   The products
are then formed by elimination of the proton or Lewis acid site.

$$R-CH_2-CH=CH_2 \xrightarrow{\overset{+}{H}-Mont} R-CH_2-\overset{+}{C}H-CH_3 \longleftrightarrow R-\overset{+}{C}H-CH_2-CH_3$$

$$\downarrow \text{(i) R'XH} \qquad\qquad\qquad \downarrow \text{(i) R'XH}$$

$$\downarrow \text{(ii) } -H^+ \qquad\qquad\qquad \downarrow \text{(ii) } -H^+$$

$$R-CH_2-CH(XR')-CH_3 \qquad R-CH(XR')-CH_2-CH_3$$

Where X = O, S or O-CO, but not NH.

In this way water can add to alkenes to give secondary alcohols and di-alk-2-yl ethers [60] over Cu(II) exchanged montmorillonite.

Alcohols can add on to alkenes in the presence of montmorillonites to give ethers. The reaction with alk-1-enes is rather slow and gives mixtures of alk-2-yl and alk-3-yl ethers [42] whereas the reaction with 2-methylpropene gives high yields of the tertiary ether at much lower temperatures [61,62].

$$Me_2C=CH_2 \;+\; MeOH \xrightarrow{\overset{+}{H}-Mont} Me_3C-OMe \quad 75\%$$

Hydrogen sulphide and thiols can add to alkenes in the presence of acid treated and Al(III) exchanged montmorillonites at 200° to give good yields of thiols [63,64] and dialkyl sulphides [65,66].

$$R-CH=CH_2 \;+\; R'-SH \xrightarrow[200^{\circ}]{Al(III)-Mont} R-CH(SR')-CH_3 \quad 40\%$$

Ammonia can add on to ethyne over acid montmorillonite to yield both pyridine and piperidine [67]. However no addition reactions have been reported with alkenes.

Carboxylic acids can add to alkenes over Al(III)-exchanged or proton-exchanged montmorillonites to give esters. Ethene and ethanoic acid acid yield ethyl ethanoate in good yield in both batch and flow conditions [68,69].

$$CH_2=CH_2 \;+\; CH_3CO_2H \xrightarrow[200^{\circ}]{Al(III)-Mont} CH_3CO_2H \quad 80\%$$

Hex-1-ene and ethanoic acid yield a mixture of hex-2-yl ethanoate and hex-3-yl ethanoate [68]. Intramolecular addition of a carboxylic acid group to an alkene takes place in cyclooctene-5-carboxylic acid over Al(III)-exchanged montmorillonite at 120° to furnish a mixture of lactones [27].

Acetals can also add on to enol ethers using acid treated montmorillonite to give a triethoxyalkane which on hydrolysis with formate buffer gives high yields of α,β-unsaturated aldehydes [70].

$$CH_3-CH_2-CH(OEt)_2 + CH_2=CH-OEt \xrightarrow{\text{H}^+\text{-Mont}} CH_3-CH_2-CH(OEt)-CH_2-CH(OEt)_2$$

$$\downarrow HCO_2Na/HCO_2H$$

$$CH_3-CH_2-CH=CH-CHO \quad 83\%$$

2-Methoxyethanol has been added to ketene in the presence of acid treated montmorillonite to give the acetate in very high yield [71].

$$MeO-CH_2-CH_2-OH + CH_2=C=O \xrightarrow{\text{H}^+\text{-Mont}} MeO-CH_2-CH_2-OCOCH_3$$
$$98\%$$

Tertiary alcohols can add to nitriles under the influence of Al(III)-exchanged montmorillonite in a reaction analagous to the Ritter reaction to give substituted amides [72]. A similar reaction can be effected with 2-bromo-2-methylpropane and a nitrile using "clayfen" to produce an intermediate, which on subsequent hydrolysis gives the substituted amide [73].

$$CH_2=CH-CN + Me_3C-OH \xrightarrow[150^\circ]{\text{Al(III)-Mont}} CH_2=CH-CO-NH-CMe_3 \quad 36\%$$

Epoxides and cyclic ethers undergo a number of ring opening addition reactions in the presence of montmorillonite catalysts. Ethylene and propylene oxides undergo easy polymerisation to give high yields of polyalkylene glycols at 35° [74].   Water [1], alcohols [75,76] and carboxylic acids [76] can all be added very efficiently to ethylene oxide at moderate temperatures under the influence of montmorillonites to give the appropriate derivative of ethane-1,2-diol.

$$\triangleright\!\!O + R-OH \xrightarrow[40^\circ]{\text{H}^+\text{-Mont}} HO-CH_2-CH_2-OR$$

Where R = H, alkyl or acyl

Tetrahydrofuran has been converted to polybutyleneglycol ethanoate by treatment with ethanoic anhydride and acid treated montmorillonite at 130° [77]. Cyclic acetals such as butane-1,4-diol formal have been polymerised using acid activated montmorillonite at 65° to give 80% of the mixed polyether (BASF) [78].

CLAYS AS CATALYSTS FOR AROMATIC ELECTROPHYLIC SUBSTITUTION REACTIONS.

Acid treated montmorillonites have been used for many years as catalysts for the alkylation of aromatic compounds with alkenes or alcohols in Friedel-Crafts type reactions. Some of the early alkylations involved toluene and propene at 500° (Phillips Petroleum) [79], benzene with propene at 80° to give cumene (Aries) [80], phenol with propene at 130° (General Aniline & Film Co.) [81], and aniline with propene at 300° (Sud-Chemie) [73,82]. The products were mainly isomeric mixtures of the isopropyl substituted products and some selectivity was observed compared with non-clay catalysed reactions.

Montmorillonites have also been used to catalyse nitrations of toluene to give a preferential selectivity for para-notrotoluene compared with normal electrophyilc nitrations [83], and to catalyse the chlorination of toluene without any unusual selectivity [84].

41%   3%   56%

Acid treated montmorillonite has been used to catalyse the electrophilic substitution of protonated formaldehyde to aromatic amines [85] and phenols [86] to give diarylmethane derivatives of commercial interest.

Montmorillonite has been used in a highly selective electrophilic substitution, to catalyse the synthesis of unsymmetrical dipyrromethanes from the coupling of α-free pyrroles with α-acetoxymethylpyrroles. The usual problem of self reaction of the α-acetoxymethylpyrrole to give symmetrical dipyrromethanes has been completely eliminated by the use of the montmorillonite catalyst [87].

Acid treated montmorillonites have the potential of trans-alkylating mixtures of alkyl benzenes and alkyl phenols to give equilibrium mixtures with different substitution patterns. Mixed alkyl products from the cumene process can be trans-alkylated to give high yields of m- and p-di-isopropylbenzene [88]. Mixtures of o- and p-alkylphenols can be trans-alkylated to give preferentially m-alkylphenols [89,90]. Recently Al_{13} pillared montmorillonite has been used to prepare high yields of 1,2,4,5,-tetramethylbenzene by the disproportionation of 1,2,4-trimethylbenzene [91].

CLAYS AS SUPPORTS FOR OXIDATION CATALYSTS

The development by Cornelis and Laszlo of "clayfen", an acid montmorillonite doped with Fe(III) nitrate, has permitted a number of oxidation processes to be catalysed by clays. Secondary alcohols can be converted in high yields to ketones [92,93] and thiols are converted to disulphides [94] with this supported reagent .

Oxidative thallation of carbonyl compounds and alkenes takes place with thallium(III) nitrate supported on K-10 montmorillonite at room temperature to give excellent yields of acetals and esters from styrene derivatives and aromatic ketones following rearrangement of the intermediates [95]. The reactions often fail completely in the absence of the montmorillonite.

$$\xrightarrow[20^{\circ}]{\text{TTN/K-10}}$$

92%

A special montmorillonite catalyst containing Sb,V and Ti oxides has been developed for the oxidative amination of 2-methylpyrazine with ammonia and oxygen to give 2-cyanopyrazine in high yield [96].

Cumene hydroperoxide, which is formed by air oxidation of cumene, can be cleaved easily by acid treated montmorillonite in an important industrial process yielding phenols and propanone [97].

$$\xrightarrow[]{H^{+}\text{-Mont }40^{\circ}}$$

$+ \; Me_2CO$

95%

DISCUSSION

Montmorillonite clays have been found to be useful as heterogeneous catalysts for a surprisingly wide range of different types of organic reactions. It will be obvious that most of the reactions discussed in this survey can be classified as normal Bronsted acid catalysed reactions. However there is evidence in clay catalysed reactions of Lewis acid activity in some of the aromatic substitution reactions and free-radical mechanisms in the reactions of several sulphur compounds and in the polymerisation of aromatic compounds.

The source of the acidity in layered clays has been very well documented and it seems clear that much of the protonic activity is due to the complexed water in the hydration shells of certain multivalent interlamellar cations, such as Al(III) and Cr(III), undergoing hydrolysis to release protons. These cationic-exchanged montmorillonites have similar catalytic activity to cold proton exchanged material.

It is possible to carry out many of the reactions discussed above with aqueous protic catalysts, but several of the reactions are unique to the clay systems.

For any given reaction, there seems to be an optimal catalyst and an optimum temperature and one has to "tune" the montmorillonite to achieve the best result with the desired selectivity.

Clay catalysts have some distinct advantages over homogeneous protic catalysts. The contact time between the products and the active acidic site is limited in a batch process by the diffusion processes which remove the product to the exterior liquid phase. In this way secondary reactions are minimised and cleaner products generally result. The contact time can be further reduced if a flow reactor is employed.

Separation of the catalyst during workup from a batch reactor is also very easy, requiring only filtration or centrifugation, with none of the solvent extraction and alkali washing procedures required to get rid of homogeneous acid catalysts.

The recent discovery of pillaring techniques, where "robust" inorganic cations have been exchanged into montmorillonites, and then calcined to deposit metal oxide pillars or clusters in the interlamellar region, has created considerable worldwide interest in the research laboratories of the petroleum industry. The resultant cross-linked clays do not swell in solvents, have larger pore sizes than many zeolites, and hence show potential for shape selective organic reactions.

An additional physical property of these pillared materials is that the cross-linking also occurs between clay granules, so that the clay lumps do not "sand" in contact with solvents. These pillared materials are therefore very suitable for reactions in flow systems where the catalyst is enclosed in a porous cage.

The major advantage of the pillared catalysts is their much improved thermal stability over the layered clays and this has induced a vast interest in their catalytic potential for petroleum processing.

In our investigations of alumina and zirconia pillared
montmorillonites and organic pillared montmorillonites, we have shown
that several of the reactions which are described in this survey can
be carried out with pillared clays, but that in almost all cases the
catalytic activity has been reduced, sometimes dramatically, over
that achieved with Al(III)-exchanged layered montmorillonite under
the same conditions.   However because of their distinct physical
advantages over layered clays, much effort is taking place worldwide
in attempts to improve their catalytic properties by chemical and
structural modifications.

REFERENCES

1.   W. Franz, P Gunther & C.E. Hofstadt, Proceedings of 5th. World
     Petroleum Congress in New York, June 1959.
2.   C.L. Thomas, J. Hickey & G. Strecker, Ind.Eng.Chem.,
     1950, 42, 866.
3.   G.A. Mills, J. Holmes & E.B. Cornelius, J.physic.Colloid Chem.,
     1950, 54, 1170.
4.   N. Lahav, U. Shani & J. Shabtai, Clays Clay Miner.,
     1978, 26, 107.
5.   D.E.W. Vaughan & P.J. Lussier, preprints, 5th. Int. Conference
     on Zeolites, Naples, 2-6 June, 1980.
6.   T. Endo, M.M. Mortland & T.J. Pinnavaia, Clays Clay Miner.,
     1980, 28, 105.
7.   S.P. Christiano, J. Wang & T.J. Pinnavaia, abstracts, 19th.
     Annual Meeting of the Clay Minerals Society, Hilo, Hawaii, 1982.
8.   Chevron Research Co., U.S. Patent, US 3655798 (1972).
9.   Chevron Research Co., U.S. Patent, US 3766292 (1973).
10. Shell Research, European Patent, EP79091 (1983).
11. C.L. Thomas, Ind.Eng.Chem., 1949, 41, 2564.
12. A. Grenall, Ind.Eng.Chem., 1948, 40, 2148;   1949, 41, 1485.
13. Toyo Soda MFG KK, Japanese Patent, J5 8083635 (1983).
14. Agency of Ind. Sci. Tech., Japanese Patent, J5 8067340 (1983).
15. Y. Morikawa, T. Goto, Y. Moro-oka & T. Ikawa, Chem.Lett.,
     1982, 1667.
16. Gulf Research & Dev. Co., U.S. Patent, US 3947483 (1976).
17. NL Industries Inc., U.S. Patent, US 4217295 (1977).
18. Shell Int. Research, Belgian Patent, BE 891410 (1980).
19. Gulf Research & Dev. Co., U.S. Patent, US 4492774 (1985).
20. Dow Chemical Co., U.S. Patent, US 3432571 (1969).
21. Gulf Research & Development Co. N.L. Patent, NL 7503 581 (1974).
22. C. Kato, K. Kuroda & H. Takahara, Clays Clay Miner.,
     1981, 29,294.
23. O. Zigo & I. Horvath, Silikaty, 1983, 27, 131.
24. Monsanto Co., U.S. Patent, US 691925 (1976).
25. Chem. Werke Huls AG, German Patent, DT 2906294 (1979).
26. J.M. Adams, S.H. Graham, P.I. Reid & J.M. Thomas,
     J.Chem.Soc.,Chem.Commun., 1977, 67.
27. J.M. Adams, S.E. Davies, S.H. Graham & J.M. Thomas,
     J.Catal., 1982, 78,197.

28. M.J. Tricker, D.T.B. Tennakoon, J.M. Thomas and S.H. Graham, Nature, 1975, 253, 110.
29. M.J. Tricker, D.T.B. Tennakoon, J.M. Thomas & J Heald, Clays Clay Miner., 1975 23, 77.
30. M.M. Mortland & L.J. Halloran, Soil Sci.Soc.Am.J., 1976, 40, 367.
31. Y. Soma, M. Soma & I Harada, Chem.Phys.Lett., 1983, 99, 153.
32. Y. Soma, M. Soma & I Harada, J.Phys.Chem., 1985, 89, 738.
33. P.J. Isaacson & B.L. Sawhney, Clay Miner., 1983, 18, 253.
34. B.L. Sawhney, R.K. Kozlaski, P.J. Isaacson & M.P.N. Gent, Clays Clay Miner., 1984, 32, 108.
35. A. Morale, P. Cloos & C. Badot, Clay Miner., 1985, 20, 29.
36. Shell Oil Co., U.S. Patent, US 4125483 (1977).
37. R. Baumann, Ang.Chem.Int.Ed.Eng.., 1981, 20, 1014.
38. P. Laszlo & J. Lucchetti, Tetrahedron Lett., 1984, 25, 1567.
39. P. Laszlo & J. Lucchetti, Tetrahedron Lett., 1984, 25, 2147.
40. P. Laszlo & J. Lucchetti, Tetrahedron Lett., 1984 25, 4387.
41. J.F. Roudier & A. Foucaud, Tetrahedron Lett., 1984, 25, 4375.
42. J.A. Ballantine, M. Davies, I. Patel, J.H. Purnell, M. Rayanakorn, K.J. Williams & J.M. Thomas, J.Mol.Catal., 1984, 26, 37.
43. Gulf Oil Canada Ltd., U.S. Patent, US 693579 (1976).
44. J.A. Ballantine, J.H. Purnell, M. Rayanakorn, K.J. Williams & J.M. Thomas, J.Mol.Catal., 1985, 30, 373.
45. J.A. Ballantine, R.P. Galvin, R.M. O'Neil, J.H. Purnell, M. Rayanakorn & J.M. Thomas, J.Chem.Soc.,Chem.Commun., 1981, 695.
46. B.A.S.F. AG, German Patent, 1211643 (1966).
47. S. Kessaissia, B Siffert & J.B. Donnet, Clay Miner., 1980, 15, 383.
48. A. Katchalsky & G Ailam, Biochem.Biophys.Acta, 1967, 140,1.
49. M. Paecht-Horowitz, Biosystems, 1977, 9, 93.
50. M. Paecht-Horowitz, J. Berger & A Katchalsky, Nature, 1970, 228, 636.
51. M. Paecht-Horowitz, Biosystems, 1977, 9, 93.
52. J.Y. Conan, A. Natat & D Privolet, Bull.Soc.Chim.Fr., 1976, 11-12, 1935.
53. T. Vu Moc, H. Petit & P. Maitte, Bull.Soc.Chim.Fr., 1979, 15, 264.
54. Sud-Chemie AG, German Patent, DBP 1086241 (1961)
55. E.C. Taylor & C. Chaing, Synthesis, 1977, 467.
56. A. Cornelis & P. Laszlo, Synthesis, 1982, 162.
57. Farbwerke Hoechst AG, German Patent, 1146871 (1963).
58. U. Schafer, Synthesis, 1981, 794.
59. S. Hunig, K. Hubner & E Benzing, Chem.Ber.,1962, 95, 931.
60. J.M. Adams, J.A. Ballantine, S.H. Graham, R.J. Laub, J.H. Purnell, P.I. Reid, W.Y.M. Shaman & J.M. Thomas, J. Catal., 1979, 58, 238.
61. Gulf Oil Canada, U.S. Patent, US 4042633 (1980).
62. A. Bylina, J.M. Adams, S.H. Graham & J.M. Thomas, J.Chem.Soc.,Chem.Commun., 1980, 1003.

63. Phillips Petroleum Co. U.S. Patent, US 096345 (1979).
64. D.H. Kubicek, Belgian Patent, BE 886261 (1982).
65. R.P. Galvin, University of Wales Ph.D. Thesis, 1983.
66. Phillips Petroleum Co. U.S. Patent, US 2610981 (1947).
67. Tashkent Poly., Soviet Patent, SU 388788 (1971).
68. J.A. Ballantine, M. Davies, R.M. O'Neil, I Patel, J.H. Purnell, M. Rayanakorn, K.J. Williams & J.M. Thomas, J.Mol.Catal., 1984, 26, 57.
69. R. Gregory, D.J.H. Smith & D.J. Westlake, Clay Miner., 1983, 18, 431.
70. D.F. Ishman, J.T. Klug & A Shani, Synthesis, 1981, 137.
71. B.A.S.F. AG, German Patent, DP 1643712 (1971).
72. R.M. O'Neil, U.C.Swansea Post-doctoral Fellowship Report, 1982.
73. E. Polla, Bull.Soc.Chim.Belge., 1985, 94, 81.
74. Fabrenfabriken Bayer, German Patent, Ger 1049103 (1959).
75. Jpn. Kokai Tokkyo Koho, 80 49,332 (1980).
76. M.E. Davies, University of Wales Ph.D. Thesis , 1982.
77. H. Mueller & O.H. Muchler, U.S. Patent, US 4243799 (1981).
78. B.A.S.F. AG, German Patent, Ger 1185376 (1965).
79. Phillips Petroleum Co., U.S. Patent, US 65971 (1948).
80. R.S. Aries, U.S. Patent, 2930820 (1960).
81. General Aniline & Film Co., U.S. Patent, US 3287422 (1966).
82. Sud-Chemie AG, German Patent, DBP 1051271 (1961).
83. J.M. Bakke & J. Liaskan, German Patent, DP 2826433 (1979).
84. J.M. Bakke, J. Liaskan & G.B. Loretzen, J.Prakt.Chem., 1982, 324, 488.
85. Mitsubishi K.K., European Patent, EP 27330 (1981).
86. Sud-Chemie AG, German Patent, DBP 1089108 (1961).
87. A.H. Jackson, R.K. Pandey, K.R.N. Rao & E. Roberts, Tetrahedron. Lett., 1985, 26, 793.
88. Hitachi Chemical KK, Japanese Patent, JA 059478 (1977).
89. Chevron Research Co., U.S. Patent, US 3655780 (1972).
90. Chevron Research Co., U.S. Patent, US 3655778 (1972).
91. E. Kikuchi, T. Matsuda, H. Fujiki & Y. Morita, Applied Catal., 1984, 11, 331.
92. A. Cornelis & P. Laszlo, Synthesis, 1980, 849.
93. A. Cornelis, P.Y. Harze & P. Laszlo, Tetrahedron Lett., 1982, 23, 5035
94. A. Cornelis, N. Depaye, A. Gerstmans & P. Laszlo, Tetrahedron Lett., 1983, 24, 3103.
95. E.C. Taylor, C. Chiang, A. McKillop & J.F. White, J.Amer.Chem.Soc., 1976, 98, 6750.
96. Degussa AG, Belgian Patent, BE 892310 (1982).
97. Phenolchemie GMBH, German Patent, Ger 1136713 (1962).

PREPARATIVE ORGANIC CHEMISTRY USING CLAYS

André Cornélis and Pierre Laszlo
Institut de Chimie Organique
Université de Liège au Sart-Tilman
B-4000 Liège, Belgium

ABSTRACT. Application of clays, modified clays or clay-supported reagents to chemistry is receiving increasing interest. This lecture will focus on recent applications to preparative organic chemistry designed and developed in our group :
A.- Metallic nitrato complexes, with covalent metal-ligand bonds, stabilized by impregnation upon a clay, offer versatile applications to organic synthesis. As a source of NO^+ nitrosonium cations, they perform a number of functional group interconversions. They effect also the regioselective mononitration of phenols.
B.- The Diels-Alder reaction is catalysed by radical cations generated by phenols in the presence of clays or by the protic or the Lewis acidic sites of montmorillonites. Clay catalysts also achieve stereo-specificities comparable to those, of cycloadditions performed in aqueous media.
C.- Saturated hydrocarbons, in the presence of acidic clays, react with carbon tetrachloride or with aromatic hydrocarbons without prior activation. These reactions have been tested on adamantane as a probe substrate.

1. INTRODUCTION.

In preparative organic chemistry there has recently been considerable growth of "supported reagents" techniques. In these, the reagents are deposited on various inorganic solid supports. The main advantages of supported reagents over conventional homogeneous solution techniques are : easier experimental set-up and work-up; milder conditions; and, often, substantial improvements of yields and/or of selectivity. Posner[1], McKillop and Young[2], Bram, d'Incan and Loupy[3] provide detailed reviews of these methodologies.

R. Setton (ed.), Chemical Reactions in Organic and Inorganic Constrained Systems, 213–228.
© 1986 by D. Reidel Publishing Company.

2.   THE DESIGN OF CLAY-SUPPORTED REAGENTS : THE EXAMPLES OF "CLAYFEN"
     AND "CLAYCOP".

We have investigated the reactivities of ferric and cupric nitrates
deposited from their acetonic solution on the acid treated bentonite
K10[4]. These two reagents were termed "clayfen" and "claycop" respectively. We shall detail here, on the example of "clayfen", its wide
spectrum of reactivity.
     As part of our search for new preparative procedures in organic
chemistry[5], we have developed this new reagent. It is simple,
effective and inexpensive.
     We shall now outline the principles followed in the rational
design of this reagent.
     In metallic nitrates, the oxygens of the nitrate groups can bind
to the metal by three different coordination modes : as unidentate,
bidentate or bridging ligands. Bidentate nitrates, in the anhydrous
state, are covalent[6,7]. The coexistence of a bidentate covalent
coordination, with the availability of lower intermediate oxidation
states for the metal, is the dual condition for high reactivity of a
metallic nitrate towards organic substrates[8]. Known for over 20
years[8,9], this exceptional reactivity had yet to be domesticated.
     Iron is abundant, cheap and non toxic ; ferric nitrate would be a
first class candidate, were it not for an unfortunate characteristic :
it is available only as the nonahydrate and is unknown to occur
without water of hydration.This hydration precludes any covalent
bonding of the nitrato group to the metal. However, in 1904, Naumann
described an anhydrous acetonic solution of ferric nitrate[10]. More
recently, the existence of complex salts of the type $(FeL_4(NO_3)_2)^+$
$(Fe(NO_3)_4)^-$, where L is a neutral ligand and where the tetranitrato-
ferrate anion is a bidentate covalent nitrate was established by
Naldini[11].
     We thus decided to investigate the reactivity of acetonic solva-
tes of ferric nitrate. Because we wanted the preparation to be easy
and inexpensive, the Naumann process, metathesis between ferric
chloride and silver nitrate in dry acetone was precluded. We concentrated, under vacuum, an acetonic solution of the commercial nonahydrate ferric nitrate. This affords a highly unstable, dark red oil. Its
analysis, performed before decomposition, shows a substantial amount
of organic material, responsible for a ca. 10% C content. In its
infrared spectrum, bands at 1220 and 1630 cm^{-1} are consentaneous with
a bidentate nitrato group on iron[12]. This solvate decomposes spontaneously in a vigorous exothermic reaction, which precludes any further
study or use without stabilization. Following the example of Alfred
Nobel, when he made dynamite, we stabilized ferric nitrate by deposition on a solid. Supported reagents have indeed the following assets :
a) activation or stabilization of the reagents;
b) promotion of selective modes of reaction;
c) ease of work-up by immobilization of by-products or of toxic
   chemicals.
These are all important for our purpose.

After comparison with other solid supports, we selected the inexpensive acidic industrial clay catalyst K10.The name "clayfen" was coined for the resulting impregnated material.

"Clayfen" (for "clay-supported ferric nitrate") is a free flowing floury powder. Left exposed to the atmosphere at room temperature, it retains its activity for only a few hours. This aging can be greatly slowed by covering the "clayfen" with n-pentane immediately after preparation, and evaporating the hydrocarbon just prior to use. Because "clayfen" loses about 40% of its reactivity when exposed to the atmosphere for four hours, or when under n-pentane for 24 hours, it is nevertheless best to use only the freshly prepared reagent. Differential calorimetric analysis shows that, above 59°C, "clayfen" decomposes with an enthalpy release of about 20 cal/g. This is a first order process, with calculated half-lives of 53 min at 69°C (in boiling n-hexane) and 14 min. at 80°C (boiling benzene temperature). In our experience, the latter temperature is the upper limit for practical use of "clayfen".

"Clayfen" is a nitrosating agent, and this property is the foundation of most of its applications.The main pattern of reactivity of "clayfen" is as a nitrosonium ion source, as in most of the examples below.

We have also studied the kindred reagent "claycop" (clay-supported copper(II) nitrate, also derived from the acetone solvate of a nitrate salt which forms covalent bidentate complexes upon dehydration) was developed from similar arguments[14]. "Claycop" shows no loss of reactivity even after standing in an open powder box for one month.

3.   APPLICATIONS OF CLAY-SUPPORTED NITRATES.

3.1. Oxidation of alcohols

Because there is no universally applicable procedure for the conversion of alcohols to aldehydes and ketones, the need for simple, selective and mild new methods still exists. "Clayfen" meets these requirements for producing ketones from secondary alcohols, or for producing aromatic aldehydes from the corresponding primary aromatic alcohols, in n-pentane or n-hexane solutions, under gentle reflux[15] (Table 1). The reaction avoids overoxidation of aromatic aldehydes into acids.

Nitrous esters are intermediates in these oxidations[16]. Running the reaction with chiral alcohols, or with conformationally anchored cyclohexanols, demonstrates that the nitrites retain the configuration of the alcohols, as is expected from reaction involving nitrosonium ions NO^+ as intermediates. In these alcohols oxidations, the supported reagent is necessary to produce the nitrite, while only its acidic K10 clay component suffices to promote decomposition of this nitrite intermediate into carbonyl product, presumably by the acid-catalysed pathway[17] of scheme I.

TABLE 1

Carbonyl compounds from oxidations of alcohols by "clayfen"[15]

compound	yield(%)
benzaldehyde[a]	85
acetophenone[a]	89
2-octanone[b]	98
4-ter-butylcyclohexanone[b,c]	74
camphor[b,d]	92
fenchone[b,d]	89
benzophenone[b]	100

a) in refluxing n-pentane   b) in refluxing n-hexane
c) from the cis/trans commercial mixture of alcohols
d) from the endo/exo commercial mixture of alcohols

SCHEME 1

$$R^1R^2CHONO + H^{\oplus} \rightleftarrows R^1R^2\overset{\oplus}{C}HONO \rightarrow R^1R^2CHOH + NO^{\oplus}$$
$$\overset{|}{H}$$

$$R^1R^2CHONO + NO^{\oplus} \rightarrow R^1R^2C \overset{O}{\underset{H}{\diagup}} O \overset{\oplus}{N} \overset{\overset{O}{\|}}{\underset{O}{\overset{\|}{N}}} \rightarrow R^1R^2C=O + H^+ + 2NO$$

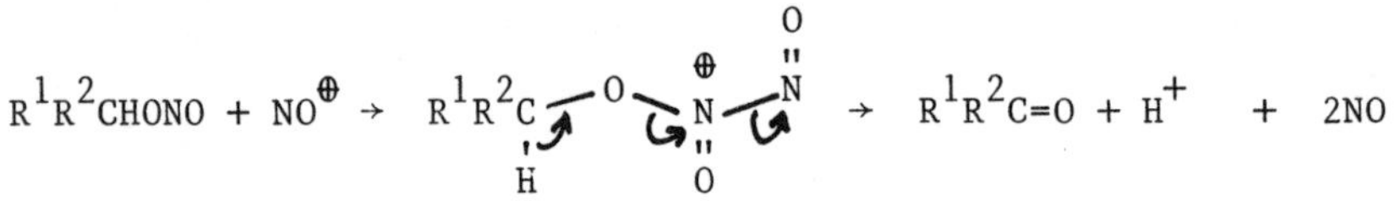

TABLE II

Compared yields of the "clayfen" and of the $Yb(NO_3)_3$ procedures for oxidation of benzoins to benzils.

R^1	R^2	Isolated yields(%)	
		"clayfen"	$Yb(NO_3)_3$
H-	H-	95	95
H-	CH_3O-	85	85
CH_3-	CH_3-	94	93
CH_3O-	CH_3O-	97	92
H-	CH_3-	96	92

Likewise, "clayfen" oxidizes benzoins into benzils[18], and the yields achieved with this dirt-cheap reagent (Table II) sustain comparison with those from the Kagan's procedure which uses the more expensive Yb nitrate as a catalyst[19].

3.2. Oxidative coupling of thiols into disulfides.

$$2RSH \xrightarrow{\text{"clayfen"}} R-S-S-R$$

This coupling by "clayfen" of thiols into symmetrical disulfides[20] differs from the classical and well-documented transition metal catalysed reaction. Here, an intermediate thionitrite (another example of a "clayfen" nitrosation) decomposes into thiyl radicals which then couple into disulfides :

$$R-SH \xrightarrow{\text{"clayfen"}} R-S-NO$$

$$R-S-NO \rightarrow RS^{\cdot} \rightarrow R-S-S-R$$

This offers a spectacular and easy test of "clayfen" activity. Freshly prepared "clayfen" reacts with thiophenol, at room temperature, in a very short time (30 s to 1 min), with dramatic color changes and sudden release of abundant nitrous fumes. The reaction is over in one or two minutes. Aged, or uncorrectly prepared reagent requires longer reaction times - about 10 min or more. "Claycop" also performs this same coupling of thiophenol, in high yield (>96%), at room temperature, in equally short times. Both reagents offer a convenient activation of thiols by thionitrite formation.

3.3. Carbonyl group deprotections.

3.3.1. Hydrolytic cleavage of thioacetals. To achieve this deprotection, which is very important but remains a rather tricky step in organic synthesis, soft cationic species (nitrosonium cations NO^+ or the Cl^+ cation) are to be preferred to transition metal ions[21,22].

As both "clayfen" and "claycop" are mild and convenient sources of nitrosonium cations, and as both are innocuous toward carbonyl groups (even in benzaldehydes) , both are reagents of choice for removal of thioacetals masking groups[14]. With cyclic thioacetals, a functionality frequently used in synthesis, they offer the extra advantage of polymerization of the difunctional sulfur residue, subsequently eliminated together with the clay by filtration. A transient and intense color suggests intermediacy of thionitrite-related species. In most cases (Table III), quantitative regeneration of the carbonyl group is achieved. Let us emphasize the 97% recovery of cholestanone from its ethylene dithioketal using "claycop" : a process in which benzeneselenic anhydride affords only a 72% yield[23], and photochemical deprotection a 77% yield[24].

TABLE III

Dethioacetalization by "clayfen" or "claycop"

carbonyl	sulfur moiety	Yield	
		"clayfen"	"claycop"
benzaldehyde	$-S(CH_2)_3S-$	100	100
benzophenone	$(-SC_2H_5)_2$	99.4	99.5
	$-S(CH_2)_2S-$	100	100
	$-S(CH_2)_3S-$	98.3	97.3
2-naphtaldehyde	$-S(CH_2)_2S-$	98.7	99.6
n-decanal	$(-SC_2H_5)_2$	86.7	87.5
	$-S(CH_2)_3S-$	99.1	91.2
2-n-undecanone	$-S(CH_2)_3S-$	98.5	98.1
5-n-nonanone	$-S(CH_2)_3S-$	98.0	98.4
5α-cholestan-3-one	$(-SC_2H_5)_2$	61.7	60.6
	$-S(CH_2)_2S-$	71.8	96.8

3.3.2. Conversion of imino derivatives to carbonyl compounds. Most of the numerous methods for unmasking carbonyl groups engaged in imino derivatives require strongly oxidative or reducing conditions, and/or basic or acidic media, and also often involve tedious procedures or expensive reagents. "Clayfen" offers an inexpensive alternative to other nitrosating reagents such as nitrosonium tetrafluoroborate[25]. Its reaction with N,N-dimethylhydrazones is fast and exothermic, and gives the expected carbonyl compounds in yields ranging from 70 to 90%[26]. Equally good results[27] are obtained with other derivatives of the imine family (semicarbazones, tosylhydrazones, phenylhydrazones, 2,4-dinitrophenylhydrazones).

3.4. Synthesis of azides and of iminophosphoranes.

While toxic and unstable, azides are nevertheless essential intermediates in organic synthesis. Nitrosonium NO^+ cations, generated by "clayfen" produce a variety of azides in fair yield (Table IV) from hydrazines by the nitrosation route, in methylene chloride suspension[28].

TABLE IV.

Conversion of hydrazines into azides

Starting hydrazine	field
p-toluenesulfonylhydrazine	83
phenylhydrazine	63
benzoylhydrazine	58
t-butylcarbazate	55-71

This easy and inexpensive access to azides was a stimulus to the preparation of iminophosphoranes, nitrogenous cousins of Wittig ylids, from the corresponding hydrazines[29] (Scheme II), (Table V).

SCHEME II

$$R-NH-NH_2 \xrightarrow[CH_2Cl_2]{\text{"clayfen"}} RN_3$$

$$RN_3 + Y_3P \xrightarrow{\text{ether or benzene}} \left[RN_3 \cdot PY_3 \right]$$

$$\left[RN_3 \cdot PY_3 \right] \longrightarrow R-\overline{N}=PY_3 + N_2 \nearrow$$

$$\theta \updownarrow \oplus$$

$$R-\overline{N}-PY_3$$

$$(Y = C_6H_5^- ; C_6H_5O^- \text{ or } C_2H_5O^-)$$

TABLE V
Preparation of iminophosphoranes according to scheme III

R	Y	Yield(%)[a]
phenylsulfonyl-	phenyl-	52
phenylsulfonyl-	phenoxy-	51
phenylsulfonyl-	ethoxy-	67
p-toluenesulfonyl-	phenyl-	50
p-toluenesulfonyl-	phenoxy-	52
p-toluenesulfonyl-	ethoxy-	64
benzoyl-	phenyl-	47-50
phenyl-	phenyl-	46-48

a : starting from the hydrazine

3.5. Aromatization of 1,4-dihydropyridines.

Most of the numerous methods available for the title reaction suffer
from rather rough conditions. As nitrosonium cation releasing reagents
are known to effect it efficiently, it was tempting to try "clayfen"
and "claycop". Applied to the preparation of diethyl 4-substituted
2,6-dimethyl-3,5-pyridinedicarboxylates, our reagents afford yields
comparable to those of previously reported procedures. Here, "claycop", while requiring longer reaction times, is more efficient than
"clayfen" and affords yields exceeding 90% in most cases[30].

3.6. Conversion of thiocarbonyls into carbonyls.

Among the many ways to turn thiocarbonyl into carbonyl compounds, two
recent methods make attractive and efficient recourse to the nitrosonium cation[31-32]. With bis-aromatic thioketones, "clayfen" offers a
much less costly alternative to salts such as nitrosonium tetrafluoroborate. Nearly quantitative conversions are then achieved[33]. The
reaction using "clayfen" also applies, but with less satisfactory
yields, to non-aromatic substrates.
        Interestingly, the reaction does not proceed via a trithiapentane
such as A, but via a dioxodithiahexane B[34].

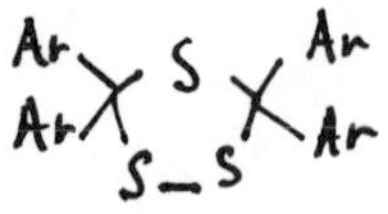

A

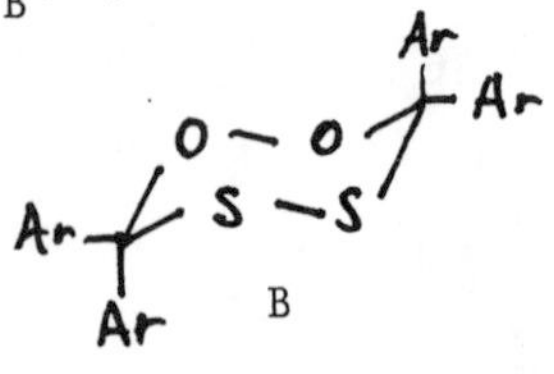

B

3.7. Regioselective nitration of phenols.

The central role of the nitrosonium cation is the mechanistic feature
that unifies all the examples described so far. In the nitration of
phenols, the role of NO^+ in the reaction mechanism is less clear-cut.[41]
We believe (based upon its electron affinity, well matched to the
first ionization potential of aromatic electrons) that it serves as an
electron acceptor in a single electron transfer step. As shown in
Table VI, "clayfen" nitrates phenols with very satisfactory results,
leading exclusively to mononitrated products[12].

  These reactions, which are all run at room temperature, take
place without any evolution of gaseous oxides of nitrogen. They can be
performed in ether, in tetrahydrofuran, or in toluene. We prefer the
latter, because it prevents contamination of the products by ferric
residues. The criterion for the solvent is that it must dissolve the
nitrated products; the solubility of the starting phenol is less
important.

TABLE VI
Phenols nitration by "clayfen".

| Substituent of | yield(%) | |
phenol	o-product	p-product
–	39	41
4-CH$_3$	58	–
3-CH$_3$	20[b]	34
4-Cl	88	–
4-F	69	–
β-naphtol	63[b]	–
4-t-Bu	92	–
3-OH	–	50[c]
4-CHO	93	–
4-CN	88	–
estrone	55[d]	–

a)	2-NO$_2$-5-CH$_3$	b)	1-NO$_2$-2-OH
c)	3-OH-4-NO$_2$	d)	2-NO$_2$-3-OH

  This is a nicely regioselective reaction : no <u>meta</u> nitration
product could be detected. With phenol itself, the <u>para</u> isomer is

favored by reference to a statistical distribution, whereas with <u>meta</u> substituted phenols, one of the <u>ortho</u> positions is clearly favored with respect to the other. The cyano and the aldehyde groups are perfectly stable under these nitration conditions.

The 55% yield of 2-nitroestrone obtained selectively with "clay-fen"[35] <u>is the best performance ever reported</u> for the preparation of <u>this gateway</u> to estrogens derivatized in ring A.

Our knowledge of the mechanism of these nitration reactions is far from being comprehensive. The available evidence[12] cannot accommodate the classical nitronium ion mechanism, nor the nitrosation route to nitration, but is compatible with the recently proposed radical[36,37] cation pathway (Scheme III).

SCHEME III

General radical cation pathway for nitration of aromatics

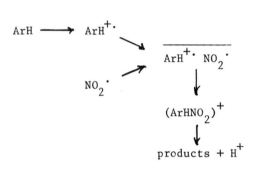

In our case, the postulated intermediate phenol radical cation could be produced from electron transfer to the charge-balancing cations of the clay, a well-documented phenomenon[38,39,40].

Nitrogen dioxide is a key component of this proposed reaction pathway. There is a dense network of interrelationships between all the oxygen combinations of nitrogen. For example, simple combination of NO^+ with NO_3^-, both available in "clayfen", produces nitrogen dioxide, according to

$$NO^+ + NO_3^- \rightleftarrows N_2O_4 \rightleftarrows 2NO_2$$

NO_2 can also result from the decomposition of an hypothetical unidentate nitrato complex of iron on the clay. In any case, the system is, from all available evidence, able to produce it : "clayfen" spontaneously decomposes with formation of NO_2 !

4. CATALYSIS OF THE DIELS ALDER REACTION IN THE PRESENCE OF CLAYS.

The Diels Alder reaction challenges the chemists : its "classical" practice often requires drastic conditions (high pressures, high temperatures, prolonged reaction times), hence, much recent work has tried to make this process fast and stereoselective. These efforts have led to the catalysis of Diels Alder reactions by Brönsted and Lewis acids, or by radical –cations, and to the use of aqueous media[42]. The acidity of clays, their ability to generate radical – cations, and the hydration of their interlamellar spaces was an invitation to test them as catalysts. I shall only briefly mention some of the results obtained in our lab :

1)   phenols generate radical –cations when adsorbed on clay surfaces. This phenol-clay system catalyses the Diels–Alder reaction of unactivated dienophiles[43], (we use K10 clay, doped with Fe(III) and 4-t-butylphenol, in dichloromethane suspension at 0°C). The efficiency of these mild conditions is obvious when the results are compared to those obtained with non activated partners under classical conditions[44].
(Table VII).

TABLE VII

Diels–Alder dimerization of 1,3-cyclohexadiene

conditions	yield(%)	endo/exo
a) 200°C, 20h.	30	4 : 1
b) K10-Fe(lII), CH_2Cl_2, 0°C, 1h.	49	4 : 1
c) K10-Fe(III), 10% 4-t-butylphenol CH_2Cl_2, 0°C, 1h.	77	4 : 1

2)   in organic solvents, the same Fe(III)-doped K10 clay, devoid   of any added phenol, promotes reactions faster than in water, with comparable stereoselectivities[45] (Table VIII).

This catalyst also overcomes the reluctance of furans to react as dienes, but, in such applications, it is not necessarily superior to Lewis acids such as $AlCl_3$[46].

TABLE VIII.

Diels Alder reaction between cyclopentadiene and methylvinylketone.

conditions[a]	isolated yield(%)	endo/exo
H_2O, 20°C, 0.3 h	95	19:1[b]
H_2O, 0°C, 2 h	91	19:1
CH_2Cl_2, 20°C, 0.3 h	97	9:1
CH_2Cl_2, 0°C, 2 h	98	13.5:1
CH_2Cl_2, -24°C, 4 h	96	21:1
C_2H_5OH, 20°C, 0.3 h	95	14:1
C_2H_5OH, 0°C, 1 h	75	25:1
C_2H_5OH, -24°C, 4 h	75	24:1

a) in all cases, 15 mmoles of cyclopentadiene, 15 mmoles of MVK, 2,5g of K10-Fe(III) and 50 ml of solvent are used.
b) the reaction performed in water alone for 3.0 hours leads to 80% yield of cycloadduct with an endo/exo ratio of 21.4:1

5. CLAY-CATALYSED CHLORINATION AND FRIEDEL AND CRAFTS ARYLATION OF ADAMANTANE.

Much recent research has attempted to circumvent the chemical inertness of paraffins. We have tried to overcome their reluctance towards substitution by means of K10 supported Fe(III) salts, selecting adamantane as the trial substrate.
    We elected this $FeCl_3$/K10 association from consideration of the various results above : ferric salts, deposited on or doping the K10 support, have given rise to an interesting spectrum of chemical reactivity, manifestly linked to the dehydration of the ferric cation.

SCHEME IV
Doping of K10 clay with ferric chloride under dehydrating conditions

$$MMT^{p-}H_p^+ \cdot nH_2O \rightarrow MMT^{p-}H_p^+ + nH_2O \nearrow$$

$$FeCl_3 \cdot 6H_2O \rightarrow FeCl_3 + 6 H_2O \nearrow$$

$$MMT^{p-} H_p^+ + FeCl_3 \rightarrow MMT^{p-}H_{(p-i)}^+ + (FeCl_{(3-i)})^{i+} + iHCl \nearrow$$

When $FeCl_3.6H_2O$, is deposited on K10 clay, and is dehydrated with the CCl_4 azeotrope, vigorous evolution of hydrochloric acid takes place at the end of dehydration : Scheme IV suggests a plausible hypothesis for this sequence of events.

Adamantane (1), admixed to the above mixture prior to dessication, leads to 1-chloro (2) and 1,3-dichloroadamantane (3), in relative proportions controled by the ratio of the hydrocarbon to the clay-ferric chloride reagent.

This suggests the intermediacy of a 1-adamantyl tertiary cation (Scheme V).

SCHEME V

$$(FeCl_{(3-i)})^{i+} + R\text{-}H \rightarrow (FeCl_{(3-i)}H)^{(i-1)+} + R^+$$

$$R^+ + CCl_4 \rightarrow R\text{-}Cl + CCl_3^+$$

$$CCl_3^+ + (FeCl_{(3-i)}H)^{(i-1)+} \rightarrow CHCl_3 + (FeCl_{(3-i)})^{i+}$$

(R = 1-Adamantyl)

The necessity of the use of carbon tetrachloride (under similar conditions, no chlorination is observed in chloroform) is consistent with this hypothesis, as well as the observed production of chloroform as the reaction proceeds; and the Mössbauer analysis of the inorganic residue at the end of the process, which shows that all the iron is still in the (+3) oxidation state. To test the formation of such an intermediate adamantyl cation, we have engaged it in Friedel-Crafts arylation : dehydration of a mixture of the K10/FeCl$_3$ reagent and adamantane, dispersed in benzene, resulted likewise in a mixture of 1-phenyl (4) and 1,3-diphenyladamantanes (5), in yield and proportions of the reagent controled by the amounts of the various components. (Table IX).

TABLE IX

Direct phenylation of adamantane by K10/FeCl$_3$.

Adamantane (mmol)	K10 clay (g)	FeCl$_3$ (mmol)	Phenyl adamantane	diphenyl adamantane
14.6	24	90	51	47
14.6	40	146	17	83

This is, to our knowledge, the first reported example of direct arylation of adamantane. The method can be extended to some other

aromatic compounds, but is unfortunately not altogether general :
aromatics bearing highly electron-donating or -withdrawing substituents either are unreactive or react by various other pathways.

     With toluene as aromatic partner, five products are obtained :
the monosubstituted meta-(6) and para-(7)tolyladamantanes, and the
disubstituted meta,meta-(8), meta,para-(9) and para,para-(10) ditolyl-
adamantanes. Ortho attack is not observed, presumably due to steric
hindrance. Equiprobability of aromatic electrophilic substitution at
the meta and para positions is conspicuous from the statistical
product ratio (6)/(7) of almost exactly 2 (34.6% and 18% respectively). The disubstituted products are likewise distributed according
to statistic prediction. We also observed that, while in most Friedel-
Crafts reactions, competition between toluene and benzene favors the
former[48], in our case, the distribution of products is purely
statistical, indicating equivalent reactivities of benzene and toluene
towards the bulky adamantyl moiety : ratios of 4 : 6 : 7 = 6 : 2 : 1
are expected, and values of 5.8 : 1.9 : 1 are indeed found.

     However, mechanistic inferences from selectivities in aromatic
Friedel-Crafts reactions, are much debated topics. Could we be dealing
with a radical rather than an electrophilic aromatic substitution
mechanism ? The observed lack of selectivity is not sufficient to
argue in favor of a radical pathway, from which one would also expect
formation of 2-aryl substituted adamantanes that could not be found.
We favor a mechanism in which a carbocation, generated from adamantane
in a "hot" state, attacks indiscriminately any aromatic neighbouring
molecule, in any position not sterically-excluded.

6.   CONCLUSIONS.

Claims that clays have served as a matrix for the appearance of the
earliest organic molecules and biopolymers have been reiterated
recently. One of us addresses elsewhere this question of the possible
role of clays in the origin of life. In any case, clays can be tailo-
red to perform efficiently, under mild conditions, useful organic
chemical transformations.

ACKNOWLEDGMENTS.

This research was generously supported by Programmation de la Politi-
que Scientifique, Brussels ("Action Concertée 82/87-34"). Support by
Fonds National de la Recherche Scientifique, Brussels, in the purchase
of major nmr and HPLC instrumentation, is gratefully acknowledged. We
are indebted to Professeur F. Grandjean (Liège) for the Mössbauer
analysis and to MM. Janin, Lamy and Tanguy (Rhône-Poulenc Recherches,
Declines, France) for calorimetric investigation of the stability of
"clayfen". We thank especially all of our co-workers listed in the
references for their unfailing enthusiastic hard work.

BIBLIOGRAPHY

(1) G.H. Posner, Angew. Chem.Int. Ed. Engl., 1978, 17, 487.
(2) A. McKillop, D.W. Young, Synthesis, 1979, 401, 481.
(3) G. Bram, E. d'Incan, A. Loupy, Nouv. J. Chim., 1982, 6, 689.
(4) A. Cornélis, P. Laszlo, Synthesis, in press.
(5) A. Cornélis, P. Laszlo, Nachr. Chem. Tech. Lab., 1985, 33, 202.
(6) C.C. Addison, Prog. Inorg. Chem., 1967, 8, 195.
(7) C.C. Addison, N. Logan, S.C. Wallwork, C.D. Garner, Q. Rev.
     Chem. Soc., 1971, 25, 289.
(8) C.C. Addison, Coord. Chem. Rev., 1966, 1, 58.
(9) C.C. Addison, Adv. Chem. Ser., 1962, 36, 131.
(10) A. Naumann, Ber. Deutch. Chem. Ges., 1904, 37, 4328.
(11) L. Naldini, Gazz. Chim. Ital., 1960, 90, 1231.
(12) A. Cornélis, P. Laszlo, P.Pennetreau, Bull. Soc. Chim. Belg.,
     1984, 93, 961.
(13) K10 clay is manufactured by Süd Chemie AG, München.
(14) M. Balogh, A. Cornélis, P. Laszlo, Tetrahedron Lett., 1984, 25,
     3313.
(15) A. Cornélis, P. Laszlo, Synthesis, 1980, 849.
(16) A. Cornélis, P.Y. Herzé, P. Laszlo, Tetrahedron Lett., 1982,
     23, 5035.
(17) D.H.R. Barton, G.C. Ramsay, D.Wege, J. Chem. Soc., 1967, 1915.
(18) M. Besemann, A. Cornélis, P.Laszlo, Compt. Rend. Acad.Sci.
     Paris, Ser II, 1984, 299, 427.
(19) P. Girard, H.B. Kagan, Teterahedron Lett., 1975, 4513.
(20) A.Cornélis, N. Depaye, A. Gerstmans, P. Laszlo, Tetrahedron
     Letters, 1983, 24, 3103.
(21) K. Fuji, K. Ichikawa, E.Fujita, Tetrahedron Lett., 1978, 3561.
(22) M.T.M. El-Wassimy, K.A. Jørgensen, S.O. Lawesson, J. Chem.Soc.
     Perkin Trans. 1, 1983, 2201.
(23) D.H.R. Barton, N.J. Cussans, S.V. Ley, J. Chem. Soc.Commun.,
     1977, 751.
(24) T.T.Takahashi, C.Y. Nakamura, J.Y. Satoh, J.Chem. Soc.Chem.Com-
     mun., 1977, 680.
(25) G.A. Olah, T.-L. Ho, Synthesis, 1976, 610.
(26) P. Laszlo, E. Polla, Tetrahedron Lett., 1984, 25, 3309.
(27) P. Laszlo, E. Polla, in press.
(28) P. Laszlo, E. Polla, Tetrahedron Lett., 1984, 25, 3701.
(29) P. Laszlo, E. Polla, Tetrahedron Lett., 1984, 25, 4651.
(30) M. Balogh, I. Hermecz, Z. Mészáros, P. Laszlo, Helv. Chim.Acta,
     1984, 67, 2270.
(31) K.A. Jørgensen, A.-B. A.G. Ghattas, S.-O. Lawesson, Tetrahedron,
     1982, 38, 1163.
(32) G.A. Olah, M. Arvanaghi, L. Ohannesian, G.K.S. Prakash,
     Synthesis, 1984, 785.
(33) S. Chalais, A. Cornélis, P. Laszlo, A. Mathy, Tetrahedron Lett.,
     1985, 19, 2327.
(34) J. Baran, Y. Houbrechts, P. Laszlo, submitted for publication.

(35) A. Cornélis, P. Laszlo, P. Pennetreau, J. Org. Chem., 1983, 48, 4771.

(36) V.D. Pokhodenko, V.A. Khizhnyi, V.G. Koshechko, O.I. Shkrebtii, Zh. Org. Khim., 1975, 11, 1873.

(37) C.L.Perrin , J. Amer. Chem. Soc., 1977, 99, 5516.

(38) T.J. Pinnavaia, P.L. Hall, S.S. Cady, M.M. Mortland, J. Phys. Chem., 1974, 78, 994.

(39) M.M. Mortland, L.J. Halloran, Soil. Sci. Soc. Amer. J., 1976, 40, 367.

(40) P.J. Isaacson, B.L. Shawney, Clay Miner., 1983, 18, 253.

(41) W.D. Reents, jr., B.S. Freiser, J. Amer. Chem. Soc., 1981, 103, 2791.

(42) P. Laszlo, J. Lucchetti, L'Actualité Chimique, 1984 (octobre), 42.

(43) P. Laszlo, J. Lucchetti, Tetrahedron Lett., 1984, 25, 1567.

(44) D. Valentine, N.J.J. Turro, G.S. Hammond, J. Amer. Chem. Soc., 1964, 86, 5202.

(45) P. Laszlo, J. Lucchetti, Tetrahedron Lett, 1984, 25, 2147.

(46) P. Laszlo, J. Lucchetti, Tetrahedron Lett, 1984, 25, 4387.

(47) S. Chalais, A. Cornélis, A. Gerstmans, W. Kołodziejski, P. Laszlo, A. Mathy, P. Métra, Helv. Chim. Acta, in press.

(48) G.A. Olah, "Friedel-Crafts Chemistry", Wiley, New York, 1973, p.217.

CLAY CATALYZED ENE-REACTIONS. SYNTHESIS OF γ-LACTONES

J.F. Roudier, A. Foucaud
Groupe de Physicochimie Structurale, U.A. C.N.R.S. 704,
Université de Rennes, Campus de Beaulieu
35042 Rennes Cédex,
France.

ABSTRACT. Ene-reactions occur generally at high temperature. Lewis acid catalysis allows ene-reactions under mild conditions. The ene-reaction of diethyloxomalonate, which is a very reactive enophile with olefins is catalyzed by kaolin, montmorillonite or H^+ doped montmorillonite. The ene-reaction products are converted into γ-lactones, especially by acidic montmorillonite. The conversion of citronellal into isopulegol is catalyzed by montmorillonite.

Many reactions are catalyzed at clay mineral surfaces or take place between the individual silicates sheets or involve the intercalation of the reactants.[1] Clays may catalyze chemical reactions by acting as Bronsted acids or Lewis acids. Unusual chemical conversions of organic molecules using silicate intercalates have been reported [2] : isomerizations,[3] acetal formation,[4] dimerizations, ester and ether formations from alkene, intermolecular hydrogen exchange and conversion of primary amines into secondary amines. The plausible mechanisms of these reactions involve protonated intermediates.[5] Clay supported reagents have been used for oxidation or Diels-Alder reactions.[6]

The reaction of an alkene having an allylic hydrogen (an "ene") with a compound containing a double bond (enophile) to form a new carbon-carbon bond is referred to as the ene-reaction [7] :

Ene-reactions occur at high temperatures, which has limited the synthetic use of this reaction. Since the enophile should be electron deficient, complexation of Lewis acids to enophiles containing basic groups should accelerate the ene-reaction. In fact, Lewis acid catalyzed ene-reactions appear to have general synthetic utility.[8]

Since clay minerals are selective acid catalysts, we have examined

229

R. Setton (ed.), Chemical Reactions in Organic and Inorganic Constrained Systems, 229–235.
© *1986 by D. Reidel Publishing Company.*

the catalytic activity of two types of sheet silicates, kaolinite, montmorillonites and doped montmorillonites, for the ene-reactions.

The intramolecular ene-reaction have been tested on citronellal,
as probe substrate. When d-citronellal 1 was heated at 180°C, the isopulegol
isomers were formed. When d-citronellal 1 was treated with acidic K10-
montmorillonite, an exothermic reaction occured, which gave a complexe
product mixture from which isopulegol 2 (8-10 %) were isolated.

In order to avoid the rise of temperature, K10 montmorillonite was
added to a solution of d-citronellal in CH_2Cl_2 at 15°C. After 15 min, the
organic layer, filtered, gave l-isopulegol 2a (56 %) and other isomers ;
d-neoisopulegol 2b (18 %) is the main component of the other three possible isomers. The other two isomers are only present in trace amounts.

Similar product distribution have been reported for the conversion
of d-citronellal into l-isopulegol, using $SnCl_4$ as Lewis acid.[10] However,
a more stereoselective preparation of l-isopulegol has been performed,
with $ZnBr_2$ as catalyst.[10]

The intermolecular ene-reaction have been tested with diethyloxomalonate 3, which is a very reactive enophile. Olefins afforded ene adducts
with 3 upon heating at 145-180°C, for 1-3 days.[11a] This ene-reaction has
been catalyzed with $SnCl_4$.[11a,b]

Diethyloxomalonate 3 reacted with 2-methyl 2-butene 5 in the presence of kaolin, at 70-90°C, to give the ene products 6, 7 and the lactone 8
(one diastereoisomer). This lactone 8 has been isolated, as by product,[12]
when the ene-reaction is catalyzed by an excess of $SnCl_4$. Compound 7
was formed by a secondary ene-reaction with the ene-product 6.

The crystalline diethyloxomalonate hydrate 4 gave the same reaction
than 3.

No reaction was observed when 3 and 5 were heated at 90°C for
48 h, in the absence of clay.

The ene adduct 7 was isolated by simple short-path distillation from
the ene-reaction mixture. Then, the ene adduct 6 and the lactone 8 were
separated by flash chromatography on silica gel (petroleum ether/ether
1/1 as eluent).

Structural assignments of 6, 7 and 8 were based on spectral data
(^1H, ^{13}C NMR, mass spectrometry). The structure of the lactone 8 was
established by ^1H NMR spectroscopy, by addition of the shift reagent
$Eu(dpm)_3$ to a solution of 8 in CCl_4.

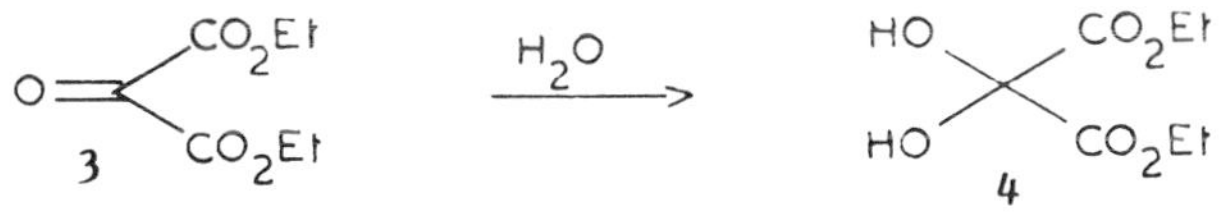

Scheme I - Reaction of **5** with **3** and kaolin (E = CO_2Et)

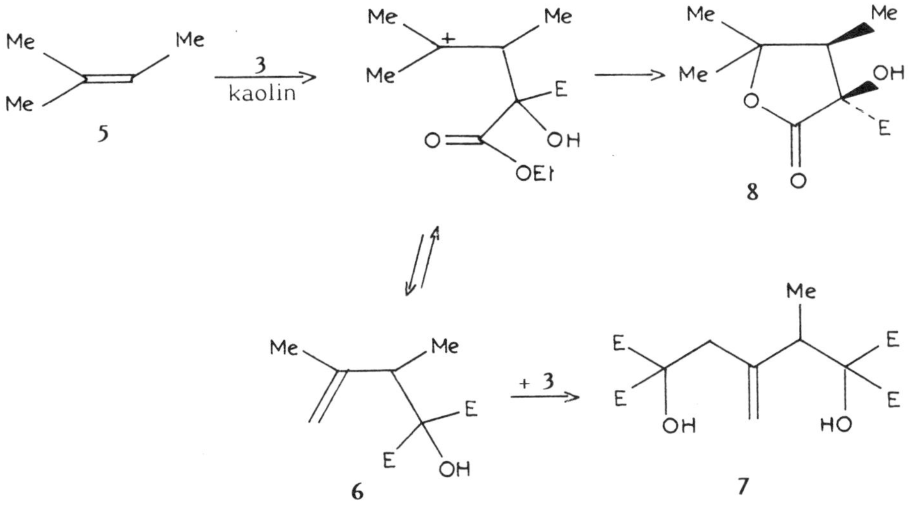

Scheme II - Reaction of **5** with **3** and montmorillonite (E = CO_2Et)

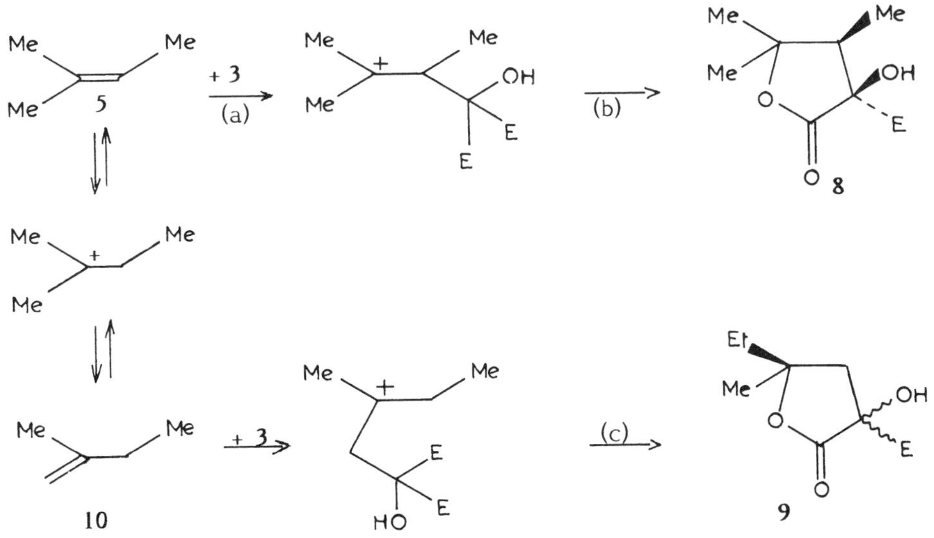

The greater downfield shifts were observed for the 4-H and the O-CH$_2$, which can be consistent with the structure **8**, if the complexation occurs predominantly both at the hydroxy group [13] and at the oxygen atom of the 3-carbonyl group.

Table I - Reaction of 2-methyl 2-butene with diethyl oxomalonate.

Reactions conditions[a]			Yield[b]	Product distribution, % [c]				
Catalyst	temp.°C	time,h	%	**4**	**6**	**7**	**8**	**9**
Kaolin [d]	72	19	82	59	37	1	3	
Kaolin [d]	125	19	82	29	63	2	5	
Kaolin [d]	90	120	78	26	5	19	49	
Mont. Camp Berteau	78	24	70	45			18	36
Mont. Camp Berteau	98	48	76	13		5	64	18
Mont. K10[e]	80	72	80	7			49	44
Al^{3+}-mont.	80	8	70	30			38	32
Cu^{2+}-mont.	80	8	60	37			40	23
Cr^{3+}-mont.	80	8	70	5			68	36
Fe^{3+}-mont.[f]	80	8	65	9			60	30
H^{+}-mont.[g]	20	8	94	10			60	30
H^{+}-mont.[g]	20	8	98	4			90	6
Bu$_4$N^{+}-mont.[h]	80	48	95	95			< 5	< 5

[a] A mixture of 2 g of catalyst, 15 mmol of **5**, 10 mmol of **3** was heated without solvent, for the time indicated.
[b] Sum of isolated yields of all components based on **3**.
[c] Determined by GLC and ^{1}H NMR.
[d] From Prolabo.
[e] From Fluka A.G..
[f] FeCl$_3$ doped K10 montmorillonite (0.6 mmol FeCl$_3$/1 g K10).
[g] H$_2$SO$_4$ doped K10 montmorillonite (3 mmol H$_2$SO$_4$/1 g K10).
[h] ref. 15.

The ene-reaction catalyzed by acidic montmorillonite gave solely the lactones **8** (only one diastereoisomer) and **9** (two diastereoisomers, in the ratio 50:50). The lactones **9** appears to be formed by the ene-reaction of **3** with the 2-methyl 1-butene (**10**), which arises from montmorillonite catalyzed isomerisation of **5**. The lactones **8** and **9** were easily separated by flash chromatography (silica gel, with ether/petroleum ether 1/1 as eluent). Structural assignment of **9** were based on spectral data (^{1}H and ^{13}C NMR, mass spectrometry).

Marked differences appear between catalysts (table I)[14]. The ratio **8/9** increased with the acidity of the montmorillonite[14] and with the temperature. Then, the rate of the step a (scheme II) increases faster than the rate of the isomerization with the acidity of the catalyst. The rates of the lactonization (steps b and c, scheme II) are very fast. The ene products are not observed except with the montmorillonite from Camp Berteau. The catalytic activity of the montmorillonite was very low when Bu_4N^+ ions occupy the exchange sites of the silicate structure.[15] Al^{3+}, Cu^{2+} and Cr^{3+} exchanged montmorillonite and K-10 gave similar product distribution.

Kaolin did not lead to double bond migration under the conditions of the table I. The ene-reactions are regiospecific and the formation of the lactone **8** is stereoselective.

This synthesis of lactones can be extended to simple alkenes. The K-10 catalyzed ene-reaction of 1-methylcyclohexene with **3** at 80°C, for 48 h, gave the lactones **10** (37 %) and **11** (23 %). In the same conditions, cyclohexene gave the lactones **12** (two diastereoisomers, 25 %).

The reaction of 1-methylcyclohexene with **3** at 80° for 17 h, in the presence of kaolin, gave **10** (20 %) and the alkenes **13** and **14** (20 %).

The 2-methyl 1-pentene **15** was treated with **3** in the presence of K-10 at 90°C for 24 h to yield the lactones **16** (one diastereoisomer, structure confirmed by ^{1}H NMR and use of shift reagent $Eu(dpm)_3$) and **17** (two diastereoisomers 50:50), the product distribution was **16** : **17** : **4** = 24 : 70 : 6. The lactones **16** and **17** were separated by chromatography on silica gel (petroleum ether/ether 1/1 as eluent).

Scheme III (E = CO$_2$Et)

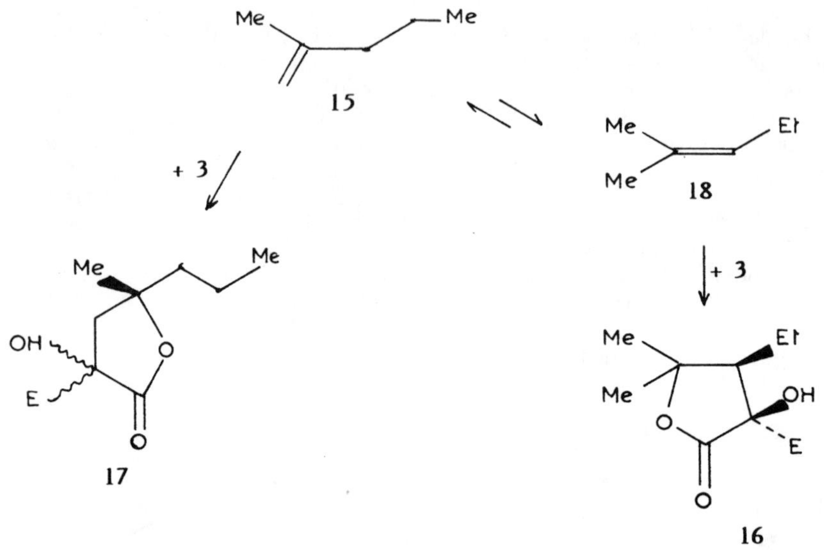

Scheme IV

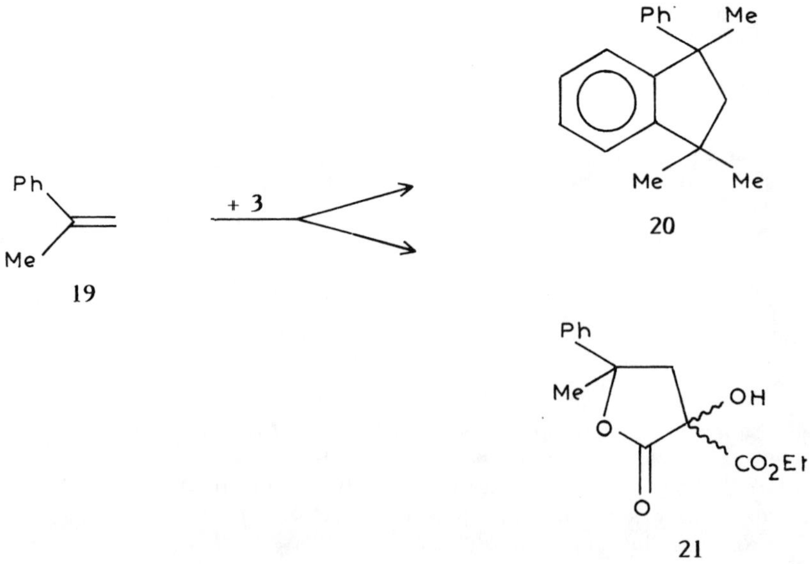

The α-méthylstyrene **19**, reacted with diethyloxomalonate in the presence of montmorillonite K10, at 80°C, affording a mixture of 1-phenyl 1,3,3-trimethylindane **20** (26 % yield) and lactones **21** (36 % yield)[17] which are separated by distillation (scheme IV).

In these reactions, clays act essentially as Bronsted acid.[16] The montmorillonite may catalyze the isomerisation of the alkene. This reaction is in competition with the ene-reaction. The montmorillonite collapses by addition of concentrated sulphuric acid. Then, only the ene-reaction is practically observed. Therefore, the isomerisation takes place mainly in the interlayer space. The ene-reaction must take place on the external surface of the clay particules and, probably, in the interlayer space.

These reactions constitute a simple procedure for the one pot synthesis of lactones. The major advantage is the relative ease which this procedure is conducted, without dry solvent and with easily available catalyst.

Acknowledgement - We thank Dr. J. Esteoule for helpful discussions and for generously providing sample of montmorillonite from Camp-Berteau.

References :
1 - B.K.C. THENG, 'The Chemistry of Clay-Organic Reactions', Hilger, London, 1974.
2 - J.M. THOMAS, 'Intercalation Chemistry', M.S. WHITTINGHAM and A.J. JACOBSON, Ed. Academic Press, New York, 1982, chapter 3.
3 - M. FRENKEL and L. HELLER KALLAI, Clays and Clay mineral, **31**, 92 (1983).
4 - E.C. TAYLOR and C.S. CHIANG, Synthesis, 467 (1977).
    V.M. THUY and P. MAITTE, Bull. Soc. Chim. Fr., 264 (1978).
5 - A. CORNELIS, P. LASZLO and P. PENNETREAU, Clay Miner. **18**, 437 (1983) and cited references.
6 - P. LASZLO and J. LUCHETTI, Tetrahedron Lett., **25**, 1567, 2147 (1984).
7 - H.M.R. HOFFMANN, Angew Chem. Int. Ed. Engl. **8**, 556 (1969).
    W. OPPOLZER and V. SNIECKUS, Angew. Chem. Int. Ed. Engl. **17**, 476 (1978).
    J.M. CONIA and P. LE PERCHEC, Synthesis, 1 (1975).
8 - B.B. SNIDER, Acc. Chem. Res. **13**, 426 (1980).
9 - K.H. SCHULTE-ELTE and G. OHLOFF, Helv. Chim. Acta, **21**, 153 (1967).
10 - Y. NAKATANI and K. KAWASHIMA, Synthesis, 147 (1978).
11 - M.F. SALOMON, S.N. PARDO and R.G. SALOMON, J. Am. Chem. Soc. (a) **102**, 2473 (1980) (b) **106**, 3797 (1984).
12 - M.F. SALOMON, S.N. PARDO and R.G. SALOMON, J. Org. Chem., **49**, 2446 (1984).
13 - R.H. WIGHTMAN, J. Chem. Soc. Chem. Comm., 818 (1979).
14 - M. FRENKEL, 'Clays and Clay Minerals', **22**, 435 (1974).
15 - S.S. CADY and T.J. PINNAVAIA, Inorg. Chem. **17**, 1501 (1978).
16 - J.M. ADAMS, D.E. CLEMENT and S.H. GRAHAM, Clays and Clay Minerals , **31**, 129 (1983).
17 - A. FOUCAUD and D. SAIB, unpublished observations.

PHOTOINDUCED CHARGE SEPARATION AT INORGANIC AND ORGANIC INTERFACES

Deh-Ying Chu, Rodney Stramel, Shuichi Hashimoto, Takashi
Nakamura, Jim Murtagh, J. Kuczynski, B. Milosavljevic, J.
Wheeler and J. K. Thomas
Chemistry Department, University of Notre Dame,
Notre Dame, IN 46556

ABSTRACT. The effects of various interfaces, inorganic or organic
colloids, on photoinduced charge separation is discussed. In
particular the nature of the charge separation, i.e., whether charge
separation occurs on contact of the two reactants or whether electron
tunneling produces two separate ions is discussed in some detail, and
also the nature of the partial screening of the reactants at the
interfaces as distinct from homogeneous solution. Finally, unique
effects in colloidal inorganic semiconductors such as cadmium sulphide
are discussed where charge separation is observed, but where the
photophysical properties of the semiconductor dramatically change when
small particles are created.

INTRODUCTION

There has been considerable interest in photoinduced charged
separation[2] for possible use in the storage of solar energy. Various
chemical systems have been suggested, some of which appear to be quite
practical and some which are more on the modeling side. This paper
discusses photochemical systems in unique organized assemblies which
produce special effects on photoinduced charged separation. It was
noted, quite early on, that charge separation, as in photoionisation
of aromatic molecules or in charge transfer between excited aromatic
molecules, was greatly enhanced in micellar systems and that the ions
had significantly longer lifetimes than those produced in homogeneous
solution. Thus further reactions of the ions could take place to
produce useful products, such as hydrogen. The immediate question
that may be asked is, whether the photoinduced charge separation is
due to a reaction of the two species as they collide on the surface
thus producing an ion pair that subsequently dissociates, or whether
the initial excitation act leads to electron tunneling while the
reactants are still separated, thus leading to an ion pair which are
already separated, which might enhance further separation.

A feature in these systems is the fact that reactant molecules
are embedded at an interface, where they experience a degree of
shielding from one another that is different from that experienced in

237

R. Setton (ed.), Chemical Reactions in Organic and Inorganic Constrained Systems, 237–252.
© *1986 by D. Reidel Publishing Company.*

homogeneous solution. Usually one side of the interface is rather
rigid in nature, where the molecule is embedded to varying degrees
depending on the system used. Experiments are described which are at
least initial investigations of such effects on photochemical
reactions.

Inorganic systems have also played a large role in modeling of
solar energy devices. They are useful as they are more resistant to
chemical corrosion than the organic systems. In particular titanium
dioxide has been a prime candidate for electron transfer to an
acceptor or electron transfer to the colloid. The main disadvantage
of such a colloid is that its adsorption is in the ultraviolet and so
it is only a model for solar energy systems. Other semiconductors, in
particular cadmium sulphide, have shown strong absorptions in the
visible part of the spectrum. This colloid has been shown to go
through many electron transfer reactions to suitable electron
acceptors which are adsorbed at the colloid surface. An additional
point of interest is that this colloid, due to its small dimensions,
shows significantly different photophysical effects compared to large
bulk semiconductors. Colloid chemistry provides a useful method of
preparing important material, such as semiconductors, with varying
architecture in order to investigate the nature of the architecture on
important photoelectrical properties of the system. Finally, the use
of rigid systems is considered where chemical reactions, such as
electron tunneling reactions, may be investigated under constrained
conditions or where the formation of a colloid is restricted, thus
forcing it to take a particular shape, which produces unique
photophysical properties, or enables the aqueous environment to be
removed from the colloid. These systems enable the experimenter to
investigate the nature of the environment of the colloid surface, and
its photophysical and photo electron transfer properties in some
detail. References are given only to reviews and to work discussed.

EXPERIMENTAL

The experimental techniques used are laser flash photolysis with a
response of less than 10^{-9} seconds, standard spectrofluorimetry and
absorption spectroscopy. Colloid particle sizes were measured by
electron microscopy or by dynamic light scattering. Other techniques
are conventional to normal chemical and physical practices. The
references refer to these experimental techniques in more detail.[3]

RESULTS AND DISCUSSION

Micellar Systems.

Micellar systems, which are small particles with highly charged
surfaces, significantly effect and promote photoinduced charge
separation quite often leading to the formation of stable or
relatively stable radical ions, which carry out useful subsequent
chemistry. A question asked has been whether the electron transfer

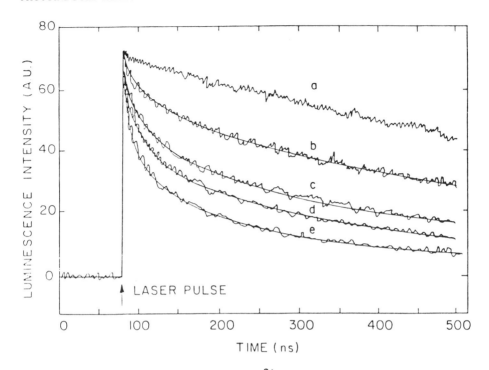

Figure 1.   Decay profiles of Ru(bpy)$_3^{2+}$ luminescence monitored at 610
            nm in the presence of various methylviologen concentrations
            (M/L):   a = 0, b = 0.025, c = 0.05, d = 0.075, e = 0.1.
            Smooth lines represent the best fit by equation 3.

occurs over a distance while the species are held together on the
surface of the assembly. Another feature is whether there is some
screening effect or geometric effects on the electron transfer
reactions due to the rigid micellar surface.   In order to solve the
first problem we have carried out reactions in rigid cellulose systems
which in may cases simulate the effects of micelles, yet do not allow
diffusion of the reactants. Any photoinduced reaction must occur
while the reactants are separated. To solve the second problem we
have carried out photoinduced electron transfer reactions in
cyclodextrins with and without added surfactants which tend to screen
the guest molecule from the attacking molecule which is in the aqueous
phase. The effects of geometric constraints on photoinduced electron
transfer reaction may be then ascertained.

Photochemistry in Cellulose Systems

It is well established in a variety of liquid solvents that the
luminescence of excited tris(2-2'-bipyridine)ruthenium(II), Ru(bpy)$_3^{2+}$,
is quenched by methylviologen, MV^{2+}, via an electron transfer
mechanism.

$$*Ru(bpy)_3^{2+} + MV^{2+} \longrightarrow Ru(bpy)_3^{3+} + MV^+ \tag{1}$$

$$k_{H_2O} = 5 \times 10^8 \text{ mol dm}^{-3} s^{-1}$$

This reaction is followed by a very fast back reaction

$$Ru(bpy)_3^{3+} + MV^+ \longrightarrow Ru(bpy)_3^{2+} + MV^{2+} \tag{2}$$

$$k_{H_2O} = 3 \times 10^9 \text{ mol dm}^{-3} s^{-1}$$

Pulsed laser photolysis studies of the above $Ru(II)/MV^{2+}$ system when located in the solid medium of cellophane film reveal a transient absorption at 400 nm which corresponds to MV^+ absorption, showing an initial build up of reduced methylyiologen which occurs within ~ 2 μs, followed by a slow decay (~ 5 ms).[4]

Figure 1 shows a comparison plot for $Ru(bpy)_3^{2+}$ luminescence intensity decays in the presence of various concentrations of methylviologen with calculated curves.

It can be observed that the luminescence decays in a non-exponential fashion (except with no quencher present) indicating a "dynamic" type of quenching.

The transient emission data were interpreted by the model originally proposed by Miller to explain the decay of trapped electrons in glassy media in the presence of scavengers. The model, which is based on a tunnelling hypothesis that an electron in a square potential well tunnels to randomly distributed acceptors, assumes that the rate constant depends exponentially on the separation of reactants:

$$k(r) = \nu \exp(-r/a) \tag{3}$$

where ν and a are associated with the vibrational frequency in the trap and the attenuation length of the wavefunction. In later work, when the problem was treated more stringently, it was shown that the model also describes other types of interactions, i.e. it could be applied to non-adiabatic electron transfer between molecules, although in such cases the meaning of the coefficients is not completely clear. The probability that the donor will survive at time t is given by:

$$P(t) = \exp[-(c/co) g(\nu t)], \text{ with } co = 3/4\pi a^3 \tag{4}$$

In the original work the function of $g(\nu t)$ is approximated by $(\ln \nu t)^3$ which can cause an error of two orders of magnitude in ν factor if the frequency factor is of the order of magnitude of $10^{10} s^{-1}$. Therefore the "exact model" derived by Tachiya and Mozumder was used. The exact model leads to the equations

$$g(\nu t) = \ln^3(\nu t) + h_1 \ln^2(\nu t) + h_2 \ln(\nu) + h_3 \quad (5)$$

where h_1 h_2 and h_3 are coefficients related to derivatives of the gamma function

$$h_1 = -3\Gamma'(1) = 1.73164699$$
$$h_2 = 3\Gamma''(1) = 5.93433597 \quad (6)$$
$$h_3 = -\Gamma'''(1) = 5.44487446$$

$$k(r) = 1.0 \times 10^{10} \exp(-0.46\ r)\ s^{-1} \quad (7)$$

In the case of an excited donor the luminescence decay, which in the absence of acceptor is given by the rate constant ko, or variation of emission intensity with time is described by:

$$I = Io\ \exp\{-kot - A\ [\ln^3(\nu t) + h_1 \ln^2(\nu t) + h_2 \ln(\nu t) + h_3]\} \quad (8)$$

where A is a factor which depends linearly on the quencher concentration:

$$A = [Q]/[(Ro/a)^3\ co] \quad (9)$$

where Ro is "a critical transfer distance" in the sense that, for an isolated donor – acceptor pair separated by Ro, the electron transfer occurs with the same rate as the spontaneous deactivation in the donor.

The average separation of RuII and MV^{2+} on a micelle is about 20 Å. The time of diffusion over this distance is given by $t = x^2/D$ where x is 2×10^{-7} cm and D is 3×10^{-6} cm^2/sec, i.e., t = 1.2×10^{-8} secs. The time for e$^-$ tunnelling over this distance as calculated from equation 7 is 10^{-6} sec. In the many systems studied the rate of e$^-$ tunnelling on micelles is much slower than the encounter time via diffusion. It is concluded, in these systems, that e$^-$ transfer takes place on contact of the reactants.

Cyclodextrin Systems

Amphiphilic molecules interact with pyrene-β-cyclodextrin (Py-β·CD) complexes leading to an extremely hydrophobic environment for Pyrene (Py), in aqueous solution.[2,5] The three component systems give rise to a 1:1:1 complex of Py, β·CD and the surfactant. The binding constant of Py and β·CD increases significantly in the presence of the surfactants, which suggests an improvement in the solubility of Py in aqueous β·CD systems. Larger binding constants of Py and β·CD are obtained in the presence of shorter chain amphiphiles between C_4- and

C_{16}-surfactants. Fluorescence quenching of Py in Py-$\beta \cdot$CD-pyridinium surfactants ($C_nPd^+X^-$) systems obeys first order kinetics, which were independent of the concentration of $C_nPd^+X^-$ above a certain concentration, while the quenching rate constant is markedly affected by the chain length of the pyridinium surfactants. Smaller rate constants are obtained for longer chain surfactants. The observed kinetics are explained in terms of a 1:1:1 complex formation of Py, $\beta \cdot$CD and $C_nPd^+X^-$, and the chain length dependent rate constants are interpreted by assuming a "diffusion controlled reaction within limited space". In this model, reactants A and B are confined in a sphere of radius b. The first-order quenching constant k_q is given by:

$$k_q = D\alpha_1^2 \qquad\qquad (10)$$

where D is the diffusion coefficient of B and α_1 is the smallest positive root of eq. 11:

$$\tan\{(b-a)\alpha\} = b\alpha \qquad\qquad (10)$$

in which a is the reaction radius. Assuming that the chain-length of the pyridinium surfactants is proportional to b, the present chain-length dependent quenching constant can be qualitatively understood by Tachiya's model, if D is 10^{-7} $cm^2 sec^{-1}$. This low value of D indicates a great restriction of movement of reactants.

Stern-Volmer kinetics were observed for Py fluorescence quenching in the Py-b$\cdot$CD-$C_{16}C_2V^{2+}$(1-ethyl-1'-hexadecyl-4,4'-bipyridinium ion) system. This is ascribed to the long-range nature of the reaction in Py(S_1^*)-viologen group systems compared to that in Py(S_1^*)-pyridinium group systems. Quenchers such as oxygen, nitromethane, copper (II) ion, thallium (I) ion etc. which reside in the aqueous phase also quench excited Py in $\beta \cdot$CD. The influence of CD with and without surfactant on the rate depends on the nature of the quenching reaction, and on the degree of screening by the host system on the guest molecule.

A marked protective effect of $\beta \cdot$CD was observed for the Py quenching of fluorescence by oxygen, nitromethane, Tl^+, and Cu^{2+} as shown in Table I.

The quenching in Table I may be divided into two groups, those that require contact of excited Py and quencher and, those that do not. The latter category includes Cu^{2+} and MV^{2+}. Quenching by MV^{2+} is via electron transfer and is only slightly affected by $\beta \cdot$CD, as the electron may be transferred over the distance larger than the collisional distance of Py and MV^{2+}. A similar mechanism applies to Cu^{2+} although the $\beta \cdot$CD does reduce the efficiency of reaction to a larger extent than MV^{2+}. Reaction of CH_3NO_2 and $C_2Pd^+Br^-$ with Py are also electron transfer in nature but are markedly affected by $\beta \cdot$CD. This indicates that a close approach of the reactants is required for reaction.

Thallium (I) ions cause intersystem crossing of excited Py and require close contact of the two reactants. This mechanism is clearly

Table I Bimolecular quenching rate constants $(M^{-1}s^{-1})$ in $\beta \cdot CD$ systems.

quencher	H_2O	$\beta \cdot CD$	medium $\beta \cdot CD$ + NaLS[a]	$\beta \cdot CD$ + CTAB[b]
O_2	1.1×10^{10}	1.2×10^{9}	6×10^{8}	6×10^{8}
CH_3NO_2	8.1×10^{9}	5.6×10^{8}	2.5×10^{7}	1.7×10^{7}
Tl^{+}	6.3×10^{9}	2×10^{7}	$<10^{7}$	$<10^{7}$
$C_2Pd^{+}Br^{-}$	7.9×10^{9}	7×10^{7}	5×10^{7}	3.5×10^{7}
Cu^{2+}	4.5×10^{9}	1.7×10^{8}	6.5×10^{8}	2.2×10^{8}
MV^{2+}	7.8×10^{9}	5×10^{9}	4.0×10^{9}	6.8×10^{8}
$C_{16}C_2V^{2+}$	7.6×10^{9}			
$C_{16}Pd^{+}$	3.6×10^{9}			

a, $[\beta \cdot CD]$= 13.7mM; [NaLS]= 15mM

b, $[\beta \cdot CD]$= 13.7mM; [CTAB]= 10mM

reflected in the dramatic reduction in reaction efficiency by $\beta \cdot CD$. Oxygen also causes intersystem crossing of excited Py, but the relatively small effects of the host molecule on the reaction efficiency indicate that close contact is not required for this reaction.

Although larger rate constants are obtained in the NaLS-CD system for positively charged quenchers, the magnitude of constants are not so different from those in surfactant-free $\beta \cdot CD$ system. It was also noticed that the quenching constant was significantly reduced for the system where complex formation is necessary for the quenching mechanism, namely Tl^{+} quenching, however, this was not the case for the Cu^{2+} and MV^{2+} quenching systems which undergo electron-transfer.

Significant steric effects on several reactions are exhibited in these systems; effects which must also operate to some degree in micellar systems.

Polyelectrolytes.

The architecture of polyelectrolyte carboxylic acids in aqueous
solution varies with pH, being tightly coiled at low pH where the
polymer is unionised, and open structure at higher pH where the
polymer is fully ionised. The polymer resembles a micelle structure
at low pH and solubilises hydrophobic molecules; these are ejected
into the aqueous phase at higher pH.[6] Covalent binding of the
hydrophobic probe molecule to the polymer enables the effect of
environment on reactions of the probe to be studied. The change in
the environment of bound pyrene on neutralization and uncoiling of the
polymer can be demonstrated by the restrictive effects that the
polymer imposes on the approach of quencher molecules such as thallium
or iodide ions to pyrene solubilized in or bound to the polymer.
Earlier studies showed that it was indeed very difficult for ionic
quenchers to traverse the polymer and react with pyrene located in the
polymer coil. Similar behavior is found for bound pyrene, a sharp
decrease occurs in the fluorescence of pyrene bound to
poly(methacrylic acid) vs. pH in the presence of Tl^+ ions. As the
polymer coil starts to open, the effectiveness of the quencher Tl^+ in
quenching bound pyrene increases and leads to a decrease in the
intensity of pyrene fluorescence. The data show a more marked
decrease in fluorescence yield than the data in the absence of Tl^+.
The same kinetic picture, that the hydrophobic environment of the
polymer chain denies entry of H_2O, thallium, or iodide to the pyrene
bound to the polymer, may be used for both systems. However, as the
chain is expanded the pyrene environment becomes progressively more
polar as water penetrates to the probe, a situation that also allows
the ionic quenchers to more readily approach pyrene. In the case of
thallium, the negative charge of the polymer increases the local
concentration of these cations around the polymer, which leads to a
dramatic increase in quenching with increasing pH and increasing
negative charge of PMA. In the case of iodide the effect is not as
remarkable as the thallium, even in the uncoiled case, as iodide tends
to be repelled by the negative charge.
     Table II lists data on the quenching of bound pyrene in
methacrylic acid-1-pyreneacrylic acid copolymer for various quenchers
at pH 2.5 and around pH 11.8, i.e., where the polymer coil is
completely closed and where it is completely open. The quenching
values tend to increase as the polymer chain opens. For the neutral
quencher oxygen, there is a slight increase, a factor of 6, on opening
of the polymer to expose pyrene to the aqueous phase. The effect is
larger in the case of the other neutral quencher, nitromethane, where
an increase of about 13 is observed, which again shows greater access
of the quencher to pyrene when there is an aqueous phase surrounding
it. These data show that O_2 tends to penetrate the polymer coil to a
much greater extend than CH_3NO_2. However, some quenchers such as
cetylpyridinium chloride (CPC) show no particular effect on opening or
closing of the polymer. This is associated with a binding of this
cationic surfactant to both the closed- and open-chain polymer, a
situation where its relative effectiveness in quenching would be

Table II

Quenching of Bound Pyrene in Methacrylic
Acid-1-Pyreneacrylic Acid Copolymer for Various Quenchers

quencher	pH	$k_q (M^{-1}s^{-1})$	A^b
$TlNO_3$	2.5	4.3×10^7	5×10^2
	11.8	2.4×10^{10}	
$CuSO_4$	2.2	1.1×10^8	1×10^2
	7.2	1.2×10^{10}	
O_2	2.5	5.4×10^8	6.3
	11.8	3.4×10^9	
NaI	2.5	7.2×10^6	6.8
	11.8	5.2×10^7	
CH_3NO_2	4.0	5.4×10^8	13
	7.0	7.0×10^9	
CPC^a	2.2	1.6×10^{10}	1.2
	7.2	1.9×10^{10}	

a Cetylpyridinium chloride.
b $= K_q(uncoiling)/K_q(coiling)$.

similar in both cases. The quenching with CPC is not a dynamic quenching, as observed with Tl^+, O_2, and CH_3NO_2, but static. The data for CPC quenching do not exhibit a change in fluorescence lifetime with increasing quenching concentration, but merely a decrease in fluorescence intensity. The quencher CPC binds close to the pyrene at all polymer configurations and quenching is extremely rapid compared to the fluorescence lifetime. Other positively charged quenchers such as Tl^+ and Cu^{2+} show a much larger quenching effect in the ionised polymer as they are bound to the polymer in close proximity to the probe.

Vesicles.

Vesicles feature the possibility of building two compartments to separate photo-induced ions, the inner compartment of water and the exterior bulk phase. A major problem is leakage of the photo products across the lipid barrier. Some attempt has been made to polymer specialized vesicle surfactants. Leakage is minimized in these systems, however, synthesis of the vesicle material is difficult. Recent work has indicated that swelling a surfactant vesicle with divinyl benzene, followed by free radical polymerization, leads to a vesicle held together by crosslinked polymer. Internal diffusion of large lipids is greatly decreased in such a system, and could help to enhance charge separation and stablization.

Colloidal Clays

Colloidal silica and clays show many properties that are reminiscent of anionic micellar solutions.[8] The fluorescent probe tris(2,2'-bipyridine)ruthenium(II), RuII, has been used to investigate the nature of the surfaces of the following clays: laponite, which is a synthetic clay, and natural hectorite and montmorillonite. RuII is adsorbed completely by the clay by ion exchange and on excitation gives rise to a luminescence spectrum in the red part of the spectrum with a lifetime of about half a microsecond. The lifetime, quantum yield, and the nature of the absorption spectrum are dependent on whether the RuII is adsorbed in layers as in the natural clays, or whether it is adsorbed on the surface as with laponite. At low concentrations of laponite, RuII is adsorbed on outer layers and is in contact with the aqueous phase. However, at higher clay concentrations or in the presence of calcium chloride, layering of the clay occurs and the probe molecule is placed progressively between the layers where its photophysics is altered. The casting of a film from the laponite-RuII exhibits maximum spectral change as associated with maximum colloid layer formation. Such changes are not as readily observed with hectorite or montmorillonite, and this indicates that for the most part, these systems exist as layered colloids and that RuII is already adsorbed between the layers. Other molecules such as Cu^{2+}, dimethylaniline, and nitrobenzene react with excited RuII through electron transfer reactions, and are also adsorbed to varying extents on the clay surface. Cu^{2+} is adsorbed strongly and the kinetics are simplified due to the strong adsorption. Here, Stern-Volmer type kinetics are observed and a quenching rate constant is obtained which is lower than that in the aqueous solution, which gives an estimate of the degree of movement of cupric ions on the clay surface. Dimethylaniline and nitrobenzene are adsorbed weakly on the clay. However, the clay catalyses the reaction of the RuII with these quenchers as both are adsorbed in a small volume, i.e., the clay surface. The kinetics that describe these latter reactions are of the Poisson form, and the kinetics indicate that the reactive quencher molecules are adsorbed around the RuII, in a zone like effect, rather than being adsorbed randomly throughout the system. This tends to indicate that the sites of adsorption are not uniform on the clay surface but occur in regions.

The model used in the clay system indicates that RuII adsorbed on clay surfaces do not change their position within the lifetime of RuII*. The quenchers, such as nitrobenzene and dimethylaniline are weakly bound in the vicinity of RuII, indicating that the movement of the quencher is limited to the vicinity of these cations. In other words, RuII* is quenched by nitrobenzene or dimethylaniline which diffuses in a limited reaction space. This model is identical mathmatically with that of micellar systems. In the present clay system it indicates that the binding sites for quenchers dimethylaniline and nitrobenzene are not randomly distributed in the clay surface but occur in zones on the particles. Analysis of the data gives the number of sites per gram of clay as 2×10^{-5} M i.e.,

2.5% of the C.E.C. of the clay. The diffusion constants for Cu^{2+} on laponite, hectorite, and gel white are, 3.3×10^{-6}, 10^{-6}, and $1.3 \times 10^{-6} cm^2/sec$ respectively.

Iron Oxide

Colloidal solutions of metal oxides has been used extensively for promoting various photochemical reactions. Ferric oxide sols have an inherent interest because of the wide distribution of this material in nature.

The photochemistry of two aqueous colloids of iron oxides (Hematite, and Amorphous), have been investigated by laser flash photolysis and steady state techniques.[9] Excited ruthenium tris bipyridine, $*Ru(bpy)_3^{2+}$, is quenched by the amorphous colloid provided polyacrylic acid, PAA, is used to stabilize the colloid and promote adsorption of $Ru(bpy)_3^{2+}$ onto the oxide surface. No net chemical change takes place and the quenching effect is most marked at pH 7.

Methyl viologen, MV^{2+} and PAA form complexes with properties which are typical of those reported for other anions, i.e. an absorption spectrum develops with a tail which extends into the visible, which is not present in either spectra of MV^{2+} or PAA alone. On irradation with UV light (308 nm) $MV^{+\cdot}$ is produced via e^- transfer from the $-COO^-$ group of the PAA to MV^{2+}. $MV^{+\cdot}$ is a stable radical with a well defined redox potential ($-.441V$) and its adsorption spectrum (λ max = 605 nm, $\varepsilon_{605} = 1.2 \times 10^4 M^{-1} cm^{-1}$) is conveniently located in the optical window of iron oxide.

Pulsed laser excitation of MV^{2+}-PAA-Fe_2O_3 colloids with light of wavelength 308 nm leads to the formation of $MV^{+\cdot}$ which decays via reaction with Fe_2O_3. Figure II shows the kinetic traces of the decay of $MV^{+\cdot}$ in the α-Fe_2O_3-PAA system under various conditions. At pH 7.0, $MV^{+\cdot}$ decays fully to the baseline, the decay being exponential with time. At pH 10.0 $MV^{+\cdot}$ only decays part-way to the base line, and a equilibrium between iron oxide and $MV^{+\cdot}$ is operative. Increasing $[Fe_2O_3]$ leads to a more rapid initial rate of decay of $MV^{+\cdot}$ and to a larger extent of decay, i.e., an increased degree of e^- transfer from $MV^{+\cdot}$ to Fe_2O_3, and a shift in the equilibrium to the Fe_2O_3.

The measured rates from Figure II are 9.3×10^9 and 6.3×10^9 L $M^{-1} s^{-1}$ for $Fe_2O_3 \cdot H_2O$ (am) (30 Å diameter), and α-Fe_2O_3 (67 Å diameter) respectively. The diffusion controlled rate is given by:

$$k = 4 \pi (r_1 + r_2) \frac{kT}{6\pi\eta} \times 6 \times 10^{20} LM^{-1} sec^{-1}$$

and is $3.2 \times 10^{10} M^{-1} S^{-1}$ using $r_1 = 15$ Å, $r_2 = 5$ Å, $\eta = 10^{-2}$ poise, T = 298K and 1.19×10^{11} using $r = 33$ Å, $r_2 = 5$ Å. The measured rates are a factor of ten less than those calculated, indicating inefficient e^- transfer from $MV^{+\cdot}$ to Fe_2O_3.

The steady state irradiation of MV^{2+}-$Fe_2O_3 \cdot H_2O$ (am)-PAA at pH 7 leads to a decrease in the Fe_2O_3 adsorption spectrum due to the production of Fe(II) as monitored by o-phenanthroline.

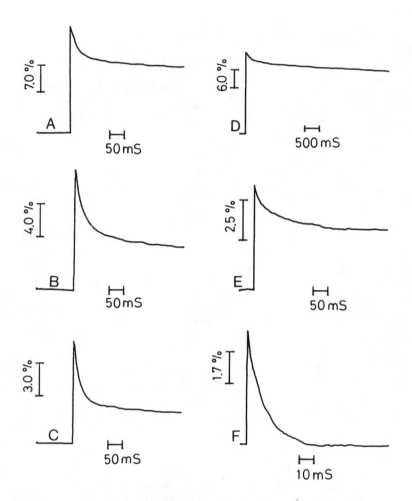

Figure 2. Reaction of MV+· with crystalline iron oxide. Left column shows the equilibrium shift at pH 9.0 with increasing particle concentration

A) 3.2×10^{-7} M particles
B) 6.4×10^{-7} M particles
C) 9.6×10^{-7} M particles

Right column shows the effect of pH on the equilibrium $[\alpha\text{-}Fe_2O_3] = 8.5 \times 10^{-7}$ particles

D) pH 11.0
E) pH 9.5
F) pH 7.1

Photophysics of polymer bound colloids

PAA stabilizes several colloids, Fe_2O_3, TiO_2, WO_3, ZnO, CoO at pH 7.0 indicating that the polymer binds to the surface of the colloid. Photolysis of these solutions should lead to CO_2 via oxidation of the $RCOO^-$ group to $RCOO^{\bullet}$ and e^- transfer to the oxide, followed by decomposition of the $RCOO^{\bullet}$ radical. The oxidation potential of the above semiconductor oxides on excitation at pH 2 are given as 2.5, 3.2, 3.4 and 3.3 eV for Fe_2O_3, TiO_2, WO_3, and ZnO respectively. The energy required to remove e^- from $RCOO^-$ is 2.1 ev. Hence, all oxides should oxidize $RCOO^-$, and CO_2 was found on irradiation of TiO_2, ZnO, WO_3 with adsorbed polymer. However, no CO_2 was found on irradiation of the $\alpha-Fe_2O_3$-PAA system. Apparently Fe_2O_3 does not oxidize adsorbed anions which may be due to the low lying surface states of this oxide, a feature which causes low photocurrent yields in $\alpha-Fe_2O_3$ electrodes.

Attempts were made to react Fe_2O_3 with other molecules by either direct irradiation of the molecule or the iron oxide, no photochemical reactions were observed with iron oxide and PVA, with triethylamine, dimethylaniline, tetramethylbenzidene, I^-, $Fe(CN)_6^{4-}$, $Fe(CN)_6^{3-}$, S^-, or EDTA.

We have not been successful in promoting chemical reaction by irradiating directly into the Fe_2O_3 absorption band. Data from previous studies on photoelectrodes indicate the poor photoresponse of Fe_2O_3 arises from surface states that efficiently mediate e^-/h^+ pair recombination.

Cadmium Sulfide

During the past two years there has been a significant increase in interest in the photochemistry of small particles of cadmium sulfide (CdS) usually in the colloidal form.[10] Much of this work stems from the search for photosystems that can be excited in the visible part of the spectrum, thereby initiating electron-transfer reactions. In colloidal CdS, photoexcitation of this molecule generates electron-hole pairs, which may migrate to the colloid surface and across the semiconductor-water interface where they may carry out the appropriate redox reactions.

Both steady-state and pulsed-laser photoexcitation techniques have been used to investigate photoinduced processes at the surface of CdS colloids. The colloid surfaces have been modified by use of adsorbed reactants and surfactants, the latter conveniently changing the charge of the colloid and lending greater stability to the system. Luminescence of CdS is observed, the spectral quality of which depends on the excitation light intensity, and the nature of the adsorbed species. The CdS luminescence is quenched rapidly ($\tau \ll 10^{-9}$) by various additives either via positive hole or e^- capture. The kinetics of the former process are Perrin, while those of the latter are Stern-Volmer in nature. Photoinduced e^- transfer arises from e^- involved with the luminescence processes and also from e^- which do not give rise to luminescence; these two processes are quite separate. Pulsed studies show that reduced products of photoinduced e^- transfer readily move on the colloid surface to react with other adsorbed species. They remain on the particle surface for extended periods of time which are greater than 260 μs.

Ethylene diamine tetra acetate, EDTA, greatly increases the efficiency of photo-induced electron transfer from aqueous colloidal cadmium sulfate to methyl viologen MV^{2+}. This effect is very apparent in colloids stabilised by cetyl trimethyl ammonium bromide, CTAB, a cationic surfactant that imparts a positive charge to the CdS surface. Experiments are reported that show that the crucial event in the system is the formation of a complex of EDTA and MV^{2+} with a resultant negative charge. This complex is electrostatically bound to the cationic CdS surface where photo-induced electron transfer occurs. Subsequent break up of the EDTA − reduced methyl viologen complex leads to MV^+ which is repelled away from the cationic CdS surface. The above effects are reversed on preparing a CdS colloid with a negatively charged surface by using sodium lauryl sulfate as a stabiliser.

Some of the recent work presented on the production of small particles of CdS indicate that the colloid size has a dramatic effect on the semiconductor properties of these colloids as compared to those of a single crystal or a large particle of CdS. This is usually achieved by a unique preparation of the CdS colloid in a nonaqueous environment. Typically, the absorption spectrum characteristic of a single crystal of CdS (with a sharp onset of about 520 nm) decreases in intensity. Also, the spectral onset shifts from 520 nm to shorter wavelengths and some indication of structure in the spectrum is observed below 500 nm, possibly the result of size on the quantum mechanical effects which normally affect the energies of the conduction band.

Several workers have also indicated that CdS may be conveniently prepared in solid matrices such as Nafion and cellulose. Indeed, it has been shown that CdS shows enhanced and unique properties in such systems. For example, in these systems small particles of CdS can be formed which possess luminescent lifetimes in excess of 1 μ sec. Furthermore, it is possible to keep them in unique form even upon removal of the water while the polymer matrix keeps the particles from coagulating. Therefore, the effect of water on the surface properties of small CdS particles can be conveniently studied. In the preparation of colloids, various stabilizers, such as surfactants, polyvinylalcohol, and large silica particles, are used to stabilize the CdS particles.

Small stable particles of cadmium sulfide (CdS) are formed by precipitation in porous Vycor glass, and as in the polymer matrices no stabilizers are needed. The known physical properties of this material, such as the dimensions of the porous channels, indicate that the CdS form must have one dimension much smaller than 40Å, i.e., the diameter of the cylinders or pores in the silica. However, the spectroscopic properties of CdS are indicative of bulk material, and quite unlike those associated with small (< 100Å radius) CdS colloids. Various other guest molecules such as methyl viologen (MV^{2+}) or water can penetrate into the CdS-Vycor system and modify the photophysics of CdS.

References

1. The authors wish to thank the National Science Foundation and the
    Army Research Office for support of this work.

2. See the following texts for reviews of the type of photochemistry
    in organized assembles.
    a. Fendler, J. H. "Membrane Mimetic Chemistry"; Wiley: New
        York, 1983.
    b. Turro, N.; Braun, A.; Grätzel, M.; Agnew Chem. Int. Ed. Eng.
        1980, 19, 675.
    c. Thomas, J. K. ACS Monogr. No. 181, 1984.

Fuller references to details of the work are given here.

3. Photoinduced Electron Transfer in Organized Assemblies
    S. S. Atik and J. K. Thomas
    J. Am. Chem. Soc. 103, 3550-3555 (1981)

4. Photochemistry of Ruthenium Tris Bipyridyl in Cellulose
    B. Miloslavovic & J. K. Thomas
    J. Phys. Chem. 87, 00 (1983)

    Photochemistry of Compounds Adsorbed in Cellulose. Methyl
    Viologen Redox Reactions.
    B. Milosavljevic & J. K. Thomas
    Int. J. Radiation Chemistry & Physics 23, 237 (1984)

    Photochemistry of Compounds Adsorbed into Cellulose. Solid-State
    Reduction of Methylviologen Photosensitized by Tris(2,2'-
    bipyridine)ruthenium(II)
    B. Milosavljevic and J. K. Thomas
    J. Phys. Chem. 89, 1830-1835 (1985)

    Photochemistry of Compounds Adsorbed into Cellulose. Diffusion-
    Controlled Mechanism of Ru(bpy)$_3^{2+}$ Luminescence Quenching by
    Copper(II)
    B. Milosavljevic & J. T. Thomas
    J. Chem. Soc., Faraday Trans. 1, 735-744 (1985)

5. S. Hashimoto and J. K. Thomas. J.A.C.S., in press.

6. Influence of the Conformational State of Polymethacrylic Acid on
    the Photophysical Properties of Pyrene in Aqueous Solution: A
    Fluorescent Probe and Laser Photolysis Study
    T. S. Chen and J. K. Thomas
    J. Polym. Sco., Polym. Chem. Ed. 17, 1103-1116 (1979)

    Photochemistry of a Water Soluble Copolymer of Methacryclic Acid
    and 1-Pyrene Acrylic Acid.

D. Chu & J. K. Thomas
Macromolecules, 17, 2142 (1984)

7. J. Murtagh and J. K. Thomas, unpublished data.

8. Photochemical Probing of Colloidal Clay Solutions
    R. DellaGuardia & J. . Thomas
    J. Phys. Chem. 87, 990 (1983)

    Photochemistry on Colloidal Clays. II.
    R. DellaGuardia & J. K. Thomas
    J. Phys. Chem. 87, 3550 (1983)

    Photoprocesses on Colloidal Clay Systems. III.
    R. DellaGuardia & J.K. Thomas
    J. Phys. Chem. 88, 964 (1984)

    Reactions of Radical Cation of Tetramethyl Benzidine with
    Colloidal Clays.
    L. Kovar, R. DellGuardia & J. K. Thomas
    J. Phys. Chem. 88, 964 (1984)

    T. Nakamura and J. K. Thomas. Langmuir, in press.

9. R. Stramel and J.K. Thomas. J. Colloid and Interface Science,
    In press.

10. a) E. F. Sneva, G. R. Olin, and M. Hair, J. Chem. Soc. Chem.
        Comm. Vol 401 (1980).
    b) K. Kalyanasundaram, E. Borgarello, and M. Grätzel. Helv.
        Chem. Acta. Vol 64, p. 382 (1982).
        M. Gratzel, JACS. Vol 104, p 2977 (1982).
    c) R. Darwent and G. Porter. Chem. Comm. Vol 4, p 145 (1981).
        Y. Nakato, A. Tsamura, H. Tsubomana, Chem. Phys. Letts.
        Vol 85, p 387 (1982).
    d) B. Kraeutler, A. Bard, JACS. Vol 100, p 4317 (1978).
        I. Isumi, F. F. Fan, A. Bard. J. Phys. Chem. Vol 85, p 218
        (1981).
        A. Bard, Ibid, Vol 86, p 172 (1982).
    e) R. Rosetti, J. L. Ellison, J.M. Gibson, and L. E. Brus.
        J. Chem. Phys. Vol 80, p4464 (1984).
    f) A. Fojtik, H. Weller, U. Koch and A. Henglein
        B. Bunsenges, Phys. Chem. Vol 88, p 967 (1984).
    g) J.P.Kuczynski, B.A. Milosavljevic and J.K. Thomas. J. Phys.
        Phys. Chem. Vol 88, p 980 (1984).
    h) M. Krishran, J.R. White, M.A. Fox, and A.J. Bard. JACS.
        Vol 105, p 7002 (1983).
    i) J.P. Kuczynski and J.K. Thomas, Langmuir Vol 1, 158 (1985).
        Ibid. J. Phys. Chem. Vol. 87, 5498 (1983).
        Ibid. J. Phys. Chem., in press.

A COMPARISON OF THE FORWARD AND BACK PHOTOELECTRON TRANSFER IN DIRECT AND REVERSE MICELLES.

M. P. Pileni[1,2], P. Brochette[2] and B. Lerebours Pigeonnière[1,2]
-1- Université P. et M. Curie
Laboratoire de chimie-physique
11 rue P. et M. Curie 75005 Paris
-2- CEN Saclay DPC 91191 Gif sur Yvette

ABSTRACT. The photoelectron transfer from porphyrins to various viologens occurs from the triplet state of the sensitizer in oil in water cetyltrimethylammonium chloride (CTAC) and in water in oil Aerosol OT micelles. In direct micelles, DM, very important charge separation is due to the entrance of reduced viologen into the micellar core. In reverse micelles, RM, the charge separation is strongly incrased by a difference in the mechanism between the forward and the back reaction.

INTRODUCTION. It si well known that micellar solutions and microemulsions, because of their microheterogeneity privilige some particular reactions taking place from part to part of the interface. Taking advantage of the increase of hydrophobicity of reactants upon redox reaction, we used micelles as interfacial barriers which allow very good charge separation (1-5).

Separation between hydrophobic and hydrophylic products of a photo-sensitized electron-transfer has been monitored when changing the environment. We used, that for, direct and reverse micelles. In one case, the reactives (hydrosoluble) are in the bulk aqueous phase (D.M.), in the other one, they are intrapped in micelles (R.M.).
We emphasize the change of charge separation and the mechanisms involved with identical reactives in the two different types of micellar solutions.

MATERIALS AND METHODS

Materials

The CTAC surfactant Eastman kodak was purified by repeated recrystallization from acetone. The AOT surfactant was purchassed from Sigma and used without further purifications. The sensitizer were Zinc tetramethylpyridylporphyrin, $ZnTMPyP^{4+}$, Zinc tetraphenylporphyrin, ZnTPP and chlorophyll a, Chl. The latters were produced by Sigma.

R. Setton (ed.), Chemical Reactions in Organic and Inorganic Constrained Systems, 253–261.

The electron acceptors were dialkylviologens $(C_x)_2V^{2+}$ with $1 < x < 12$ or propylviologen sulfonate, PVS.

Apparatus

The laser photolysis experiments were performed with Nd laser (2ns) and flash photolysis experiments with conventional flash photolysis apparatus.

RESULTS

In direct micelles, DM, and in reverse micelles, RM, we have studied the same photoelectron transfer from $ZnTMPyP^{4+}$ to viologens. In R.M we have also studied the photoelectron transfer from ZnTPP or from Chl to viologens.

1. FORWARD PHOTOELECTRON TRANSFER

In DM and RM, the triplet state of ZnTPP and $ZnTMPyP^{4+}$, formed by laser or flash photolysis, are characterized by their absorption spectrum respectively centered at 460 nm and 470 nm (6). In RM, Chl triplet state is characterized by an absorption centered at 750 nm (7). The lifetime of the triplet state of sensitizer is given table I :

Table I : rate constant of triplet decay of sensitizer :

	$ZnTMPyP^{4+}$	ZnTPP	Chl
k_o	$4 \ 10^2 \ (s^{-1})$ $1.5 \ 10^3 \ (s^{-1})$	(8) unmeasurable	(9) $5 \ 10^3 \ (s^{-1})$

With ZnTPP a multiexponential triplet decay is observed which is probably due to various location of ZnTPP in micellar core and in the hydrocarbon bulk.

The triplet state of the various sensitizer used is quenched by viologens addition. The disappearance of the triplet state is most of the time followed by the appearance of new species attributed to the reduced viologen and to the oxidized sensitizer.

Table II : maximum absorption of various compounds :

V^+	$ZnTMPyP^{5+}$	$ZnTPP^+$	Chl^+
395, 605 nm	700 nm	400 nm	850 nm

The triplet quenching rate constant, k_q, was determined by monitoring the quenching of the triplet of the sensitizer at various viologen concentrations.

In direct CTAC micelles, DM :
The kq values obtained with the water soluble porphyrin, ZnTMPyP^{4+}, and the dialkylviologens, $(C_x)V^{2+}$, are similar to that obtained in aqueous solution (table III) indicating that the forward photoelectron transfer occurs in the bulk water phase and is equal to 2.5 10^6 M^{-1} s^{-1} (10).

Table III : Triplet quenching rate constant of ZnTMPyP^{4+} :

k_q x 10^{-6} (M^{-1} s^{-1})	$(C_1)_2V^{2+}$	$(C_3)_2V^{2+}$	$(C_6)_2V^{2+}$	$(C_8)_2V^{2+}$	$(C_{12})_2V^{2+}$	PVS
aqueous solutions	2	3	3	2	2	200
DM	3	3	2.5	2	3	80

In reverse AOT micelles, RM ; the triplet decay of the sensitizers follows a complex rate law (figure 1):

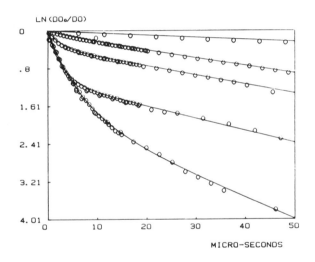

Figure 1. Kinetic treatment assuming a Poisson's law distribution : experimental points (0) and fitted curves (full line). [AOT] = 0.1 M, [Ch1] = 10^{-5} M, [PVS] = 5.10^{-4} M.

To determine k_q, it is estimated that the reactants are distributed amongst the water pools according to a Poisson distribution law (11-16). The kinetic treatment produces the following time dependence for the triplet of the sensitizer :

$$Ln \ [^3P^*]/[^3P_0^*] = - (k_o + ke \ [V^{2+}])t - \bar{n}(1-exp(-k_q \ t))$$

where $^3P^*$ and $^3P_0^*$ are the sensitizer triplet concentrations at time t and time zero respectively, k_o is the first order rate constant governing the triplet sensitizer decay in the absence of viologen, kq is the intramicellar quenching rate constant, k_e is the bimolecular exchange rate constant involving collisions between the water pools and $\overline{n} = [V^{2+}] / [WP]$ where $[V^{2+}]$ and $[WP]$ are respectively the viologen and the water pool concentrations. Table IV gives the values obtained with the various sensitizers and with the various viologens for a water content $w = [H_2O] / [AOT] = 20$.

Table IV : kinetic rate constants obtained from Poisson's distribution law : k_q (s^{-1}) k_e $(M^{-1} s^{-1})$

Sensitizer		PVS	$(C_1)_2V^{2+}$	$(C_3)_2V^{2+}$	$(C_6)_2V^{2+}$	$(C_8)_2V^{2+}$
	k_q	$3\ 10^5$		$3\ 10^5$	$2.6\ 10^5$	$4\ 10^5$
Chl	k_e	$3\ 10^7$	–	–	$3.2\ 10^7$	$1.1\ 10^7$
ZnTMPyP^{4+}	k_q	$6\ 10^8$	$1.5\ 10^8$	–	–	–
	k_e	$1.6\ 10^7$	$4.6\ 10^6$	–	–	–

with ZnTPP no kinetic studies are possible because of the various locations of ZnTPP in micelles and in the bulk.

2.   BACK REACTION

In direct micelles :
The results obtained in DM can be splitted in two parts :
- dialkylviologens with less than eight carbon atoms per side chain or with PVS :
With such viologens, the back reaction is similar to that observed in aqueous solution and the half lives, $\tau_{1/2}$, of the back reaction for all the reduced viologens is about 2.5 ms (table V). These results indicate that, in these cases, micelles do not perturb the back reaction.
- dialkylviologens with more than eight carbon atoms per side chains :
In micellar solution, the back reaction is considerably retarded whereas in aqueous solution no changes have been observed. Table V shows that the delay of the back reaction increase with the chain length of the viologens :

Table V : half lives of reduced viologen :

$\tau_{1/2}$ (ms)	$(C_1)_2V^{2+}$	$(C_3)_2V^{2+}$	$(C_6)_2V^{2+}$	$(C_8)_2V^{2+}$	$(C_{12})_2V^{2+}$
aqueous solution	2.5	2.4	2.3	2.6	2.5
DM	2.5	2.6	7	48	123

Such delay in the back reaction observed with $(C_8)_2V^{2+}$ and $(C_{12})_2V^{2+}$ is obtained at CTAC concentrations above the critical micellar concentration whereas below c.m.c concentration, the half lives obtained are similar to those obtained in aqueous solution. From NMR experiments it has been shown that the reduced dioctyl and dodecyl viologens are mainly located in the hydrocarbon core (17).

In reverse micelles, RM :
The results obtained with the various sensitizers and viologens
strongly differ :
-1- With ZnTMPyP^{4+} as a sensitizer, the triplet state is quenched by
viologen addition but no reduced viologen formation is observed after
the laser pulse. To observe reduced viologen, a water soluble electron
acceptor such as NADH, has to be added to the solution. Hence NADH
reacts with the porphyrin cation and induced a stabilization of reduced viologen. The formation of reduced viologen in such case and the
quenching of ZnTMPyP^{4+} triplet state by viologen addition indicates
that the forward electron transfer reaction takes place. The failure
to observe reduced viologen by laser photolysis can be explained in
terms of the location of species : the forward reaction is followed
by a very fast back reaction which is due to the high proximity of the
reactive species in the water pool droplet.
-2- With Chl, the back electron transfer reaction follows a complex
kinetics decay (figure 2). Which is unchanged with the various

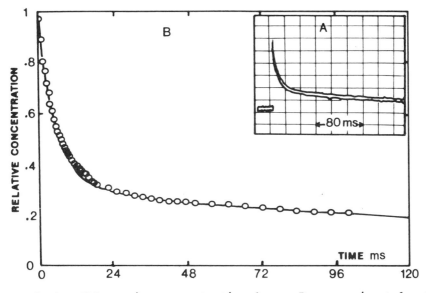

Figure 2. A : Chl cation concentration decay. B : experimental points
and simulated curve of Chl cation concentration.

dialkylviologens. To explain such decays we assume that the short time
part of the decay curve is due to the intramicellar back reaction
whereas the long time kinetic decay is due to the exchange of photo-
lytic products from one droplet to another, which could induce a delay
in the back reaction. These processes could be represented by the
following :

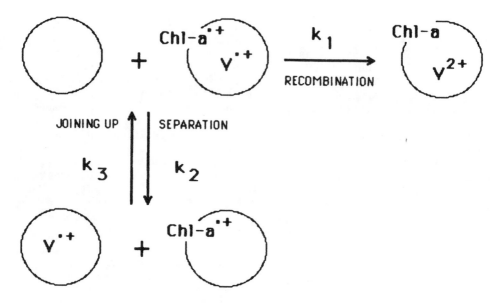

where the circle represent the micelles.

From the kinetic treatment the various rate constants of the processes are determined (18) (table VI).

Table IV : kinetic rate constant :

k_1	k_2	k_3
$108 \ s^{-1}$	$1.6 \ 10^5 \ M^{-1} \ s^{-1}$	$3.10^7 \ M^{-1} \ s^{-1}$

Discussion

In RM and in DM, the forward photoelectron transfer occurs from the triplet state of the sensitizer. However the data obtained in RM and in DM are strongly differents :

In direct micelles :
The quenching rate of the $ZnTMPyP^{4+}$ triplet state by the various dialkylviologens is similar in the presence and in the absence of micelles (table II). This indicate that the forward electron transfer occurs in the bulk aqueous phase. Similarly, with dialkylviologens having less than eight carbon atoms per side chain, the unchanged values of the back electron rate constant obtained in aqueous and micellar solutions indicate than the back reaction also occurs in the bulk phase. The strong increase of the half lives of reduced viologen having eight or more carbon atoms per side chain, in micellar solution in comparison to those obtained in aqueous solution (table V) and their changes below the critical micellar concentration indicate a micellar effect and can be explain by the entrance of reduced viologen inside the micellar core. This is confirmed by the increase of hydrophobicity

of reduced viologen by increasing the chain length (19) and by the
changes of CTAC NMR spectrum (17).

In reverse micelles :
The forward and the back electron transfer strongly depend on
the relative location of the sensitizer and the electron acceptor :

-1- The electron acceptor and the sensitizer are located in the same
water pool : in this case the forward and the back electron rate constant are very fast and no reduced viologen is detected. This is probably due to the high proximity of the sensitizer and the acceptor.

-2- The electron acceptor and the sensitizer are located in the same
droplet, on both sides of the interface.
The forward electron transfer is slower than the previous case
and the back reaction is strongly delayed. The forward electron transfer rate constant is about 3.10^5 s^{-1} and the back reaction is about
100 s^{-1}. This delay is attributed to changes of the electron configuration of the solutes after the photoelectron transfer through the
interface.
From kinetic treatment of the triplet state of the sensitizer
at various viologens concentration, the micellar exchange rate constant is determined and is equal to 3.10^7 M^{-1} s^{-1}. This value is in
good agreement to that given previously in litterature (20).
The rate constant determined from the kinetic treatment of the
back reaction indicates that the exchange of reduced viologen from one
droplet to another is less efficient (10^{-5} M^{-1} s^{-1}) when the two photo-
products are located in the same droplet than when they are in two
different droplets (3.10^7 M^{-1} s^{-1}).

Conclusion:

In the present paper we have shown that the back reaction
induced by photoelectron transfer can be strongly delayed using
micellar solutions. However the reasons of such delay depend on the
organized media used. In oil in water micelles (DM), the back electron
transfer is prevented by the hydrophobic character of the reduced
dialkylviologen when in water in oil micelles (RM) the delay is
for one part due to changes in the configuration of chlorphyll cation
and reduced viologen and for another part to the micellar exchange
processes.

REFERENCES

1. J. Kiwi, K. Kalyanasundaram, M. Gratzel, in *"Structure and bonding ; 49"* Ed. C.K. Jorgensen, Springer-Verlag, Berlin Heidelberg, 1982.

2. P. A. Brugger, P. P. Infelta, A. M. Braun, M. Gratzel, *J. Am. Chem. Soc.*, 103, 320, 1981.

3. P. P. Infelta, P. A. Brugger, *Chem. Phys. Lett.*, 82, 3, 462, 1981.

4. T. Nagamura, T. Kurihara, T. Matsuo, M. Sumitani, K. Yoshihara, *J. Phys. Chem.*, 86, 4368, 1982.

5. M. P. Piléni, B. Lerebours, P. Brochette, Y. Chevalier, *J. Photochem.*, 28, 273, 1985.

6. K. Kalyanasundaram, M. Neumann-Spallant, *J. Phys. Chem.*, 86, 5163, 1982.
   A. Harriman, *J. Chem. Soc. Faraday Trans. II*, 77, 1281, 1981.

7. P. Mathis, P. Sétif, *J. Israel Chem.*, 21, 316, 1981.

8. M. P. Pileni, *Chem. Phys. Lett.*, 75, 3, 540, 1980.

9. J. Kiwi, M. Gratzel, *J. Phys. Chem.*, 84, 1503, 1980.

10. S. Chevalier, B. Lerebours, M. P. Pileni, *J. Photochem.*, 27, 301, 1984.

11. P. P. Infelta, M. Gratzel, J. K. Thomas, *J. Phys. Chem.*, 78, 190, 1974.

12. M. Tachiya, *Chem. Phys. Lett.*, 33, 289, 1975.

13. S. S. Atik, J. K. Thomas, *J. Am. Chem. Soc.*, 103, 3543, 1981.

14. M. P. Piléni, P. Brochette, B. Hickel, B. Lerebours, *J. Coll. Int. Sci.*, 98, 549, 1984.

15. Y. Moroï, *J. Phys. Chem.*, 84, 2186, 1980.

16. J. C. Dederen, M. Van der Aumeraer, F. C. de Schryver, *Chem. Phys. Lett.*, 68, 213, 451, 1979.

17. B. Lerebours, Y. Chevalier, M. P. Pileni, *Chem. Phys. Lett.*, 117, 89 (1985).

18. P. Brochette, T. Zemb, P. Mathis, M. P. Pileni, *submitted to Helvetica Chimica Acta*.

19. I. Tabushi, 9th *IUPAC Symp. on Photochemistry*, Pau, 1982.
    I. Tabushi, *Pure Appl. Chem.*, <u>54</u>, 1733, 1984.

20. P. D. I. Fletcher, Howe, B. H. Robinson, Steyler in *"Reverse micel-les, (Proc. Eur. Sci. Found. Workshop, Rigi-Kaldbalt, Switz., 1982)"*
    Ed. P. L. Luisi, and B. E. Straub, *Plenum-press*, 1984.

    S. S. Atik, J. K. Thomas, *Chem. Phys. Lett.*, <u>79</u>, 351, 1981.

EFFECT OF MICROEMULSIONS ON PHOTOCHEMICAL REACTIONS

Monique Rivière
Laboratoire I.M.R.C.P.- UA CNRS 470
Université Paul Sabatier
118 route de Narbonne
F 31062 Toulouse cedex
France

ABSTRACT. Three photochemical reactions have been studied in both anionic and cationic microemulsions.
- E $\rightleftarrows$ Z isomerization of O-methyl oxime ethers of aromatic ketones;
- photocycloaddition of α-cyclenones;
- photodimerization of cinnamic derivatives.
Our results showed that the magnitude of the microemulsion effect depends on :
- the substrate localization;
- the reaction mechanism and more specially the involvement of an excimer;
- the structure of the microemulsion.
We have also demonstrated the important part played by the interface (localization, effect of proximity of substrate molecules, high local concentrations, selective extraction and sensitivity to salts),leading to high yields, regio and stereoselectivities.
The photoreactivity of cinnamic derivatives in microemulsions shows that it is possible to obtain in these fluid media dimerization reactions similar to those observed in the solid state.

1. INTRODUCTION

The results presented in this talk are an overview of a study carried out in our laboratory on the effect of organized media on photoreactivity in organic chemistry. Interactions in organized media play an important part in biological processes taking place in microscopically heterogeneous environments.

Micellar solutions and microemulsions have been widely used as models for complex bioaggregates such as membranes and other cell organelles (1). However the solubilization ability and the stability are better for microemulsions than for micellar solutions. So, in most cases we used microemulsions.

Microemulsions (2-7) are thermodynamically stable and transparent media which are microscopically heterogeneous. They are generally made up of four components (water, oil, surfactant and cosurfactant), with

R. Setton (ed.), Chemical Reactions in Organic and Inorganic Constrained Systems, 263–282.

a structure depending on their relative proportions (fig.1).

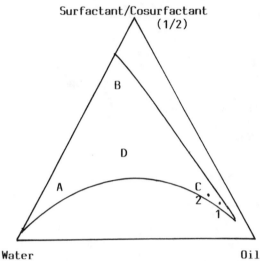

Fig.1: Phase diagram of microemulsions

The water-rich area (A) contains micelles of oil in water. These consist of droplets of oil surrounded by a mixed film of surfactant and cosurfactant. These micelles are noted S_1 according to Winsor's nomenclature (4)

The area (B) is rich in surfactant and cosurfactant, its structure is more complex, containing, for example, various lamellar organizations.

In the oil-rich area (C), micelles of water droplets (S_2) are dispersed in a mixture of oil, cosurfactant and sometimes a few molecules of water.

In the area (D), various structures have been proposed : liquid crystals, gels, lamellae, bicontinuous structures. According to De Gennes and Taupin (8), a bicontinuous structure can be described as a network of water channels in hydrocarbon, produced by the coalescence of S_1 and S_2 micelles.

The boundaries between these different areas are hard to define exactly, although they can be evaluated using physical (9) or theoretical methods (10,11).

Two photochemical reactions have been studied in anionic and cationic microemulsions:
- E $\rightleftarrows$ Z isomerization of O-methyl ethers of aromatic ketones;
- photodimerizations.
This choice has been induced by differences:
- in the solubility properties;
- in the possibility of aggregation in the ground state or in the excited state;

- in the nature of excited state.
Biological implications were also on our mind.

2. PHOTOISOMERIZATION REACTIONS

We observed in the literature (12-15) that different probes can give
conflicting results. This may be related to differences, not only in
solubilization sites but also in reactions mechanisms.
We report here a comparison of the behaviour, in anionic and cationic
microemulsions, of two isomerizable probes:
   - the 0-methyl ether of 2-acetonaphtone oxime 1;
   - the 0-methyl ether of acetone oxime 2.
   These two reactions have been studied in "classical" solvents by
Padwa (16-18).

In the case of the oxime ether 1, the data obtained by Padwa indicate that the final photostationnary state composition is dependent
on both the concentration and the solvent used (pentane or benzene).
We have completed this study and found that the concentration effects
were more marked in protic than in aprotic solvents (fig.2) and that
the photostationnary state composition was not wavelength dependent
(14-15).
   Padwa proposed the following mechanism:

monomer ──→ monomer* (^{1}S) ──→ predominantly 1Z

                              +monomer
monomer ──→ monomer* (^{1}S) ──────→ excimer ──→ predominantly 1E

The hypothesis of an excimer is supported by the fact that a new

component appears in the fluorescence of the isomer 1Z, around 500 nm
when its concentration in pentane increases from 0.1 to 1.2 M.

In the case of the oxime ether 2, Padwa found that the composition of the photostationnary state obtained by direct irradiation is
independent of the concentration and can be calculated by the following relation:

$$\frac{[2E]}{[2Z]} = \frac{\varepsilon_Z (\phi_{Z \to E})}{\varepsilon_E (\phi_{E \to Z})}$$

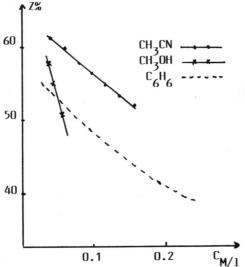

Fig.2: Isomerization of 1 in solvents

2.1. Photoisomerization of compound 1 in anionic microemulsions

The microemulsions were constituted of water-benzene-sodium dodecyl
sulfate/butanol (1/2).

We analysed the behaviour of compound 1 in the whole phase
diagram.

2.1.1. Microemulsions in the benzene-rich area (C). For this area, in
all the microemulsions, we observed a monotonic reduction of the quantity of 1Z at photostationnary equilibrium as concentration increased
(fig.3).

The slopes of the graphs of percentage of 1Z against the total
concentration were slightly less steep than for benzene.

2.1.2. <u>Microemulsions of the areas (A), (B) and (D)</u>. For all others
microemulsions, a more complex phenomenon was seen than for the microemulsions in the area (C) (fig.3).

The proportion of isomer $\underline{1}Z$ against the concentration shows a
discontinuity approximately represented by two lines intersecting at a
given concentration (noted C_s).

In all cases C_s is lower than the surfactant concentration.

Even if phenomena are quite similar for all these microemulsions,
the study of the saturation concentration C_s and of the slopes of the
curves before C_s enables three groups of microemulsions to be distinguished (fig.3).

For the microemulsions situated in the middle part of the diagram
(D), the slopes are quite high and the saturation concentration varies
linearly with the proportion of the surfactant (15). It is therefore
quite likely that, in this area, the medium has a similar structure.

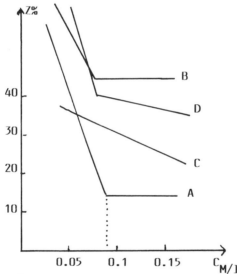

Fig.3: Isomerization of $\underline{1}$ in microémulsions (SDS)

For microemulsions of the area (B), the slopes are lower, and the
saturation concentrations C_s are not very great , relative to the high
proportion of surfactant, so the nature of the interface is then quite
different from the preceding case.

The behaviour of microemulsion in the area (A) is abnormal: it
contains a small amount of surfactant while the saturation concentration is very high; on the other hand, the proportion of the Z isomer
is very much lower than in the other microemulsions.

In this study, interfacial processes appears to play an important
role and there is a close relationship between the saturation concen-

tration and the nature of the interface. To provide further evidence
for this, we altered the nature of the interface by the addition of
salts (sodium chloride and potassium chloride).

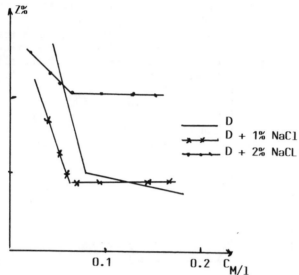

Fig.4: Salt effects in SDS microemulsions

    In all cases (fig.4), the addition of 1% of salt led to a reduc-
tion of the saturation concentration. When the salt concentration increased (to 2%), the saturation concentration did not change. So, it
is likely that the micellar area situated near the polar heads of the
surfactant can only accept a limited quantity of salt and the maximum
value is reached with a 1% salt concentration.
    In most cases, potassium chloride reduces C_s to a smaller extent
than sodium chloride. This is probably due to cationic effects at the
interface, Na^+ coming closer to the polar heads than K^+. These results suggest that the interface can only take up a limited amount of
sustrate which depends on the availability of a number of "sites",
this number increases with the proportion of SDS and is reduced by addition of salts. In all cases studied, after addition of salts, the
proportion of isomer Z does not change above the saturation
concentration. A sort of "buffer effect" on the reaction is observed.

2.2. Photoisomerization of compound 1 in cationic microemulsions

The system chosen was composed of : water, benzene, cetyltrimethyl
ammonium bromide/butanol (1/2).
    Fig.5 shows that the concentrations and salts effects are very
similar to those observed in anionic microemulsions.

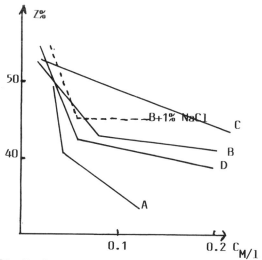

Fig.5: Isomerization of 1 in microemulsions (CTAB)

2.3. Photoisomerization of compound 2

We compared the photoisomerization of 2 in methanol, in benzene and in anionic and cationic microemulsions.

Our observations confirm the results of Padwa: the reaction is not concentration dependent. Moreover the effect of microemulsions and classical solvents are quite similar.

The different behaviour of compounds 1 and 2 in microemulsions may be due to formation of the excimer postulated by Padwa in the photoisomerization of compound 1.

On the basis of this hypothesis we decided to study the fluorescence of compound 1 in benzene, in methanol (since the interface region in microemulsions is protic) and in microemulsions.

2.4. Fluorescence of compound 1

This fluorescence study was carried out at various concentrations for 1Z and 1E.

It is important to note that, in methanol, concentrations of 1E above 0.1 M could not be studied, due to the low solubility of isomer 1E in protic solvents.

In methanol, at concentrations above 0.1 M a new component was observed (λ_{max} = 480 nm) in the fluorescence of 1Z. The emission intensity increased and became very marked for concentrations around 1 M (fig.6).

This component was not seen in benzene either for 1Z and 1E.

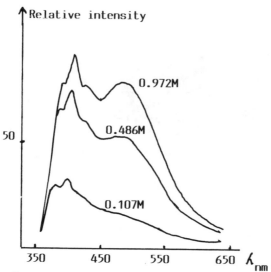

Fig.6: Fluorescence of 1Z in methanol

In microemulsions, the two isomers were soluble and the concentration effects on fluorescence could be studied for both isomers
(fig.7). We chose an anionic microemulsion in the area (D), which
shows the most favourable break in the concentration effect (fig.3).

At low concentration of 1Z (0.051 M), the normal fluorescence is
observed and a slight fluorescence band appears at longer wavelengths
(460-500 nm). This new emission became much more marked at concentrations from 0.051 to 0.512 M.

Excimer fluorescence can thus be observed in microemulsions, even
at low concentration. However the fluorescence intensity increases
slightly less rapidly in microemulsions, as concentration increases,
than in methanol.

These facts can be explained by the different localizations taken
up by the oxime ether 1 in the various regions of the organelles.

Three areas can be considered :
- the contact area between the polar heads of the surfactant and the
aqueous phase referred to as the interface;
- the area constituted by the hydrophobic tails of the surfactant, referred to as the interphase,
- the third area is formed by the benzene phase.

At low concentration, the substrate (not soluble in water) takes
up a position in the interface where excimer formation is facilitated.
When the concentration increases, a proportion of the substrate molecules is found in the interphase or in the benzene phase where excimer
formation is less favoured. This could explain why the proportion of
excimer increases less rapidly as concentration increases in a microemulsion medium than in methanol.

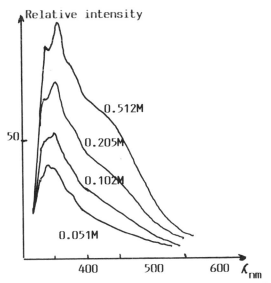

Fig.7: Fluorescence of 1Z in microemulsions

We repeated the experiments with isomer E and, as in methanol, no change of the fluorescence spectrum was seen, so isomer E probably does not form an excimer.

As well as verifying the formation of an excimer as postulated by Padwa (16-17) our results showed that only the excimer Z..Z is formed.

2.5. Discussion

We have observed that:
- in media in which a strong excimer fluorescence is observed, there is a marked concentration effect on the photoisomerization reaction (methanol, microemulsions);
- in media in which excimer fluorescence is not observed, a small concentration effect is seen. This would seem to indicate that, even if the excimer is formed, the quantity is too low to be detected.

These observations, joined to the fact that microemulsions have no effect on the isomerization of compound 2 in which an excimer is not involved, allow us to explain the differences in the behaviour of compound 1 in the different areas of the phase diagram.

For microemulsions in the benzene-rich area (C) the isomerization takes place essentially in the continuous benzene phase in which the probe is very soluble. The lower slope observed could be due to the fact that some oxime ether molecules are in the microemulsion interphase, lowering the probability of excimer formation and reducing the concentration effect.

The break in the slopes observed in the areas (A),(B) and (D) can be explained by the distribution of the substrate through the medium. Not being soluble in water and water/1-butanol mixtures, the oxime

ether can only reside in the interface, the interphase or in the benzenic region of the microemulsion. It is likely that there is a distribution of the oxime ether between these pseudo-phases with a preference
of one of the isomers for the interface region. The fluorescence results suggest a preferential localization of Z at the interface, which
is confirmed by the greater solubility of this isomer in microemulsions, as well as in protic media, in general.

- At low concentration, the substrate is preferentially located
at the interface, near the polar heads of the surfactant (SDS) : the
reactive group is situated in a protic and structured medium which favours excimer formation, leading to a marked concentration effect.

- The point of intersection of the lines (fig.3) corresponds to
the saturation of the interface by the oxime ether (C_s)

- Above this concentration, the proportion of oxime ether reacting in the other pseudo-phases increases. Several phenomena occur simultaneously :

(i)   isomerisation in the interface where excimer formation is
facilitated ;

(ii)   isomerisation in the benzenic phase;

(iii) selective extraction of the Z isomer by the interface in
contrast to the E isomer which is removed by the benzenic phase.

These phenomena suggest that the isomerisation process is sterically hindered by the presence of the surfactant hydrophobic tails for
both isomers, in the interphase region.

The relative rates of these different reactions leads to a fairly
stable proportion of isomer Z in the photostationnary state, for different macroscopic concentrations of the oxime ether above C_s.

The difference observed between the areas A, B and D can be related to the different shapes of the interfaces modifying the phenomena
described above. This is particularly true in the case of a microemulsion made up of spherical micelles in a continuous aqueous phase. The
spherical structure of the interface would favour the localization of
the oxime ether near the polar heads of the surfactant, facilitating
excimer formation.

The two salt effects observed :
       - reduced C_s
       - "buffer" effect

can also be explained in the same manner.

C_s are reduced because the additionnal amounts of cations at the
interface reduce the possibility of solubilization of 1Z at this interface and so reduce the possibility of excimer formation.

The "buffer" effect is probably observed because the salt alters
the rate of selective extraction of Z by the interface. So, a form of
equilibrium between the two sites of isomerization is obtained.

2.6. Conclusion

This study, in both anionic and cationic microemulsions, on the photoisomerization E$\rightleftharpoons$Z of the O-methyl oxime ether of 2-acetonaphtone and
acetophenone for which the mechanism in apolar solvents is known (16-
18) showed that the magnitude of the microemulsion effect depends on

- the substrate localization;
- the reaction mechanism, and more specifically the participation of an excimer;
- the structure of the microemulsion.

We have shown that the concentration dependent photoisomerization can be used as a "chemical probe" to descriminate between the different structures of the microemulsions in the monophasic area. The use of such a reaction represents a simple and quick technique which complements physical methods for the analysis of these media.

We have also demonstrated the important part played by the interface : localization of the molecules, high local concentrations, selective extraction and sensitivity to salts.

3. PHOTODIMERIZATION REACTIONS

Many reactions of photodimerization are dependent on the structure of the medium because they are influenced by molecular interactions in the ground state as well as in the excited state.

We chose to study the dimerization of three molecules which had been extensively studied in classical media :
- isophorone was chosen first as the simplest molecule to test the media;
- the coumarin moiety is included in furocoumarins (like psoralens) drugs used in photodermatology;
- the cinnamate derivatives are also used in dermatology as protection against U.V. solar radiations; they are known to dimerize mainly in the solid state.

For all these molecules the photocycloadditions showed a strong dependence on the reaction media which influence yields, rates, regioselectivities and stereoselectivities. They also present differences in solubilities and reaction mechanisms.
- Isophorone is soluble in cyclohexane but not in water. No molecular association has been established for this reaction in the ground state or in the excited state. The photodimerization involves triplet states of isophorone.
- Coumarin is neither soluble in water nor in cyclohexane. Two reactions mechanisms are possible leading to two different stereoisomers.
- Cinnamate derivatives are insoluble in water and slighly soluble in cyclohexane. They dimerize in the crystalline state under topochemical control.

3.1 Isophorone dimerization

In solvents, the irradiation of isophorone leads to the formation of three photodimers : HT, head to tail (anti>>syn); HH, head to head. Previous studies (19,20) have shown that the dimerization rates are higher in polar and protic media than in apolar solvents. This effect is important for the HH isomer, the formation of which is clearly favoured by the solvent polarity (methanol, 80% HH ; cyclohexane, 10% HH).

HT syn   HT anti   HH anti

According to Chapman et al. (19), the photodimerization involves triplet states of isophorone. These researchers consider the possibility that two triplet species lead to the two observed dimers. No molecular association has been established for this reaction, in the ground state or in the excited state.

We studied this reaction essentially in anionic micellar media and microemulsions with sodium dodecyl sulfate (SDS) as surfactant (21,22).

3.1.1. : <u>Dimerization in micellar media and in microemulsions constituted with water, SDS/butanol(1/2), cyclohexane.</u> The results summarized in table I show that isophorone dimerization is carried out more rapidly and with higher yields in micellar media and in microemulsions than in solvents. Furthermore the reaction presents a high regioselectivity.

In microemulsions, the quantity of HH isomer obtained is always substantial (50%-90%). It should be noted that even when the continuous phase is cyclohexanic, a minimum of 50% of HH regioisomer is obtained, although isophorone is soluble in cyclohexane but not in water.

In order to explain this phenomenon, we have put forward the assumption of a preferential reactivity of isophorone at the interface (21). According to this interpretation, a variation in the ratio of HH and HT isomers could be expected for such a medium if the isophorone concentration was modified.

As indicated in Table I, an increase in the HH isomer yield is observed when the isophorone concentration increases from 0.025 M to 0.2 M ; for higher concentrations the quantity of HH isomer remains constant.

These observations can be explained by the following facts.
(i) At the interface, the HH dimerization rate is higher than the HT dimerization rate :

$$(v_{HH})_i \ggg (v'_{HT})_i$$

(ii) In cyclohexane the opposite phenomenon occurs :

$$(v'_{HH})_{cyclohexane} \lll (v'_{HT})_i$$

(iii) At the interface, the rate of HH isomer formation is much higher than the rate of HT formation in cyclohexane, probably because the rate constants are different but also because local concentration effects are important in the micelles :

$$(v_{HH})_i \ggg (v'_{HT})_{cyclohexane}$$

(iv) The photodimer produced at the interface does not remain in the micelle; it is then replaced by new isophorone molecules. Isophorone is consequently extracted continuously from the cyclohexanic phase.

Table I : Dimerization of isophorone in various media

media	concentration	% isophorone conversion[a]	% dimers isoph. transformed	yield of HT(%)[b]	yield of HH(%)[c]
cyclohexane	0.1M	40	25	90	10
methanol	0.1M	30	70	20	80
micellar	0.02M	100	95	5	95
A[d]	0.1M	100	95	5	95
B[d]	0.1M	100	80	15	85
D[d]	0.1M	100	90	15	85
C$_1$[d]	0.1M	100	70	50	50
C$_2$[d]	0.025M	100	80	40	60
	0.1M	100	75	30	70
	0.2M	100	75	25	75
	0.4M	100	75	25	75
	0.6M	100	75	25	75

a : After 24h irradiation at 254 nm
b : HT anti + HT syn
c : HH anti
d : microemulsions of the different areas of the phase diagram (fig.1).

As long as the interface is not saturated, the concentration effect on the HH dimerization rate inside the micelle is predominant and the yield of HH isomer increases with the isophorone concentration.

When the interface is saturated, the local isophorone concentration and consequently the HH dimerization rate are constant in the micelle. The rate of HT isomer formation increases then in cyclohexanic phase, but this effect is balanced by the continuous extraction of isophorone towards the dispersed phase. Thus the ratio of HH and HT isomers remains almost constant.

3.1.2 <u>Dimerization in microemulsions in which isophorone replaces the non-polar oil.</u> During our study of the concentration effects, isophorone was found to be easily soluble in the microemulsions under investigations.

In microemulsion C_2 (fig. 1) it was possible to add up to 50 vol.% of isophorone and still retain a transparent stable monophasic system. This led us to consider that isophorone could replace the hydrocarbon

M. RIVIERE

in microemulsions.

We therefore studied systems containing four components : water, SDS/butanol (1/2), isophorone. Several mixtures of these four components led to stable and transparent media, apparently monophasic. The phase diagram shown in fig.8 is quite similar to that for the cyclohexanic microemulsions (fig.1).

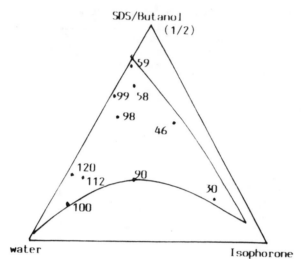

Fig.8: microemulsions with isophorone as "oil"

As this was the first time that a polar substance had been used as "oil", it was difficult to decide wether these systems were true microemulsions or cosolubilized systems.

However, for system 90 (fig.8), containing significant quantities of two non miscible liquids (water: 43%; isophorone: 29.9%) it is unlikely that cosolubilization was attained just by adding small quantities of SDS and butanol. Furthermore, we checked, for this system, that the mixture (isophorone-water-butanol) was heterogeneous and that the addition of SDS led immediately to a homogeneous medium. Mixture 30 was tested in the same way. It is difficult to explain these facts except by an organization of the medium due to the surfactant.

The results obtained by irradiation of the systems containing isophorone as "oil" in comparison with those previously obtained for cyclohexanic microemulsions could give information on their structure. Systems 30 to 120 (fig.8) in which isophorone is one of the constituents,were irradiated (254 nm) for 24h. The results are listed in Table II.

For all the systems studied, the yield of HH is much higher than that from pure isophorone or from isophorone-butanol mixtures. This shows that a significant quantity of isophorone reacts in a polar medium and probably also in a structured medium. In most cases, the

results are in fact similar to those obtained in oil-water cyclohexanic
microemulsions or in micellar media.

Table II : dimerization in microemulsions in which isophorone
           replaces the non polar oil

Solvent	Yield of HT isomer (%)	Yield of HH isomer (%)
Isophorone:butanol (90:10)	40	60
Pure isophorone	60	40
Microemulsion 30	25	75
Microemulsion 46	15	85
Microemulsion 58	10	90
Microemulsion 59	10	90
Microemulsion 90	20	80
Microemulsion 98	10	90
Microemulsion 99	10	90
Microemulsion 100	25	75
Microemulsion 112	10	90
Microemulsion 120	10	90

    For system 30 (fig. 8), in which the quantity of isophorone is
substantial, we obtained the same result as in oil-water cyclohexanic
microemulsions with an isophorone concentration superior to 0.2 M.
    Systems 90 and 100 (fig. 8), situated on the borderline, have a
particular behaviour. They are homogeneous before irradiation and become biphasic after isophorone photolysis, i.e. after the isophorone has
partly disappeared, giving rise to the dimer. Water-SDS/butanol-cyclo-
hexane microemulsions situated in the same area of the phase diagram
behave in the same way when cyclohexane is removed by evaporation.
    All these results are consistent with the existence of a micro-
structure for the water -(SDS/butanol)-isophorone medium.
    Since this work we have found that it was also possible to prepa-
re microemulsions with furan as "oil" and the structure of these media
has been confirmed by physical means.

3.1.3. Conclusion. Three results must be singled out from this study.
(i) The localization of the substrate at the interface combines the
polarity and condensation effects, leading to particularly high regioselectivities.
(ii) In microemulsions, a phenomenon of continuous extraction of reactants and products occurs during photochemical processes. Because of
this, microemulsions are found to be interesting systems which allow
high conversion rates and yields.
(iii) The formation of a microemulsion is a general phenomenon which is
not limited to surfactant - cosurfactant - water - hydrocarbon systems.
Our work has shown that a polar component, isophorone, can replace the
hydrocarbon.

3.2. - Coumarin dimerization

In the last example, isophorone was dissolved by hydrocarbons and the observed results were consistent with a competition between two reactions :
  - reaction at the interface leading to the HH (head to head) dimer;
  - reaction in the hydrocarbon phase leading to the HT (head to tail) dimer.
In order to circumvent this competitive reaction and observe a reaction occuring only at the interface, we decided to study the photo-dimerization of coumarin in organized media. This compound which is of biological interest, is neither soluble in water nor in cyclohexane, but is soluble in micellar media (solutions of sodium dodecyl sulfate : SDS) and in microemulsions (water-SDS/butanol: 2/1-cyclohexane). Coumarin is thereby localized at the interface.
The photodimerization of this compound presents another inte-resting feature. Depending on the nature of the solvent, two reaction mechanisms are possible, leading to two different stereoisomers (23-26).

The anti HH dimer (II) is the only dimer formed in non polar solvents while both the syn (I) and the anti (II) dimers are obtained in polar solvents.
However, even in polar solvents, high concentrations of coumarin are required to obtain a high syn/anti ratio eg. I/II = 13 in a 0.3 M solution of coumarin in methanol (25). From a study of the reaction

mechanism, it was deduced that the anti isomer resulted from the triplet state and the syn isomer is generated from a singlet excimer (24).

It was thought that the microstructure of the micellar medium or the microemulsions would have an influence on the orientation of coumarin molecules in this photochemical reaction.In a recent paper, K. Muthuramu et al. (27) obtained satisfactory results using a micellar reaction medium: 21% of syn HH dimer after 22h irradiation in 0.02 M aqueous solution of SDS.

We have investigated such reactions in microemulsions, from both a mechanistic and a preparative standpoint.

Coumarin was photolyzed in ethanol (concentration 0.3M), in an aqueous solution of SDS (5% by weight) and in several microemulsions (water-SDS/butanol-cyclohexane) (28).In all these organized media the concentration of coumarin was 0.034 M. The microemulsions were chosen from all the different areas of the phase diagram (fig.1).

In methanol, we obtained predominantly the syn HH dimer but the yield was low.

In organized media, irradiation led to the syn HH dimer only, [MP= 276°C, in agreement with the results of Morrison et al. (25)].

The yields of dimer I, calculated from the coumarin transformed, were over 90% in all cases. Analysis by ^{1}H NMR spectroscopy of the crude irradiated mixture indicated only small quantities of the anti HH dimer (II) and/or its decomposition products (< 5%). Even with low concentrations (1/10 compared with ethanol) the yield of dimer I was much higher in organized media.

It appears therefore, that in micellar media and in microemulsions, only one photoreaction takes place, namely, the formation of the HH dimer I through the excited singlet state of coumarin.

We conclude that the triplet reaction leading to dimer II does not take place in organized media. This probably results from the orientation of the coumarin molecules at the interface, which only allows the formation of the syn dimer. Results on bis-furocoumarin reactions (29) lend support to this explanation. The geometry of these molecules forces the syn cycloaddition.

These results show that the reaction in a microemulsion inhibits the triplet state reaction in coumarin photodimerization, and confirm the role played by interfacial organization in the reaction mechanism. Effects of the orientation and high local concentration of the substrate were thought to explain the reaction observed.

However K. Muthuramu et al.(29), using 7-alkoxy coumarins, only observed the formation of syn HT dimers in organic solvents and in micellar media, with no orientation reversal. We suggest that the critical factor in these dimerizations is the enhancement of local concentration of substrate in both micellar and microemulsion systems.

3.3. Dimerization of cinnamate derivatives

In the case of cinnamate derivatives, two photoreactions can occur: E ⇄ Z isomerization and dimerization.
The literature (31-35) shows the following results.

- With cinnamic acid there is no dimerization in solution. This dimerization occurs for some crystalline forms under topochemical control: the α crystals lead to α truxillic acid, the β crystals to β truxinic acid and the γ crystals are stable to light. In the crystalline forms, the distance between the two double bonds have to be less than 3.7 Å to allow dimerization.

- With ethyl cinnamate, dimerizations have been observed in liquid films or in glasses at low temperatures, the regioselectivities were low (36).

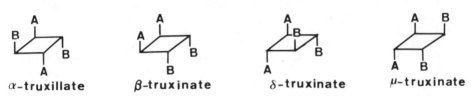

α-truxillate β-truxinate δ-truxinate μ-truxinate

We studied the photoreactivity of cinnamic acid and methyl cinnamate in cyclohexane, methanol and in microemulsions of the different areas of the phase diagram (water,SDS/butanol (1/2), cyclohexane).

In both cases we observed only isomerization in cyclohexane and in methanol.

In microemulsions, we observed not only isomerization but also dimerization; this reaction was quantitative after 6 days of irradiation (λ = 300 nm).

With cinnamic acid, we have identified the δ dimer, but the other products of reaction (one or two) have not yet been separated.

With methyl cinnamate, we obtained only two dimers: δ and μ (80/20) in all the microemulsions studied.

It therefore appears that the reaction occurs at the interface, leading only to HH dimers.

4. CONCLUSION

We can conclude from this work that in micellar media and in microemulsions the effects depend on: the localization of the probe, the reaction mechanism and the structure of the media.

The localization of molecules at the interface allows a combination of concentration, organization and proximity of substrate molecules leading to high yields, regio and stereoselectivities.

The photoreactivity of cinnamic derivatives in microemulsions shows that it is possible to obtain in these fluid media dimerization reactions similar to those observed in the solid state.

REFERENCES

(1) - N.J. Turro, M. Grätzel and A.M. Braun, Angew. Chem. Intern. Ed., **19**, 675 (1980).

(2) - J.P. Hoar and J.H. Schulman, Nature (London), **152**, 102 (1943).
(3) - T. Flaim and S.E. Friberg in "Advances in liquid crystals", Academic press, New york, pp. 137-155 (1982).
(4) - P.A. Winsor in "Solvent properties of amphiphilic compounds", Butterworth ed.(London), (1954).
(5) - J.H. Fendler, J. Phys. Chem., **83**, 1485 (1980).
(6) - C.A. Jones, L.E. Weaner and R. Mackay, J. Phys. Chem., **83**), (1980).
(7) - S.J. Gregoritch and J.K. Thomas, J. Phys. Chem., **83**, 1491, (1980).
(8) - P.G. de Gennes and C. Taupin, J. Phys. Chem., **86**, 2294 (1982)
(9) - S.L. Holt, J Dispersion Science and Technology, **1** (4), 423, (1980).
(10) - P. Bothorel, J. Biais, B. Clin P. Lalanne and P. Maelstaf, C.R. Acad. Sci. Paris, **289** (16), serie C, 409 (1979)
(11) - J. Biais, P; Bothorel, B. Clin and P. Lalanne, J. Dispersion Science and Technology, **2** (1), 67 (1981).
(12) - K.S. Schanze, and D.G. Whitten, J. Amer. Chem. Soc., **105**, 6734 (1982).
(13) - C. de Bourayne, M.T. Maurette, E. Oliveros, M. Rivière and A. Lattes, J. de chimie physique, 79 (2), 140 (1982).
(14) - I. Rico, M.T. Maurette, E. Oliveros, M. Rivière and A. Lattes, Tetrahedron Letters, 4795 (1978).
(15) - I. Rico, M.T. Maurette, E. Oliveros, M. Rivière and A. Lattes, Tetrahedron, **36**, 1779 (1980).
(16) - A. Padwa and F. Albrecht, Tetrahedron Letters, 1083 (1974).
(17) - A. Padwa and F. Albrecht, J. Org. Chem., 39(16), 2361 (1974).
(18) - A. Padwa and F. Albrecht, J. Amer. Chem. Soc., **94** (2), 1000 (1972).
(19) - O.L. Chapman, P.J. Nelson, R.W. King, D.J. Trecker and A.A. Griswold, Rec. Chem. Prog., **28**, 167 (1967).
(20) - R.E. Koning, G.J. Visser and A. Vos, Rec. Trav. Chim. Pays-Bas, **89**, 920 (1970).
(21) - R. Fargues, M.T. Maurette, E. Oliveros, M. Rivière and A. Lattes Nouveau Journal de Chimie, **3**, 487 (1979).
(22) - R. Sakellariou-Fargues, M.T. Maurette, E. Oliveros, M. Rivière and A. Lattes, J. of Photochemistry, **18**, 101 (1982).
(23) - G.S. Hammond, C.A. Stout and A.A. Lamola, J. Amer. Chem. Soc., **86**, 3103 (1964).
(24) - F. Hoffman, P. Wells and H. Morrison, J. Org. Chem., **36** (1971).
(25) - H. Morrison, H. Curtis and T. Mac Dowell, J. Amer. Chem. Soc., **88**, 5415 (1966).
(26) - G.O. Schenck, I. Von Wilucki and C.H. Krauch, Chem. Ber., **95**, 1409 (1962).
(27) - K. Muthuramu and V. Ramamurthy, J. Org. Chem., **47**, 3976 (1982).
(28) - R. Sakallariou-Fargues, M.T. Maurette, E. Oliveros, M. Rivière and A. Lattes, Tetrahedron, **40** (12), 2381 (1984).
(29) - J. Gervais and F.C. de Schryver, Photochem. and Photobiol., **21**, 71 (1975).
(30) - K. Muthuramu, N. Ramnath and V. Ramamurthy, J. Org. Chem., **48**, 1872 (1983).

(31) - H. Stobbe, Chem. Ber., **52** B, 666 (1919).
(32) - H. Stobbe and K. Bremer, J. Prakt. Chem., **123**, 1 (1929).
(33) - Solid State Photochemistry, D. Ginsburg Ed., Verlag Chemie, New York (1976) and references therein.
(34) - D.A. Ben Efraïm and B.S. Green, Tetrahedron, **30**, 2357 (1974).
(35) - M. Bolte, C. Lorain and J. Lemaire, C.R. Acad. Sci. Paris, **293** serie II, 817 (1981).
(36) - P.L. Egerton, E.M. Hyde, J. Trigg, A. Payne, P. Beynon, M.V. Mijovic and A. Reiser, J. Amer. Chem. Soc., **103**, 3859 (1981).
(37) - H. Amarouche, C. de Bourayne, M. Rivière and A. Lattes, C.R. Acad. Sci. Paris, **298** série II, 121 (1984).

ENERGETICAL AND GEOMETRICAL CONSTRAINTS ON ADSORPTION AND REACTION
KINETICS ON CLAY SURFACES.

H. VAN DAMME, P. LEVITZ and L. GATINEAU
C.N.R.S. - C.R.S.O.C.I.
45071 - ORLEANS CEDEX 2, France

ABSTRACT. The Euclidean dimension of an ideal solid surface is 2. We
show that the effective dimensionality experienced by reactant molecules
on real surfaces may be quite different. This happens, for instance, on
energetically heterogeneous surfaces where the reactant molecules spend
most of their time either in mobility *valleys* or in isolated mobility
clusters, with an effective dimensionality, at short times, close to
1.33. This considerably modifies the reaction kinetics. A cluster-
limited electron transfer reaction on clay surfaces is discussed.
Microporous surfaces may also have an effective dimensionality different from 2. In pillared clays, due to interpillar separation distances
of the order of the molecular diameter, it is close to 1.9.

1. INTRODUCTION

It is customary to assume that, with respect to homogeneous reaction
conditions, surface reactions involve the loss of at least one degree
of freedom. The loss of the first degree of freedom stems from the
decrease in dimensionality, on going from a bulk phase to an interfacial
region, whereas further losses are eventually associated with localization of the adsorbed molecules on adsorption sites, within the
adsorbed layer. This classical approach has been developed in a general
statistical thermodynamical framework, more than twenty years ago [1,2].
The aim of the present contribution is not to add more details to those
models. Our purpose is to analyze, using recent theoretical concepts,
the experimental behavior which was observed upon adsorbing and reacting
molecules in a particular type of constrained system: clay- and pillared
clay surface. However, the approach seems to be general enough to allow
for the extension of the analysis to other systems.
    Generally speaking, two types of constraints can be considered:
energetical constraints on the one hand, and *geometrical constraints* on
the other hand. The former involve all the physical and chemical interactions between the adsorbate or the reactants and the solid surface,
whereas the latter can be thought of in terms of surface roughness and
(or) pore structure. We will show that both types of constraints can lead

283

R. Setton (ed.), Chemical Reactions in Organic and Inorganic Constrained Systems, 283–304.

to a modification of the effective dimensionality of space. In the first part of the paper (section 2), we will analyze the consequences of the energetic constraints on the surface-reaction rates. The reactions which will be considered are simple light-induced electron and energy transfer reactions, which can be monitored by luminescence quenching measurements. The second part (section 3) will be devoted to an analysis of adsorption in pillared clays, and will stress the relationship between microporosity, fractal dimension and molecular sieving properties.

2. INFLUENCE OF ENERGETICAL HETEROGENEITY ON REACTION KINETICS ON CLAY SURFACES.

2.1. Reaction Rates in Homogeneous and Constrained (Inhomogeneous) Media.

Let us consider a simple reaction between A and B molecules in an homogeneous medium:

$$A + B \longrightarrow \text{Products} \tag{1}$$

The time evolution of the concentrations is described by the linear differential equation:

$$d\,[A(t)]/dt = -\,k\,[A(t)]\,[B(t)] \tag{2}$$

where k is the rate constant. According to (1), $d\,[A(t)]/dt = d\,[B(t)]/dt$. If $[A(t)] = [B(t)]$, the reaction is strictly bimolecular. If one of the reactants (B, for instance) is present in much larger concentration, the reaction becomes pseudo-unimolecular and solving eq.(2) leads to:

$$[A(t)] = [A(o)]\ \ \exp\left\{-\,k\,[B(o)]\,t\right\} \tag{3}$$

In fact, this classical way of treating bimolecular and pseudo-unimolecular (trapping) reactions implies severe assumptions about the reaction medium. In particular, the use of a time-independent rate constant, k, implies that the reaction medium is fluid and remains homogeneous, even at microscopic level, during the whole course of the reaction. It turns out that these conditions are closely approached in well-stirred solutions, but in many instances drastically different reaction conditions are found. Typical examples are enzyme pockets and artificial molecular cavities, intracrystalline spaces (zeolites, clays, ...), micellar systems, polymers or even simple solid surfaces. In all these examples, which are, in fact, constrained media, one generally expects some "anomalous" effects with respect to homogeneous kinetics, which manifest themselves by a rate "constant" k which is no longer time-independent.

It has been shown by DE GENNES [3], in a work on diffusion-controlled processes in dense polymer systems, that the differences between reaction kinetics in solution-like and constrained media can be put on a very general basis by considering the *exploration behavior* of the reactants. Two types of behavior may be distinguished: *noncompact exploration* on the one hand, and *compact exploration* on the other hand. Noncompact exploration is an exploration mode where the reactants have only a small probability to pass twice on the same site. As time goes on, the probability for a molecule to come back to a spot which was already explored decreases steadily, and finally vanishes for $t \to \infty$. Noncompact exploration also implies that the exploration clouds of A and B can penetrate each other without the reaction necessarily taking place (even though A and B can react as soon as they "meet" each other). Conversely, in compact exploration, the probability to come back to any site tends to 1 as $t \to \infty$. The reactants pass many times over the same spots, and the reaction between A and B takes place very soon after overlapping of their exploration volumes.

Mathematically, the exploration behavior is controlled by the time evolution of the rms displacement, $x(t)$, of the reactants, or, more exactly, by the time evolution of their *exploration volume*, $\Omega(t) \simeq x^d(t)$, when reaction occurs in a d-dimensional space [3]. If $\Omega(t)/t$ is an increasing function of time, exploration is noncompact. This is, for instance, the case with diffusion-controlled reactions in a three-dimensional solution: $x(t) \sim t^{1/2}$; d = 3, hence, $\Omega(t) \sim t^{3/2}$, and $\Omega(t)/t$ increases with time. In such an exploration mode, there are many unexplored "holes" left within $\Omega(t)$. Conversely, if $\Omega(t)/t$ is a decreasing function of time, exploration is compact. The whole space volume within $\Omega(t)$ is visited during time t. This would be the case, for instance, with a diffusion-controlled reaction occuring within a quasi-one dimensional pore with a diameter roughly equal to the molecular size: $x(t) \sim t^{1/2}$; d = 1; hence, $\Omega(t) \sim t^{1/2}$ and $\Omega(t)/t$ decreases with time. Overlapping of the exploration volumes of A and B *ipso facto* implies the encounter of A and B (this is of course quite obvious for a strictly one-dimensional system, but it is also true for any d < 2).

Many constrained media can be described in terms of a lattice model. The "lattice" which has to be considered is not necessarily a crystal lattice. It can be any ensemble of potential wells, cages, binding sites, ... considered as lattice "sites". The only condition is that the reaction can only occur on the lattice sites. In a lattice model, the conditions for noncompact or compact exploration can be put in a slightly different form, by referring to the number of "jumps" between lattice sites, J(t), performed during a time t. In noncompact exploration, $x^d(t) > a^d J(t)$, where a is the average distance between nearest neighbor sites (i.e. the lattice parameter), whereas, in compact exploration, $x^d(t)$ is smaller than $a^d J(t)$.

As anticipated, the fundamental difference between noncompact and compact exploration shows up very clearly in the behavior of the rate "constant", k. Only in noncompact exploration is k a true constant, i.e. a time-independent parameter. In compact exploration, it becomes a time-dependent quantity, which can be expressed as:

$$k(t) \sim x^d(t)/t \qquad \text{or} \qquad k(t) \sim S(t) \sim S(t)/J(t) \qquad (4)$$

in a continuous or a lattice model, respectively [3,4]. S(t) is the
number of *distinct* sites visited during a time t. In order to account
for the reaction rate in a given constrained system, the problem is
then to know the time dependence of x(t) or S(t). In what follows, we
will concentrate on S(t), because the lattice model is probably more
appropriate to describe motion in a constrained medium with adsorption
or binding sites accessible to the reactants, such as a heterogeneous
surface.

       In fact, it turns out that the problem of the time–dependence of
S(t) is intimaly related to the knowledge of the *effective* dimension
of the reaction medium, i.e. of the lattice of accessible sites. In
most cases, this effective dimension will *not* be the dimension of the
embedding Euclidean space, d (i.e. d = 3 for a reaction occuring in a
"volume", or d = 2 for a reaction occuring on a surface). It will be
lower than d, because the network of all the possible jumps in the
system is usually far from being space (3d or 2d)–filling. Due to the
fact that some jumps between closely located sites are impossible or
highly unlikely (because of walls, energy barriers, pore necks, ...)
the network of allowed jumps has a reduced dimensionality. Several
authors [5,6,7] have shown that the time–dependence of S(t) (and other
dynamical processes) is governed by the so-called *spectral dimension*
of the system, d_s, which characterizes the *connectedness* of the system.
A simple power-law dependence is predicted:

$$S(t) \sim t^{d_s/2} \sim J^{d_s/2}(t) \tag{5}$$

The spectral dimension is intimately related [4,5] to another characteristic dimension of the system: its *fractal dimension*, d_f, which
characterizes the distribution of sites. However, as far as motion in
the medium is concerned, it is easily understandable that the connectedness of the sites is more important than their distribution [7].

       The fractal dimension, d_f, of the lattice is defined from the
space volume dependence of the number of sites. If for a lattice
embedded in a d-dimensional space, the number of sites, $N(\ell)$, included
in a space volume of length ℓ is scaling as:

$$N(\ell) \sim \ell^{d_f}, \tag{6}$$

for regular lattices, $d_f = d$, but in general, d_f need not to be an
integer. Lattices in which the number of sites scales according to
equation (6), but with a non–integer exponent, are fractal lattices.
Fractals are dilationaly self–similar structures [8]: the general
morphology remains the same upon changing the observation scale.

       An example ot fractal lattice is shown in Figure 1A. It is the
"infinite" cluster in a two-dimensional percolating system. Consider
a regular lattice of sites which can be either occupied, with a
probability p, or empty, with a probability 1-p. When p is very low,
only isolated occupied sites exist. As p increases, an increasing
number of occupied sites form clusters of finite size. As p increases

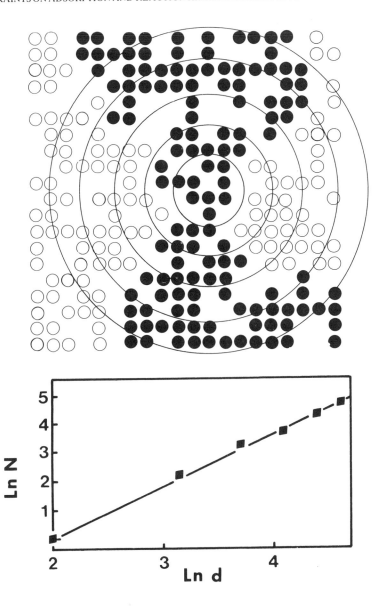

Figure 1. (A): Infinite cluster (●) in a two-dimensional site percolation system over a 20 x 20 square lattice; (B): Determination of the fractal dimension of the infinite cluster from the radius dependence of the number of sites.

further, fusion of clusters occurs and finally, for some probability
$p = p_c$ called the percolation threshold, an infinite cluster, crossing
the whole lattice on which it is mapped, is formed. For a two-dimensional
square lattice, $p_c = 0.50$ [9]. It has been shown that the infinite
clusters in percolation problems are fractals [9]. For the example
shown in Figure 1, $d_f = 1.89$. The power law dependence of $N(\ell)$ in this
system clearly shows up (Fig. 1). Remarkably, the spectral dimension
of an infinite percolation cluster at threshold is independent of the
dimensionality of the Euclidean space on which it is mapped: d_s is close
to 4/3, for all d [5]. In other words, if one would measure the number
of distinct sites, S(t), visited by a random walker on an infinite
percolation cluster at threshold, in any dimension, one would find
$S(t) \sim t^{2/3}$. This has been verified by computer simulation [10].

Returning now to the reaction rate between A and B molecules that
we considered at the beginning of this section, we can obtain the
differential rate equation for reaction in a compact medium by combining
equations 2, 4 and 5:

$$d\,[A(t)]/dt = - k_o\; t^{d_s/2-1}\; [A(t)]\; [B(t)] \tag{7}$$

In pseudo-unimolecular conditions ([A] $\ll$ [B]), the solution of this
equation is [4, 11]:

$$[A(t)] = [A(o)] \quad \exp\left\{-(2k_o/d_s)\; [B(o)]\quad t^{\;d_s/2}\right\} \tag{8}$$

For a reaction occurring on a percolation cluster ($d_s = 4/3$), one
predicts:

$$Ln\left\{[A(t)]/\; [A(o)]\right\} \sim - t^{\;2/3} \tag{9}$$

This is not only a simple example. As will be shown in the next sections,
it is closely related to reaction rates on heterogeneous solid surfaces.

2.2 Diffusion and Reaction on Energetically Heterogeneous Surfaces

Let us consider first an (almost) ideal surface on which all the
potential wells corresponding to the adsorption sites would have the
same depth, with respect to the gas or solution phase, and on which the
energy barriers between sites would be small (Langmuir model).
Diffusion on such a surface is very much like a random walk on a two-
dimensional regular lattice, at least at very low coverage, when
collisions between adsorbed molecules are unlikely. Figure 2 illustrates
the type of exploration which can be performed by a walker on a square
lattice, in one thousand steps. As expected for a random process, the
details of the walk are subject to some fluctuations, which are important for short walks. For instance, the first walk which is shown in
Figure 2 is denser than the others. The number of distinct sites, S,
visited after 1000 jumps is between 295 and 375, for the walks which
are shown. As discussed in section 2.1, the most important feature for
reaction kinetics is the S(t) vs J(t) behavior. On the average, for
two-dimensional regular lattices, $S \sim J$, (and t), neglecting logarithmic

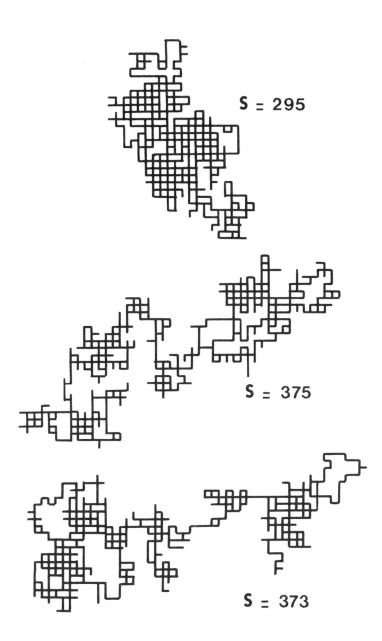

Figure 2. Three examples of a random walk performed over a two dimensional square lattice, without barriers. J = 1000 is the number of jumps, which is equivalent to "time" t = 1000. S is the number of distinct sites visited.

terms. In other words, and in first approximation a random walk on a
two-dimensional lattice is a limiting case of noncompact exploration, at
least after the very first jumps of the walker around its starting point,
and the rate "constant" $k \sim S(t)/t$ is in this case a true time-independent parameter. For instance, for the reaction of walkers with randomly
distributed traps, the survival probability of the walkers would be
given by equation (3), which is equivalent to equation (8) with
$d_s = d_f = d = 2$.

Most real surfaces differ seriously from the previous model in the
sense that they are energetically heterogeneous. The heat of adsorption
i.e. the depth of the potential wells with respect to the gas phase or
the solution phase, is not constant and there is a whole distribution
of potential energy barrier heights for diffusion. A first approach to
this complex situation is to consider only one type of disorder –over
the energy barriers for instance– and to perform random walks on such
a disordered lattice. It should be pointed out that considering only
the disorder in the distribution of barrier heights between sites leads
to a symmetrical behavior for the jump probabilities between adjacent
sites i and $j = W_{ij} = W_{ji}$. The existence of a second disorder, in the
potential wells depth, would lead to $W_{ij} \neq W_{ji}$, in general. Figure 3
shows a random walk over a square lattice with five different jump
probabilities (0.02; 0.05; 0.15; 0.33; 1.0) corresponding to five
different barrier heights distributed randomly over the bonds connecting
adjacent sites. One thousand jump trials were performed, but only 324
led to effective jumps: $J = 324$, for $t = 1000$. This is a first, trivial
consequence of the presence of barriers; the average jump probability
is (of course) lower than in the absence of barriers. A second, less
trivial effect shows up when one compares the number of distinct sites
visited, S, with respect to J. S is only 54 for $J = 324$. The S/J ratio
is 0.17, as compared to $S/J \simeq 0.35$ in the homogeneous situation
(Figure 2) (It should be noticed that S/J is not affected whether there
are no barriers, or barriers all identical to each other. The result of
introducing homogeneous barriers is merely to lengthen the time scale:
$J(t) < t$). In other words, the barrier *disorder* reduces the effectiveness of the walk: the walker passes many times over sites which have
already been visited: exploration is now compact. The S(t) vs J(t)
relationship is no longer linear.

The physical reason for this important modification shows clearly
in Figure 3, where the barrier map has been added to the walk
pattern. One can see that, most of the time, the walker just followed
the valleys along which motion is easy. Very few high barriers have
been overpassed. This suggests that diffusion on a heterogeneous surface
could be studied with an even simpler model, in which one would only
consider a random distribution of two types of barriers: infinite
barriers ($W_{ij} = W_{ji} = 0$) on the one hand, and zero barriers
($W_{ij} = W_{ji} = 1$) on the other hand. At this point, we are back to the
percolation problem discussed in the previous section, but the percolating lattice is no longer a lattice of sites, but a lattice of bonds
(the relationship between the two lattices is however quite obvious).
Depending on the relative concentrations of infinite and zero (or
finite) barriers, four different, well characterized situations can be

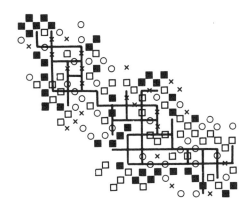

Figure 3. Random walk performed over a two dimensional square lattice with barriers between neighbouring sites. Five different barrier heights were randomly distributed over the bonds between sites. They correspond to jump probabilities of 1 ( ); 0.33 (×); 0.15 (○); 0.05 (□); and 0.02 (■). The upper part ot the Figure shows only the walk, drawn at the same scale as Figure 2. The lower part is an enlarged view of the walk, with the barrier map.

experienced by the walkers: (i) at a concentration of zero barriers
$p = p_c = 0.59$, an important fraction of the population of walkers
(adsorbed molecules), i.e. $\sim$ 50%, will be on an infinite percolation
cluster *at criticality*. For those walkers, one expects the "super-
universal" behavior $S(t) \sim t^{2/3}$; (ii) At higher concentrations of
finite barriers, $p > p_c$, an increasingly large population of walkers
will be on the growing infinite cluster. However, the infinite cluster
above criticality, is no longer a fractal in the whole dimension range.
It is still fractal (dilationaly self-similar) over short lengths
(below the correlation length), but becomes Euclidean (translationaly
self-similar) over larger distances [9]. Hence, for walkers on such
a structure, one expects the $S(t) \sim t^{2/3}$ at short time, and a crossover to an Euclidean two-dimensional noncompact regime $(S(t) \sim t)$ at
longer times. As p approaches 1, diffusion increasingly resembles
diffusion on a regular lattice, and the crossover happens at increasingly short times. This has indeed been observed by computer simulation [10]; (iii) At concentrations of finite barriers somewhat below
p_c, all the walkers will be in finite clusters, i.e. in valleys of
connected sites which no longer span the whole lattice. At long
time, the walkers will experience the finite character of their
exploration space: $S(t)$ will saturate. However, at short times, one
still expects the $S \sim t^{2/3}$ regime to be observed, because, in the
early instants of their walk, the walkers do not know that they are
in closed valleys; (iv) At very low concentrations of finite barriers,
almost all the walkers will be on single sites. The system in this
state is completely frozen.

From this, and inasmuch as the percolation model is a good
approximation for surface heterogeneity, one can predict the following
general trends for the reaction kinetics of walkers (A molecules) with
static traps (B molecules) on a heterogeneous surface. In terms of
survival probability $P = [A(t)] / [A(o)]$, one should always observe
Ln P decreasing as $- [B_o] t^{2/3}$ at short times. Unless the ratio of
high (infinite) to low (zero) energy barriers is by chance exactly
equal to the percolation threshold, this regime should not last for
very long times. If the fraction of low barriers is larger than that
of high barriers, one should observe a crossover to classical pseudo-
unimolecular kinetics (equation 3): $Ln P \sim - [B_o] t$. If the fraction
of low barriers is smaller than that of high barriers, typical cluster-
limited kinetics should be observed. The A molecules which are in
mobility valleys containing one or more B traps will very rapidly
react. On the other hand, those which are in trap-free valleys will
never react and will survive forever. Hence, on the average, the
survival probability should reach a non-zero time independent value,
P_∞, at long times. The higher the trap concentration, the lower P_∞
will be. This, as well as the exploration behavior, $S(t)$, is illustrated
in Figure 4.

2.3. Cluster-Limited Luminescence Quenching on Clay Surfaces.

Luminescence quenching measurements provide a convenient method for

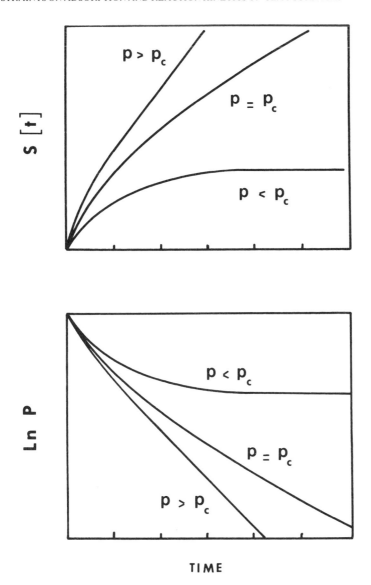

Figure 4. *Upper part:* Number of distinct sites visited, S(t), by a random walker over a lattice with either zero (or finite) or infinite barriers between sites, at increasing concentrations of zero (or finite) barriers, p. p_c is the percolation threshold. *Lower part:* Survival probability, P, of the walker in the same situations when traps are distributed over the lattice.

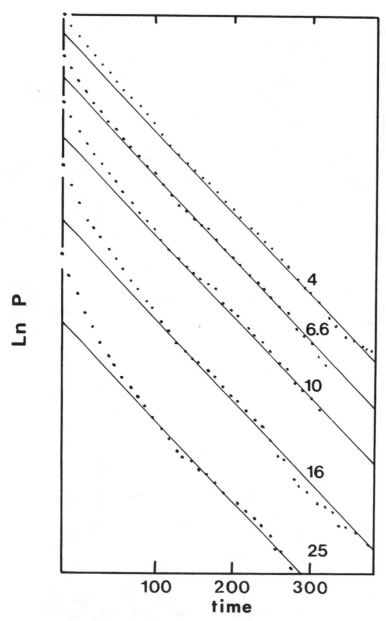

Figure 5. Simulation of the survival probability of a random walker over a lattice containing 4, 6.6, 10, 16 and 25% of traps, respectively (from top to bottom), below the percolation threshold. The spontaneous dead probability of the walker is 0.01 per unit time. The jump probabilities between neighbouring sites are either 0.36 (finite barriers) or 0 (infinite barriers). The barrier concentrations are 40% and 60%, respectively.

monitoring the efficiency of a photochemical reaction. The basis for
the method is the competition among all the relaxation modes of an
excited state. An excited state, A*, can relax via four main physical
or chemical processes: (i) luminescence; (ii) chemical decomposition;
(iii) non radiative relaxation by coupling with the thermal energy
levels of the surrounding; (iv) a quenching reaction with a specific
partner, B. The rate of three of these processes can be expressed by
simple monomolecular or pseudo-unimolecular rate equations:

$$V_\ell = k_\ell \ [A*] \tag{10}$$

$$V_{nr} = k_{nr} \ [A*] \tag{11}$$

$$V_q = k_q \ [A*] \ [B], \text{ with } [A*] \ll [B] \tag{12}$$

(A* will be assumed not to decompose). Since the total relaxation
probability of A* is necessarily equal to 1, any increase of the
quenching reaction rate (3), by increasing the B quencher concentration
for instance, will necessarily lead to a decreased probability for
the other processes. In other words, by monitoring the luminescence,
one can indirectly estimate the efficiency of the quenching reaction.
This can be done either by steady-state or time-resolved measurements.
The latter are best suited for kinetic studies. The principle is to
generate a population of excited species by a short pulse of light,
and to monitor the decrease of this population via the luminescence
decay. After the pulse, the population of A* decays according to:

$$d \ [A*]/dt = - \ k_\ell \ [A*] - k_{nr} \ [A*] - k_q \ [A*][B] \tag{13}$$

For a quenching reaction occuring on a solid surface, we know from
sections 2.1 and 2.2 that k_q will not, in general, be a time-independent
parameter, unless the surface is perfectly homogeneous. If, by chance,
this happens to be the case, one expects a classical single exponential
decay (or survival probability):

$$[A*(t)]/[A*(o)] = P = \exp \left\{ - \ (k_o + k_q \ [B_o]) \ t \right\} \tag{14}$$

where $k_o = k_\ell + k_{nr}$. Associated with equation (14) is a characteristic
lifetime $\tau = 1/(k_o + k_q \ [B_o])$. This is typically the behavior observed
in homogeneous solutions, in the absence of ground state A-B associations (complexes, ion pairs, ...).

The experimental situation which will be described here is the
exact equivalent to the walker-trap problem discussed in 2.2. The
walker is a cationic coordination compound -tris (2,2' bipyridine)
ruthenium(II)$^{2+}$- adsorbed on the surface of various clay minerals. The
fixed traps are the Fe^{3+} ions of the clay crystal lattice. Steady state
luminescence measurements show that even in this situation (i.e.
embedded in the structure of the solid), the Fe^{3+} ions are strong
quenchers for the luminescence of the probe molecules adsorbed on the
surface of the clay particles [12, 13]. There is however a significant

difference with respect to the situation of section 2.2.: even in a
trap-free environment, the walkers (the excited Ru probes) do not
survive forever. Their survival probability decays as a single
exponential function of time, with lifetime τ_o = $1/k_o$. Hence, even
at a fraction of high barriers for diffusion above p_c (i.e. in a
situation where only closed valleys exist), the survival probability
of the probes will not reach a constant value. It will decay as
$\exp(-t/\tau_o)$. Nevertheless, the clustered (closed valleys) situation
should be easily distinguishable from the Euclidean (open valleys)
situation at long time since, in the former case, the exponential
limit is independent of trap concentration, whereas in the latter case,
it becomes faster (shorter τ) as the trap concentration increases.
A simulation of the luminescence decays expected in the clustered
situation is shown in Figure 5.

Figure 6 shows the experimental results, for six different clays,
with Fe^{3+} contents spanning $\simeq$ three orders of magnitude. The relationship with the simulated decays of Figure 5 is striking, clearly showing
that the diffusion of coordination compounds like tris(2,2' bpy) Ru(II
on clay surfaces can be described in terms of a percolation model,
and that its reaction rate with fixed traps obeys a cluster-limited
diffusion-reaction scheme. The exponent derived from the short-time
behavior of the decays is $\simeq$ 0.7, in good agreement with the theoretical
value (2/3) [14].

The conclusion from the previous results should not be taken too
strictly. In other words, one should not conclude that the adsorbed
complexes will *never* escape the valleys of easy motion in which they
are confined. Actually, what the results show is that there are *two*
time scales to be considered: (i) a short time scale during which the
reactant molecules very rapidly explore, in a compact regime, a restricted number of sites on the surface, and (ii) a much longer time scale,
which is associated with macroscopic diffusion, i.e. with the crossing
of the high energy barriers which separate one fast motion valley from
the next one. If the measurements are performed on a rather short time
scale (one or two microseconds, in the example considered here), the
slower valley-to-valley motion will not be noticeable. This behavior
might well prove to be a general one for diffusion-reaction in constrained systems where the energetical disorder is large.

Finally, it should be pointed out that the barrier map is not an
intrinsic property of the surface. The barriers are created by the interactions of the reactant molecules with the surface. Hence, different
interactions will lead to different barrier heights. Thus, on the same
surface, one should be able to observe simultaneously different types
of kinetic regimes. For instance, a molecule like tris(2,2' bipyridine)
Ru(II), which is strongly retained on the clay surface, will have an
extremely slow valley-to-valley motion, whereas less strongly bound
species, like $Cu(H_2O)_6^{2+}$ for instance, will feel a much reduced energetical disorder and will much more often overpass the highest energy
barriers. In such a case, the survival probability of the excited Ru
complex in the presence of co-adsorbed Cu ions is no longer strictly
cluster-limited. A crossover to an Euclidean regime is observed at long

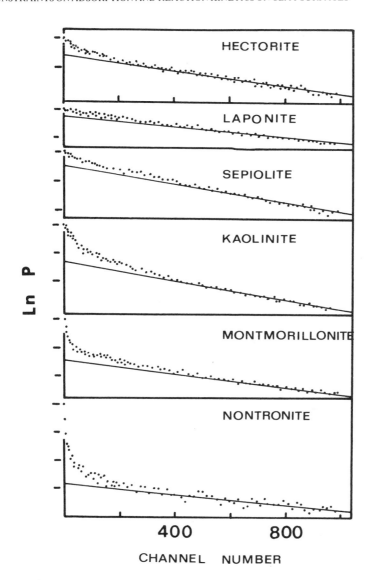

Figure 6. Experimental decay (or survival probability, P) of excited [tris (2,2' bpy) Ru II] adsorbed on various clays with increasing trap (Fe) concentration. The time scale is 1.6 nanosecond per channel.

time [14], in which Ln P decays as a single exponential with a lifetime becoming shorter as the concentration of Cu increases: $\tau = 1/(k_o+k_q [Cu])$. In fact, such a system can be described by a dual lattice model, in which one of the reactants would be in a situation corresponding to $p < p_c$, and the other reactant in a situation corresponding to $p > p_c$.

3. AN ANALYSIS OF GEOMETRICAL CONSTRAINTS IN PILLARED CLAYS IN TERMS OF FRACTAL DIMENSION

Several inorganic constrained systems (zeolites, pillared clays, silicic acids, ...) are in fact microporous solids and, due to this, impose severe geometric constraints on the molecules which react on their surface. Shape selectivity and molecular sieving are the properties first associated with this microporosity. However, microporosity alone is not a sufficient condition to induce clear-cut molecular sieving properties (we will concentrate on molecular sieving rather than on shape selectivity because shape selectivity implies chemical inter- actions as well as geometrical constraints). Many highly microporous materials are poor sieves (charcoals, silicagels, ...). In order to be a good sieve, a material has to be cristalline or quasi-cristal- line. Typical examples are zeolites. Cristallinity seems to be necessary to have well-defined pore- or, more exactly, window-sizes. On the other hand, ill-organized materials tend to be poor sieves because randomness produces irregular surfaces, which generate a whole distribution of pore-sizes. Hence, the fundamental question to be asked in order to know whether a material is potentially a good sieve or not is: how "regular" is its surface? A "regular" surface is not necessarily a flat surface. The surface of zeolites results from the interpenetration of complex shapes which can in no way be considered as flat over distances larger than a few Angströms. Yet, zeolites are good sieves. In fact, a regular surface is a surface which can be generated by the repetitive *translation* of an elementary shape of molecular size, which can be as complex as crystal chemistry may allow for. A hypothetical example is shown in Figure 7. Conversely, irregular surfaces cannot be generated by translational symmetry operations, but in many cases -as shown by AVNIR, PFEIFER and FARIN [15-19]- they might be characterized by *dilatational* symmetry. In other words, many ill-organized or defectuous materials might be *fractals* in the sense of MANDELBROT [15], and their surface irregulatities are most probably invariant over a certain range of scale transformation. Increasing or decreasing the resolution power does not modify the general morphology of the surface. An N-fold magni- fication of a surface "detail" reveals N^D smaller details morphological- ly similar to the previous one. An example is shown in Figure 7 (for a line, not for a surface). D is the fractal dimension of the surface. It can be anywhere in the range $2 < D < 3$ (or $1 < D < 2$ for a fractal line, as in Fig. 7). The higer D, the higher the wiggleness and the "space-filling" character of the surface.

The first method developped by AVNIR and PFEIFER [15] for measuring D is a surface chemical resolution analysis based upon the determination

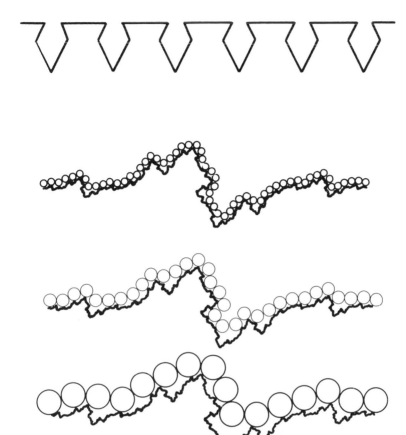

Figure 7. *Upper part:* Example of a periodic, i.e. translationaly invariant, line, which would be a good sieve. *Lower part:* Example of a aperiodic, dilationaly invariant (fractal) line. Covering this line with molecules of increasing radius r yields a power law dependence $N \sim r^{-D}$; where D, the fractal dimension is neither 1 nor 2, but 1.16.

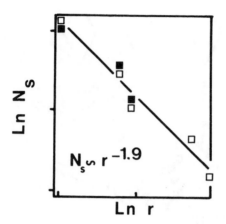

Figure 8. Plot of the amounts adsorbed at saturation, N_s, by an Al_{13}-pillared montmorillonite (data from reference 22) vs kinetic diameter of the adsorbates. (■): air dried samples. (□): freeze dried samples.

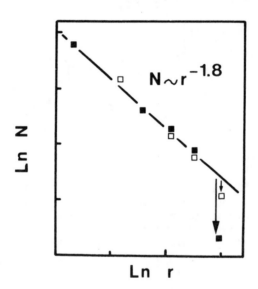

Figure 10. Plot of the amount "adsorbed" at saturation in a "pillared clay", from a simple dimensional simulation (Fig. 9). (■): quasi-crystalline distribution of pillars. (□): random distribution of pillars. The arrows indicate the sieving effect.

of monolayer coverage with different adsorbates, i.e. different
yardsticks. Indeed, on a surface of fractal dimension D, the number of
spherical molecules of radius r necessary to reach monolayer coverage
scales down with increasing r as:

$$N \sim r^{-D} \tag{15}$$

or, equivalently, as:

$$N \sim \sigma^{-D/2} \tag{16}$$

where σ is the cross section of the probe molecules. The limit D = 2
corresponds to a regular or Euclidean surface. As shown by AVNIR et al.
[17], this limit is closely approached by faujasite type zeolites. On
the other hand, with porous silicagel, porous alumina and charcoal, one
approaches D $\simeq$ 3 [15,17,18].

Pillared clays [20] are interesting materials for a fractal
analysis: on the one hand, the individual clay layers are expected to
have a rather smooth and Euclidean surface in the molecular range, but
on the other hand, the pillars in the interlayer gallery are a potential
source of randomness. If the pillars are distributed in a quasicrystalline arrangement, one expects good molecular sieving properties. If
the pillars are randomly distributed, the material should still be
highly microporous, but its quality as a sieve should be much inferior.
The question is: Is the fractal dimension, D, a useful and quantitative
parameter for characteristing the porous structure of pillared clays?

Experimentally, a broad range of properties have been observed.
For instance, BARRER and coworkers [21] found that smectite clays
pillared with large alkylammonium ions (the so-called organo clays) are
excellent sieves for separating oxygen from nitrogen at low temperature.
PINNAVAIA et al. [22] found that montmorillonites pillared with poly-
hydroxyaluminium cations, and air-dried after pillaring, are also good
sieves, but with somewhat larger size windows. They do no not adsorb
molecules with a kinetic diameter larger than 10Å. Surprisingly, the
same authors found that the same Al-pillared montmorillonites, freeze-
dried after the pillaring treatment, no longer display any sieving
properties.

VAN DAMME and FRIPIAT [23] performed a fractal analysis of adsorption in these materials. For the organo clays, the analysis was performed by using data obtained by adsorbing only one type of yardstick
molecule (nitrogen) into a series clays pillared with organo cations of
increasing size [21]. This can be done by using the relation between
N (monolayer coverage) and σ_p (the cross section of the pillar) which
can be easily derived from equation (16) [23]. The result is
D = 2.0 ± 0.04, clearly showing that the intracrystalline surface
on which the organic pillars are deposited is Euclidean. On the other
hand, for the Al-montmorillonites, the analysis was performed by using
data [22] obtained by adsorbing a series of molecules of increasing
size on the pillared clay, and using equation (16). The result is
D = 1.90 ± 0.10; close to but slightly below 2 (Figure 8). Interestingly,

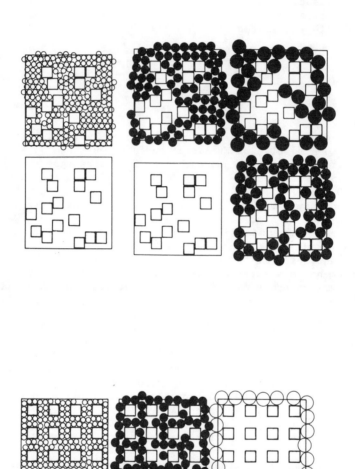

Figure 9. Two-dimensional simulation of adsorption in pillared clays with a quasicrystalline (left) or random (right) distribution of pillars. The relative size of the "molecules" with respect to the pillars was chosen in the range corresponding to the experimental situation for Al_{13}-pillared montmorillonites.

a simple two-dimensional simulation, illustrated in Figure 9, yields
D ≃ 1.8 (Figure 10), again close to but below 2. In fact, this stems
from the fact that the average interpillar distance is of the order
of the molecular diameters used to probe the porous space. Using
molecules much smaller than the interpillar separation distance leads
to D = 2.0, as it should be (the molecules do not "feel" the walls
of the pillars).

   Another interesting observation is that the same fractal dimension
holds for the freeze-dried clays, i.e. for a material which is a
molecular sieve and for a material which is not. At first sight, this
does not seem to be consistent with an eventual modification of the
pillar distribution induced by the drying treatment (and this is the
conclusion drawn in ref. 23). In fact, the two-dimensional simulation
shows that D is *not* sensitive to the type of pillar distribution
(Figure 10). Only the molecular sieving effect allows one to discriminate the random and quasicrystalline arrangement of pillars.

   This clearly shows the limits of the fractal dimension for
characterizing the geometrical constraints which can be applied by a
given system to the reactant molecules. D is certainly useful to
determine whether a system has self-similar properties or not. On one
hand D ≃ 2 seems to be a *necessary* condition to have good sieving
properties (zeolites, pillared clays). However, it is not a *sufficient*
condition. Any randomness in the system, without scale-invariant (self-
similar) character will loosen the geometrical constraints, without
modifying D.

REFERENCES

1. R.H. Fowler and E.A. Guggenheim, Statistical Thermodynamics,
   Cambridge University Press, Cambridge (1939), Chap. X.
2. T.L. Hill, Introduction to Statistical Thermodynamics, Addison-
   Wesley, Reading, Mass (1960), Chaps. 7 and 14.
3. P.G. de Gennes, J. Chem. Phys., 76, 3316 (1982).
4. (a) P. Evesque, J. Physique, 44, 1217 (1983); (b) P. Evesque and
   J. Duran, J. Chem. Phys., 80, 3016 (1984).
5. S. Alexander and R. Orbach, J. Physique Lett., 43, L-625 (1982).
6. R. Rammal and G. Toulouse, J. Physique Lett., 44, L-13 (1983).
7. J. Klafter, G. Zumofen and A. Blumen, J. Physique Lett., 45, L-49
   (1984).
8. B.B. Mandelbrot, The Fractal Geometry of Nature, Freeman,
   San Franscisco (1982).
9. D. Stauffer, Phys. Rep., 54, 1 (1979).
10. P. Argyrakis and R. Kopelman, J. Chem. Phys., 81, 1015 (1984).
11. (a) P.G. de Gennes, C.R. Acad. Sc. Paris, 296, II-881 (1983);
    (b) I. Webman, Phys. Rev. Lett., 52, 220 (1984).
12. D. Krenske, S. Abdo, H. Van Damme, M. Cruz and J.J. Fripiat, J.
    Phys. Chem., 84, 2447 (1980).
13. A. Habti, D. Keravis, P. Levitz and H. Van Damme, J. Chem. Soc.,
    Faraday Trans. 2, 80, 67 (1984).

14. H. Van Damme, P. Levitz and L. Gatineau, J. Physique Lett., to be submitted.
15. D. Avnir and P. Pfeifer, Nouv. J. Chim., 7, 71 (1983).
16. P. Pfeifer and D. Avnir, J. Chem. Phys., 79, 3558 (1983).
17. D. Avnir, D. Farin and P. Pfeifer, J. Chem. Phys., 79, 3566 (1983).
18. D. Avnir, D. Farin and P. Pfeifer, Nature, 308, 261 (1984).
19. D. Avnir, D. Farin and P. Pfeifer, J. Coll. Interf. Sci., 103,112 (1985).
20. T.J. Pinnavaia, Science, 220, 365 (1983), and this volume.
21. R.M. Barrer, Philos. Trans. R. Soc. London, A 311, 333 (1984), and references therein.
22. T.J. Pinnavaia, M.S. Tzou, S.D. Landau and R.H. Raythatha, J. Molec. Catal., 27, 195 (1984).
23. H. Van Damme and J.J. Fripiat, J. Chem. Phys., 82, 2785 (1985).

REACTIONS BETWEEN ACIDIC MICROEMULSIONS AND CALCIUM CARBONATE

J. Sjöblom, H. Söderlund and T. Wärnheim
Institute for Surface Chemistry
Box 5607
S-114 86 Stockholm
Sweden

ABSTRACT. Properties of acidic microemulsions are presented. We have investigated in some detail two model systems, i.e. water (HCl)/ butanol/cetylpyridiniumchloride (CPC)/dodecane and water (HCl)/poly-oxyethylene nonylfenol ether/dodecane. NMR-measurements providing the self-diffusion coefficients of the components in the microemulsion indicate that the system based on the cationic surfactant + butanol has a bicontinuous structure, while the nonionic system can be classified as containing more-or-less closed aggregates with a restricted water diffusion. In the reactions with calcium carbonate a retardation sequence water (HCl): cationic microemulsion (HCl): nonionic micro-emulsion (HCl)= 1:25:500 is found. The reaction rate is obviously dependent on the microemulsion structure, which can be varied by designing different systems. The possibility to monitor the variation in the reactivity of the microemulsion by measuring the self-diffusion coefficients by means of NMR is encouraging.

1. INTRODUCTION

Due to its large technical importance in oilwell stimulation, the reaction between hydrochloric acid and calcite (calcium carbonate) has been a subject of intense research (1). For technical reasons it is of interest to retard the reaction between HCl and $CaCO_3$ and, hence, to obtain a system that is capable of penetrating deeper into the formation. Traditionally, when using strong acids, the retarded systems have contained polymers as viscosity increasing agents or the acid has been emulsified. In both cases the basic idea is to lower the diffusion of hydrogen ions and, hence, to lower the reaction rate. In this paper we point at the possibility to dissolve the strong acid into a microemul-sion, i.e. a solution containing water, surfactant, co-surfactant and hydrocarbon in the same thermodynamically stable solution, and to let this acidic solution react with calcium carbonate. The acidic microemul-sion in oilwell stimulation is not beneficial only as a retarded system but also as an effective system for uniform matrix acidizing due to its low viscosity and low interfacial tension towards oil and water.

R. Setton (ed.), Chemical Reactions in Organic and Inorganic Constrained Systems, 305–313.
© 1986 by D. Reidel Publishing Company.

2. MICROEMULSION SYSTEMS

A microemulsion is a solution containing high amounts of water and
hydrocarbon. Stability is achieved with the aid of a surface active
agent. If the surfactant is ionic (cationic or anionic) it must be
combined with a co-surfactant, which is generally a medium chain alcohol, like for instance n-butanol. The basic research on the microemulsion field tries to clarify the structure of these unique solutions
possessing both polar and nonpolar properties. Several microemulsion
systems have been investigated by means of different experimental
techniques (2-5). The most promising structural information about
microemulsions has been obtained from NMR-self-diffusion measurements
(6-11). These have shown that the diffusion coefficients of water as
well as hydrocarbon are high and comparable with those of the pure
components. This experimental result has been gathered into a structural
picture of the solution. Usually one speaks about a "bicontinuous"
structure of microemulsions stabilized by surfactant and possibly co-
surfactant. In this picture water and hydrocarbon domains constitute
different regions, but the interface between these regions is considered
very flexible and dynamical in nature. This structural picture is also
in agreement with the low viscosity of the microemulsions. One reason
for the technical interest in microemulsions is the possibility to
achieve ultralow (down to $10^{-3} - 10^{-4}$ mNm^{-1}) interfacial tension towards
an excess water or oil phase. This property gives them good washing and
penetration properties. Therefore microemulsions have found large use
in cleaning system and are of considerable interest as flooding systems
in enhanced oil recovery.

2.1. Microemulsions Based on Ionic Surfactants

Generally one finds a larger interest in formulations based on anionic
surfactants than on cationics, but this is also dependent on the application as seen later. In order to stabilize a water + hydrocarbon mixture the ionic surfactant must be combined with a co-surfactant,
usually an alcohol. Normally one uses as surfactant alkyl carboxylates,
sulphates of sulphonates, where the alkyl chain has 8-16 carbon atoms.
Among the cationics pyridinium salts and amines are commonly used. The
micelle forming substances described above have a very limited capacity
to solubilize hydrocarbon in a pure aqueous solution. When, however, a
co-surfactant is added it is possible to obtain high mutual solubility
of hydrocarbon and water in the same solution. The co-surfactant is believed to modify the interfacial properties towards a lower degree of
rigidity, and to create new conditions for solubility of the polar and
nonpolar components. The co-surfactant molecule must thus itself be
highly dynamical and partition between the different regions in the
microemulsion. The isotropic microemulsion phase is sensitive to addition of electrolytes, especially divalent ions are capable of breaking
the solutions.

2.2. Microemulsions Based on Nonionic Surfactants

With the use of nonionic surfactants it is possible to achieve high
mutual solubility of oil and water with only three components. In this
case the system is extremely temperature sensitive and solubilization
of oil in aqueous micellar solutions and water in oily micellar solutions varies much as temperature changes. Optimal solubilization occurs
at a temperature called P.I.T. (Phase Inversion Temperature) (12-15).
At this temperature there occurs a so-called surfactant phase which in-
corporates large amounts of water and hydrocarbon, i.e. a microemulsion
phase. A temperature increase will change the hydrophilic/lipophilic
balance of the nonionic surfactant towards a higher degree of lipophili-
city and, hence, make the surfactant more oilsoluble. On a molecular
level this is usually attributed to the changed interaction between the
etoxygroups (EO) of the surfactant and vicinal water molecules.
Important in applications of these systems is that the isotropic microemulsion region is extremely temperature sensitive and exists only
within a narrow temperature range.

3. ACIDIC MICROEMULSION SYSTEMS

Disregarding the chemical reactivity an addition of a strong acid to
microemulsion would have to a first approximation the same consequences
on stability as an addition of an ordinary 1:1 electrolyte. One should,
thus, expect a reduction in the region of existence for ionic microemulsions and a change in P.I.T. of the nonionic microemulsions. We have
investigated two acidic microemulsion systems. One system is based on a
cationic surfactant and the other on a nonionic surfactant.

3.1. The Model System Cetylpyridiniumchloride (CPC)/Butanol/
Water (4MHCl)/Dodecane

It is obvious that a microemulsion formulation with a single ionic
surface-active agent must in this case be based on a cationic surfactant. An anionic surfactant will at these pH:s lose its surface activity. The phase diagram of the model system CPC, butanol and water is
presented in Fig. 1.

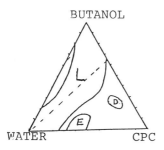

BUTANOL

WATER   CPC

Figure 1. Phase diagram for
butanol, water and cetylpyridi-
niumchloride (CPC) at 298K.
---- represents the ratio
CPC/butanol = 35:65

The ternary system consists of a large solution phase L and two anisotropic mesophases E and D. The properties of the latter phases will not
be discussed in this context. It is possible to obtain rather large
microemulsion regions, when adding dodecane, if a ratio CPC:butanol =
35:65 is used. This phase diagram is shown in Fig. 2.

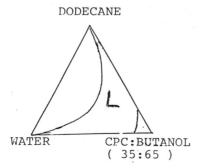

DODECANE

WATER   CPC:BUTANOL
                          (35:65)

Figure 2. A schematic phase
diagram for dodecane, water and
a fix ratio between CPC and
butanol (35:65)

The figure reveals the existence of a one phase region containing at
the same time high amounts of water and dodecane. An addition of HCl to
this system will somewhat reduce the extension of the microemulsion
region (L). In order to make a first characterization of the microemulsions within the isotropic phase, we have determined the self-diffusion
coefficients of the components by means of NMR-measurements, as the
content of water and dodecane is changed. The result is summarized in
Fig. 3.

The self-diffusion coefficient of water is proportional to the amount
of water in the microemulsion. The self-diffusion coefficient of
butanol (not given in the figure) is more or less constant over the
entire concentration range (16). This behaviour is similar to that of
other microemulsions stabilized by an ionic surfactant and butanol (8-9).
On the time-scale of the diffusion measurements ($\sim$ 100 ms) the present
system must be regarded as effectively bicontinuous.

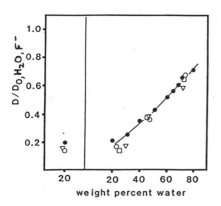

weight percent water

Figure 3. The self-diffusion of
water ($\bullet$), fluoride ($\circ$),
fluoride in presence of a nonyl-
phenol with 30 ethoxy groups ($\square$)
and 50 ethoxy groups ($\triangledown$), respectively, as a function of the
water content in the microemulsion. D_0 is the self-diffusion
coefficient of pure water and
1M KF, respectively. The two
values at 20% H_2O represent two
different contents of dodecane.

Since our main interest is to investigate what influences the transport
properties of the system, we have dissolved fluoride to be used as a
probe for the transport of water soluble compounds. According to the
data in Fig. 3 there is a proportionality between the water diffusion
rate and the fluoride diffusion rate. In an attempt to reduce the
diffusion rate of the system without increasing the viscosity of the
solution too drastically, small amounts of nonionic surfactants with a
long hydrophilic chain (30-50 ethoxy groups) were added and the fluoride
diffusion was determined. Figure 3 reveals only minor effects on the
diffusion of fluoride in presence of the hydrophilic nonionics. This
result is clearly consistent with a maintained bicontinuous structure.
Previous investigations on ternary systems containing nonionic surfac-
tant/water/hydrocarbon indicate that if the nonionic surfactant is
hydrophilic enough, there seems to be no distinct separation into hydrophilic and hydrophobic domains (10-11). Evidently, the same conclusion
applies for our system with mixtures of ionic and hydrophilic nonionic
surfactants. Summarizing we have found no indications that the microemulsion system based on CPC should or could easily be modified, to
contain well-defined aggregates with a closed structure.

3.2. The Model System Polyoxyethylene (EO:4.5-5.5) Nonylfenol
Ether/Water (4MHCl)/Dodecane

It is generally known that an addition of a 1:1 electrolyte to an
aqueous nonionic surfactant solution will lower the cloud point.
However, in the acidic solution the cloud point is increased (12).
Molecularly this can be attributed to a hydrogen bond between the ether
oxygens and the acidic protons (17). Raising the cloud point is
analogous of making the surfactant more hydrophilic. Therefore, the
P.I.T. of oil/water/nonionic surfactant is expected to increase when
adding HCl. The maximum ability to solubilize water in the L_2-phase
will then be shifted to higher temperatures in acidic microemulsions
compared to those containing pure water, as seen in Figure 4. In the
microemulsion stabilized by Triton N-42, which is a nonylfenol with
4.5 EO groups, the system containing pure water is well above the
P.I.T. at 25°C, while addition of acid makes the surfactant more hydrophilic, and raises the P.I.T. Triton N-57 (5.5 EO groups) behaves analogously, but here the optimum solubilizing capacity in the L_2-phase
occurs around room temperature in the water system and around 60°C in
the acidic system. By combining the two surfactants the acid solubilization in the temperature interval 20°C-60°C may be optimized, as seen
in Figure 5.

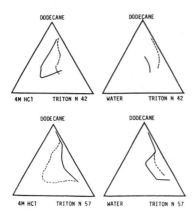

Figure 4.   Partial phase diagrams for nonionic microemulsions. The
solution phases extend from the oil-surfactant binary axis towards the
water (or acid) corner. Concentrations in weight percent. Solid line
indicate the extensions at 25°C. Hatched line indicate the extensions
at 60°C. (18)

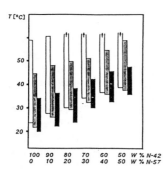

Figure 5. Temperature interval where the acid microemulsions are stable
for different mixtures of nonionic surfactants.Weight ratio dodecane:
surfactant 60:40. White area indicates 20% 4M HCl content. Grey area
indicates 30% 4M HCl content. Black area indicates 40% 4M HCl content.
No measurements were performed at higher temperature than 60°C. This
has been indicated in the figure by not drawing any upper temperature
limit for microemulsions stable at that temperature. At the lower
temperature boundary, the microemulsion is in equilibrium with a liquid
crystalline phase. At the upper temperature limit, the microemulsion
is in equilibrium with a water-rich solution phase. (18)

The water self-diffusion coefficients for the system water or 4M HCl/
dodecane/Triton N-42 are presented in Figure 6. The hydrocarbon self-
diffusion (not shown in the figure) is rapid in both systems and
fairly constant. The self-diffusion coefficient of pure water
incorporated in the microemulsion is low and decreases with increasing
water content. This indicates the presence of rather well-defined
water-in-oil aggregates (reversed micelles).

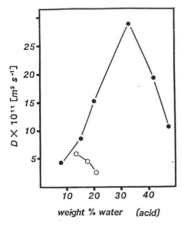

Figure 6. Water self-diffusion coefficients in microemulsions stabilized by Triton N-42. Weight ratio dodecane: surfactant 50:50.

(◐) microemulsion containing water
(●) microemulsion containing aqueous 4M HCl

After ref. (18).

In the system with added acid, the water self-diffusion coefficient first increases with increasing water concentration, and then, around 35-40% starts to decrease. Usually, the self-diffusion of all components is rapid in similar systems with large solution regions, indicating a bicontinuous structure on the timescale of the diffusion measurements, 100 ms. This behaviour is evident in the first concentration interval, where the self-diffusion of water increases rapidly. The decrease of the water self-diffusion coefficient starting at higher water contents indicates that the aggregates become more well-defined or that they grow in size (or possibly both). If we have well-defined water droplets in an oil-phase at these high concentrations of water, they must for geometrical reasons be rather large, since too small droplets will give a too large interfacial area; the surfactant will not suffice to disperse the water. A closer scrutiny of these phenomena must be performed in more well-defined systems.

4. REACTIONS WITH CALCIUM CARBONATE

It has been shown that the reaction between $CaCO_3$ and HCl is diffusion controlled (1) and, thus, provides a proper basis for our investigations with retarded microemulsion systems. Contrary to acidic aqueous solutions, the masstransport in a microemulsion of H^+ to the solid surface, is dependent of the droplet diffusion in the bulk. In this paper we do not attempt to investigate the reaction mechanisms, but to experimentally compare the reactivity between $CaCO_3$ and HCl for two microemulsion systems with different self-diffusion behaviour. The reaction between $CaCO_3$ and microemulsions in the model system CPC/H_2O (4M HCl)/butanol/dodecane is illustrated in Fig. 7.

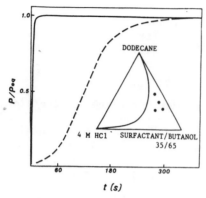

t (s)

Figure 7. The relative carbon dioxide pressure (P_{CO_2}/P_{CO_2}, equilibrium) as a function of the reaction time for a) acidic aqueous solutions (solid line), b) acidic microemuslions (dotted line) in the inserted system. Reaction $2H^+ + CaCO_3 = Ca^{2+} + CO_2 + H_2O$

The figure shows that the reaction is retarded in comparison with pure aqueous acidic solutions. An equivalent amount of HCl is in this micro-emulsion system consumed about 25 times slower than in the pure aqueous system. Small variations in microemulsion composition do not affect the reaction to any higher degree. In figure 8 we summarize the result from the reactions between the nonionic (Triton N-42) microemulsions and $CaCO_3$. In this case we find a considerable retardation of 500 times before the equivalent amount of HCl is consumed.

Summarizing the retardation sequence HCl in water: HCl in a cationic microemulsion: HCl in a nonionic microemulsion is $\approx$ 1:25:500. Qualita-tively the trend in this sequence can be understood from the structure of the microemulsions. In the cationic system we found a bicontinuous structure with a rather rapid water self-diffusion ($D_{H_2O} \propto D_{H_3O^+}$) while the nonionic system revealed for the tested microemulsion solutions a limited water self-diffusion. It is our purpose to return in a forth-coming paper to a more quantitative description of the process.

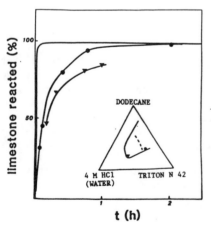

t (h)

Figure 8. The amount of reacted limestone as a function of the reaction time for a) acidic aqueous solutions (solid line), b) acidic microemulsions (dotted line) in the inserted system.
Same reaction as in Fig. 7.

5. CONCLUSIONS

In the present paper we have focused on the relations between macro-scopic diffusion properties of microemulsions, their structure and their reactivity with solid calcium carbonate. We have found that the systems with the lowest reaction rate are those, where water and water-soluble compounds have a slower self-diffusion. A slow water self-diffusion in a microemulsion is generally interpreted in terms of closed water-in-oil aggregates (6-9). We may thus conclude that we are able to design acidic microemulsion systems with varying reactivity, and to monitor this variation by measuring the self-diffusion coefficients of the components.

ACKNOWLEDGEMENT

This work was financially supported by Norges Tekniske og Naturviden-skaplige Forskningsråd, NTNF.

REFERENCES

1. Williams, B.B., Gidley, J.L. and Schechter, R.S., "Acidizing Fundamentals", SPE Monograph Series, Vol. 6, (1979)
2. K.L. Mittal, Ed, "Micellization, Solubilization and Microemulsions", Plenum Press, New York 1977
3. K.L. Mittal and E.J. Fendler. (Eds.), "Solution Behavior of Surfactants", Plenum Press, New York 1982
4. K.L. Mittal and B. Lindman, (Eds.), "Surfactants in Solution", Plenum Press, New York 1984
5. I.D- Robb, Ed, "Microemulsions", Plenum Press, New York 1981
6. B. Lindman, P. Stilbs and M.E. Moseley, J. Colloid Interface Sci., 83, 569 (1981)
7. P. Stilbs, M.E. Moseley and B. Lindman, J. Magn. Reson., 40, 401 (1980)
8. P. Stilbs, K. Rapacki and B. Lindman, J. Colloid Interface Sci., 95, 583 (1983)
9. T. Wärnheim, E. Sjöblom, U. Henriksson and P. Stilbs, J. Phys. Chem., 88, 5420 (1984)
10. P.G. Nilsson and B. Lindman, J. Phys. Chem., 86, 271 (1982)
11. P.G. Nilsson, Thesis, University of Lund, Lund 1984
12. H. Saito and K. Shinoda, J. Colloid Interface Sci., 32, 649 (1970)
13. K. Shinoda, J. Colloid Interface Sci., 34, 278 (1972)
14. K. Shinoda and H. Kuneida, J. Colloid Interface Sci., 42, 381 (1973)
15. K. Shinoda and S. Friberg, Adv. Colloid Interface Sci., 4, 281 (1975)
16. T. Wärnheim and J. Sjöblom, Tenside, in press
17. H. Schott and A.E. Royce, J. Pharm. Sci., 73, 793 (1984)
18. J. Bokström, J. Sjöblom, P. Stenius, T. Wärnheim and H.S. Fogler, Tenside, in press

PHOTOCHEMICAL REACTIONS AT THE AIR-WATER INTERFACE INVESTIGATED WITH
THE REFLECTION METHOD

Dietmar Möbius and Michel Orrit[*]
Max-Planck-Institut für biophysikalische Chemie
(Karl-Friedrich-Bonhoeffer-Institut)
Postfach 2841
D-3400 Göttingen-Nikolausberg
[*] on leave of absence from Centre de Physique Moléculaire
Optique et Hertzienne, Université de Bordeaux I et C.N.R.S.
(LA 283), F-33405 Talence, France

ABSTRACT. The reflection of light from the air-water interface is
modified in the presence of a dye monolayer in the spectral range of
the dye absorption band. The average orientation of the chromophores
can be determined by using linearly polarized light under oblique
incidence. The rate of photochemical cis-trans isomerization of a
thioindigo derivative at the air-water interface depends on the surface
pressure. The isomerization of an amphiphilic spiropyran to the
merocyanine form causes a sudden rise in surface pressure at constant
area. The propagation of the resulting compressional wave has been
studied by the reflection method.

1. INTRODUCTION

Photochemical reactions at interfaces are characterized by high local
concentrations of reactants, restricted mobility and anisotropic
intermolecular interactions. These different environmental conditions
as compared to photoreactions in homogeneous systems lead to new and
unexpected observations. At the air-water interface, the organization
of insoluble molecules can be controlled by variation of the environment in mixed monolayers and by changing the surface pressure. The
chemical reactivity, therefore, depends strongly on the structure of
the monolayers (1-3). Many photochemical reactions involve large
structural changes of the excited molecule which give rise to reorientation of molecular subunits at the interface. Such processes can be
investigated spectroscopically, e.g. by using the reflection method (4)
and the usual techniques like the measurement of surface pressure at
constant area or surface potential changes during the reaction (5).
    The reflection method (4) is based on the modification of the
light reflection from the air-water interface in the presence of a dye
monolayer. The reflection spectrum, i.e. the difference in reflectivity
of the water surface with dye monolayer and the reflectivity of the

R. Setton (ed.), Chemical Reactions in Organic and Inorganic Constrained Systems, 315–322.

pure water surface, ΔR, plotted vs the wavelength of the incident
light, follows the absorption spectrum. The difference ΔR is
proportional to the dye density in the monolayer for small dye
densities (4). A spectrum can be measured at normal incidence of light
(angle of incidence $\alpha = 0°$) only, if the transition moments of the dye
molecules have a component in the monolayer plane. Information on the
orientation of the transition moments in the spread monolayer is
obtained (6,7) when linearly polarized light is used under an angle of
incidence different from $\alpha = 0°$. In the case of an orientation of the
transition moments normal to the water surface, only p-polarized light
(electric vector oscillating in the plane of incidence) gives rise to a
change in reflectivity of the water, and ΔR is negative for $\alpha = 45°$ in
this case.

The cis-trans photoisomerization of an amphiphilic thioindigo
derivative has been studied in mixed monolayers at the air-water
interface by following the decrease in area at constant surface
pressure since the trans form has a larger area/molecule in the mixed
monolayer than the cis form (2). The reaction has also been studied in
transferred monolayer systems on glass plates (3) by measurement of the
absorption spectra (difference in transmission ΔT between section of
glass plate with dye monolayer and reference section without dye
monolayer). The dependence of the rate of photoisomerization on surface
pressure in mixed monolayers at the air-water interface was determined
using the reflection method (3).

The photoisomerization of an amphiphilic spiropyran to the
merocyanine form gives rise to a sudden increase of surface pressure at
constant area (3). If only a part of the entire monolayer area is
illuminated, a 2-dimensional compressional wave is generated which
travels across the remaining area. The monolayer of a cyanine dye in
contact with the spiropyran monolayer may undergo a phase transition
from monomer to J-aggregate when the compressional wave passes. The
different specroscopic properties of cyanine monomers and J-aggregates
allows one to detect the arrival of the compressional wave at the
observed spot by the reflection method (8). The time dependence of the
reflection signal after rapid generation of the compressional wave
contains information on the frequency-dependent damping of the wave at
the interface provided the monomer-aggregate transformation is
sufficiently fast.

2. EXPERIMENTAL

Materials: Bidistilled water prepared as described earlier was used in
most of the experiments; the spectra shown in Fig.1 were obtained from
monolayers on water taken from a Milli-Q unit (Millipore). Chloroform
(Baker analyzed reagent) used as spreading solvent was run through a 1
m column with alumina (B-Super 1, Woelm) and stabilized by addition of
1 vol% of ethanol (p.a. Merck). Arachidic acid and stearic acid (p.a.
Merck) was used after recrystallization from ethanol; octadecanol (p.a.
Merck) was used without further purification. The azo dye AZ (structure
see Fig.1) was prepared by J. Heesemann (9), the thioindigo derivative

TI (structure see Fig. 2) was supplied by D.G. Whitten (10), the spiropyran SP and the cyanine dye (CY) (structures see Fig.3) were synthesized by J. Sondermann (11,12).

Methods: Monolayers were prepared by spreading with the monolayer material on the water surface and compression of the material after solvent evaporation. The solutions used were 10^{-3} M AZ, 10^{-3} M TI, 10^{-3} M SP, 10^{-3} M CY, 5×10^{-3} M arachidic acid (C20), 5×10^{-3} M stearic acid (C18), and 5×10^{-3} M octadecanol (OD). The mixtures were prepared from these solutions according to the molar ratios given with the results. A rectangular trough provided with a filter paper Wilhelmy balance and electronic feed-back acting on the movable barrier in order to maintain constant surface pressure (6) was used for measuring the reflection spectra under 45° incidence, and details of this spectrometer will be given in a forthcoming paper (7). Reflection spectra at normal incidence were measured using a circular trough of the type described by Fromherz (13) with the reflection spectrometer described earlier (4). This setup was also used for studying the propagation of a compressional wave generated photochemically (8). The sections on the trough for the different monolayers in contact with each other (see Fig.3) were initially separated by thin films of polytetrafluorethylene during spreading and compression. These films were removed immediately before flash exposure of the spiropyran monolayer. The output of the photomultiplier tube was fed into a storage oscilloscope (Tektronix 7613) using a 1 k. resistor. A photographic flash (Metz, Mecablitz 40 CT 1) was used for exposures of the spiropyran.

2. RESULTS AND DISCUSSION

2.1 Effect of chromophore orientation on light reflection

Reflection spectra ΔR of monolayers of the azo dye AZ, measured at a surface pressure of 20 mN/m, are shown in Fig.1. The upper curve refers to an angle of incidence α=45°, s-polarization (electric vector oscillating in the plane of incidence), the curve in the middle to normal incidence (α=0°), unpolarized, and the lower curve to α=45°, p-polarization (electric vector oscillating in the plane of incidence).

The broad band observed in the case of normal incidence and of α=45°, s-polarization is attributed to monomer molecules of the azo dye lying flat on the water surface. The ratio of ΔR(44°,s)/ΔR(0°) = 2.02 ± 0.04 is found at 470 nm. According to model of reflection by a multilayer sytem of air-monolayer-water including secondary reflections (6), the ratio ΔR(44°,s)/ΔR(0°) = 1.78 is calculated for transition moments oriented statistically in the layer plane, supporting the interpretation given above.

The spectrum observed with p-polarized light shows a particular feature in the blue range. At wavelengths shorter than 450 nm, the reflection ΔR(44°,p) is negative, i.e. the reflectivity of the monolayer covered surface is smaller than that of the clean water surface. Such a phenomenon is expected for transition moments oriented normal to the water surface. Since these do not interact with normally

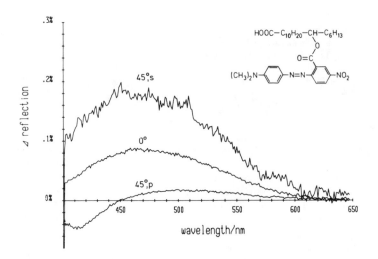

Figure 1. Reflection spectra ΔR from a water surface covered with a
monolayer of the azo dye AZ, structure given in the figure. Angle of
incidence α=45°, s-polarization of light (upper curve); α=0°,
unpolarized light (middle curve); α=45°, p-polarization (lower curve).
Broad band with maximum at 470 nm due to dye monomers oriented
statistically in the layer plane; negative band with maximum at 420 nm
attributed to aggregates with transition moment normal to the layer
plane. Surface pressure: 20 mN/m, subphase: water pH=5.6, 22°C.

incident or s-polarized light, the two other curves in Fig. 1 have no
corresponding band. The result is in agreement with results of light
transmission measurements (6,9).

This example shows that the reflection method provides information
on the orientation of transition moments in monolayers at the gas-water
interface. The interpretation of results is straightforward in contrast
to the methods based on polarized emission of light.

2.2 Photoisomerization in monolayers at the air-water interface

The surface active thioindigo derivative TI (structure see Fig. 2) can
be photoisomerized from the trans to the cis form, and monolayers of
the two isomers can be formed. The isotherms of mixed monolayers of TI
and arachidic acid (C20), molar ratio TI:C20 = 1:3, depend on the isomer:
cis-TI requires a larger area/molecule than trans-TI in the mixed
monolayer.

In monolayers at the air-water interface and in monolayer systems
on glass plates, cis-trans photoisomerization has been observed (3,10),
trans-cis photoisomerization, however, not so far. One explanation for
this behaviour may be the different spatial requirements of cis and
trans TI. It might be expected then, that the reaction should be accelerated by using a higher surface pressure during photoisomerization.
The reaction can be studied by measuring the enhanced light reflection

from the water surface in the range of the trans-TI absorption band.
Fig. 2 shows the reflection spectra of mixed monolayers of TI:C20 = 1:3,
obtained with cis-TI (c) and trans-TI (t). The cis-TI reflection
spectrum is characterized by a band with maximum at 480 nm, where the
reflection spectrum of the trans-TI has a minimum. The trans-TI
reflection spectrum has a maximum at 400 nm and a band with maximum at
550 nm. Following the reflection change at 550 nm during illumination
of the mixed cis-TI:C20 = 1:3 monolayer, the apparent rate constant for
photoisomerization can be determined. It was found, that the rate
decreases with increasing surface pressure (3).

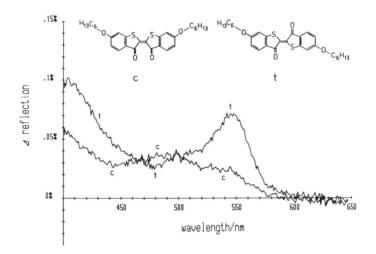

Figure 2. Reflection spectra ΔR of mixed monolayers of the thioindigo
derivative TI and arachidic acid, molar ratio TI:C20 = 1:3, obtained
with cis-TI (c) and trans-TI (t). Surface pressure: 10 mN/m, subphase:
water, pH=5.6, 22°C.

This is in contrast to the expected behaviour, although the spatial
requirement of the intermediate state after photoexcitation from which
either cis-TI or trans-TI is formed, must be taken into account instead
of the ground state molecules involved. Another important aspect should
also be considered: molecular mobility and chromophore orientation.
According to molecular models, trans-cis isomerization should not be
hindered more than cis-trans. In the compressed monolayer, however, the
cis-TI chromophore might have a particular orientation and environment
which is quite different from that of the trans-TI and cannot be
regained after relaxion of the system following the photoisomerization.
The orientation of the cis-TI chromophore is different from that of
the trans-TI, according to measurements of the reflection spectra with
plane polarized light at an angle of incidence of 45°. A more detailed
investigation including transient measurements at the air-water interface may be required in order to clarify the role of the molecular
organization.

2.3 Photochemical generation of a two-dimensional compressional wave

The cis-trans photoisomerization of the thioindigo derivative in mixed monolayers caused a decrease in area at constant surface pressure. On photoisomerization of the surface active spiropyran SP (structure see Fig. 3) to the merocyanine form in mixed monolayers with octadecanol, SP:OD = 1:5, the area of the monolayer increases at constant surface pressure (8). If the area is kept constant, the surface pressure rises rapidly on flash excitation of the spiropyran (3).

Therefore, the fast photoreaction provides a possibility to generate a surface pressure jump in a monolayer without the undesired direct mechanical excitation of the aqueous bulk phase. The transmission of the resulting compressional wave across a monolayer is of particular interest with respect to the dynamics of monolayers and membranes and the dependence of the damping of such waves on the structure. Indeed, the compressional wave should induce structural changes in a monolayer, and the kinetics of phase transitions may be investigated with this method.

On the other hand, such a structural change can be used to indicate the arrival of the compressional wave after generating it in a section of the monolayer remote from the place of detection. If the phase transition is correlated with a spectral change as in the case of surface pressure dependent dye association, the resulting change in reflectivity of the water surface can be used for detection of the compressional wave. An arrangement for the study of the transmission

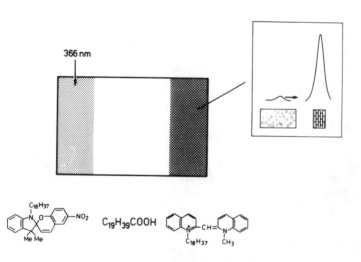

Figure 3. Arrangement for the study of the transmission of compressional waves in monolayers (schematic). Monolayer sections from left to right: generating layer spiropyran SP:OD =1:3; transmitting layer arachidic acid; detection layer cyanine CY:C18 = 1:1. The light reflection of the dye layer is increased on arrival of the compressional wave.

of compressional waves across monolayers is shown schematically in Fig. 3. On the left hand side, lightly shaded area, the compressional wave is generated by photoisomerization of the spiropyran. The wave travels across the monolayer of arachidic acid, middle section, and finally reaches the monolayer of the cyanine dye, whose light reflection is increased in the spectral range of the aggregate band on arrival of the compressional wave.

With the output of the photomultiplier tube connected via a 1 kΩ resistor to a storage oscilloscope, the change in light reflection from the detection monolayer can be recorded after flash excitation of the generating monolayer. The reponse of the system is shown in Fig. 4. The upper trace was obtained with a 5 cm wide transmission monolayer of arachidic acid, the lower trace with 20 cm distance between generating and detection monolayer. A short delay between flash and onset of the reflection change is clearly seen which is larger for the lower trace as compared to the upper trace. Further, it can be immediately seen that the change is slower and smaller in case of the lower trace. This indicates that the high frequency components of the compressional wave spectrum generated by the fast photoreaction have been damped to a larger extent than the small frequency components during transmission across 20 cm of an arachidic acid monolayer. From the different shapes of the two traces it must be concluded that at least in case of the lower trace, the kinetics of the phase transition used for optical detection is not determining the form of the recorded surface pressure jump.

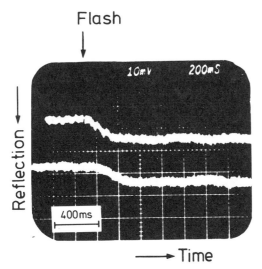

Figure 4. Oscilloscope traces representing the change in reflectivity of the detection monolayer (see Fig. 3) after photochemical generation of a compressional wave. Width of the transmission layer (arachidic acid) 5 cm, upper trace; 20 cm, lower trace. Surface pressure: 10 mN/m; subphase: water, pH=5.6, 22°C.

CONCLUSION

Photochemical reactions at the air-water interface can be investigated
with the reflection method. Using plane polarized light under oblique
incidence, the orientation of transition moments can be evaluated,
which may change in the course of a photochemical reaction. Structural
changes at the air-water interface can be used to generate surface
compressional waves without direct mechanical excitation of the bulk.
The transmission of these waves can be investigated optically by
following a surface pressure dependent phase transition associated with
a change of the absorption spectrum.

REFERENCES

(1) G. L. Gaines, Jr., Insoluble Monolayers at Liquid-Gas
     Interfaces, Interscience Publ., New York, 1966.
(2) D. G. Whitten, Angew. Chem. Intern. Ed. Engl. 18, 440 (1979).
(3) D. Möbius, "Organic Photochemical Reactions in Monolayers and
     Monolayer Systems", in M. A. Fox (Ed.), Organic Phototrans-
     formations in Non-Homogeneous Media, ACS Symposium Series 278,
     Washington 1985, p.113.
(4) H. Grüniger, D. Möbius and H. Meyer, J. Chem. Phys., 79, 3701
     (1983).
(5) D. Möbius, H. Bücher, H. Kuhn and J. Sondermann, Ber. Bunsenges.
     Phys. Chem., 73, 845 (1969).
(6) D. Möbius, M. Orrit, H. Grüniger and H. Meyer, Thin Solid Films,
     submitted.
(7) M. Orrit, D. Möbius, H. Grüniger and H. Meyer, in preparation.
(8) D. Möbius and H. Grüniger, "Signal Transduction in Monolayers
     After Photo-induced Molecular Reorganisation", in M. J. Allen and
     P. N. R. Usherwood (Eds.), Charge and Field Effects in
     Biosystems, Abacus Press, Tunbridge Wells, 1984, p.265.
(9) J. Heesemann, J. Am. Chem. Soc., 102, 2167 (1980).
(10) D. G. Whitten, Angew. Chem. Intern. Ed. Engl., 18, 440 (1979).
(11) J. Sondermann, Liebigs Ann. Chem., 749, 183 (1971).
(12) E. E. Polymeropoulos and D. Möbius, Ber. Bunsenges. Phys. Chem.,
     83, 1215 (1979).
(13) P. Fromherz, Rev. Sci. Instrum., 46, 1380 (1975).

GEOMETRICAL AND ELECTRONIC CONSTRAINTS IN REDOX INTERCALATION SYSTEMS

R. Schöllhorn
Technische Universität Berlin
Institut für Anorganische und Analytische Chemie
Strasse des 17. Juni 135
1000 Berlin 12
West Germany

ABSTRACT. The intercalation of guest species into solid host lattices
via electron/ion transfer processes at low temperatures provides an
interesting example for a strong correlation between structure, electronic properties and chemical reactivity of the host lattice and the
nature and concentration of guest species. The mutual constraints between host and guest are discussed under various aspects.

INTRODUCTION

The concept of the well-known inclusion compounds (e.g. hydrate chlath-
rates, urea inclusion compounds (1) is based on the geometrical distin-
ction between a solid molecular "host" structure and molecular "guest"
species located inside the host lattice with weak interactions between
host and guest. The guest species are localized at their lattice sites
i.e. they cannot undergo translational motion. Inclusion compounds are
usually formed by the constituent molecular species from homogenous
solution; the specific host framework is stable only in the presence
of the guest species. A host/guest concept is similarly used for the
description of the formation and structure of the so-called intercalation compounds. However, in the latter case the structure of the host
lattice is stable in the absence of guest species which can be intercalated reversibly into the bulk of the host via a system of vacant
lattice sites that now must be arranged in a way that a continuous
diffusion path is provided across the whole volume of the host structure. Intercalation processes are thus solid state reactions which
proceed at low temperatures and therefore show a series of characteristic phenomena usually not found for solid state reactions proceeding at high temperatures under equilibrium conditions.
    A most interesting group among the solid compounds able to under-
go intercalation reactions are those which exhibit electronic conductivity, since they show a particularly strong correlation between
structure, electronic properties and chemical reactivity of the host
lattice and the nature and concentration of the inserted guest species
(2,3). A simple reaction scheme for this type of intercalation proces-

R. Setton (ed.), Chemical Reactions in Organic and Inorganic Constrained Systems, 323–340.
© *1986 by D. Reidel Publishing Company.*

ses is given by eq. 1.

$$xA^+ + xe^- + Z \longrightarrow A_x^+[Z]^{x-} \tag{1}$$

Cationic guest species are entering the lattice and occupy vacant sites, while electrons, that are provided by a coupled chemical or electrochemical system are synchronously taken up filling energetically suitable levels in the band structure of the host which itself becomes a macroanion in chemical terms. These processes may be rather complex and are correlated with specific mutual constraints between host and guest under various aspects.

A particular point with respect to the uptake of potential guest species is the geometrical aspect. Basically the number of available sites and the dimensions of the diffusion channels determine the maximum possible stoichiometry and the limiting size of the guest; both factors depend upon the specific host lattice structure (Fig. 1). For framework structures with intersecting or parallel isolated lattice channels the effective minimum diameter along the diffusion path represents an upper limit for the uptake of guest species. In contrast, host lattices with layer or chain type structures may obviously adapt flexibly to the dimensions of the guest species by an increase in the interlayer or interchain spacing, respectively. Interlayer spacings up to 5000 pm have been measured for two-dimensional host structures. The sites occupied depend on the nature and size of the guest cation, the guest ion concentration and the ionicity of the host lattice (4). The interaction host/guest in terms of chemical bonding may vary from essentially ionic to predominantly covalent. Strongly electropositive guest ions (e.g. alkali ions) tend to reside on octahedral or trigonal prismatic sites in structurally simple host lattices, while highly polarizable ions (e.g. Cu, Ag) prefer tetrahedral sites. Site changes e.g. octahedral $\rightleftharpoons$ trigonal prismatic or octahedral $\rightleftharpoons$ tetrahedral can be observed with changes in stoichiometry. Similarly temperature dependent transitions order $\rightleftharpoons$ disorder in the guest ion sublattice are frequently found.

In the course of the intercalation rather small dimensional changes are observed for the host lattice matrix with changes in guest concentration; in the case of low dimensional host lattices, however, significant overall changes of certain lattice parameters may appear due to changes in interlayer or interchain distances depending upon the size and arrangement of the guest species.

Characteristic geometrical transformations are found for layered systems (Fig. 2). Stacking changes, i.e. alterations in the ordering sequence of the layer matrix units perpendicular to the basal planes, may occur either initially upon intercalation or at higher critical concentrations of guest species resulting in changes of the coordination geometry of the guest ions. At high interlayer spacings stacking changes may lead to total one-dimensional disorder of layered systems and finally to colloid states /3,5,6/. A second specific transformation of layered hosts in the course of intercalation reactions via electron/ ion transfer is the appearance of staging phenomena. At low concentrations of guest ions a homogenous distribution of guest ions over all

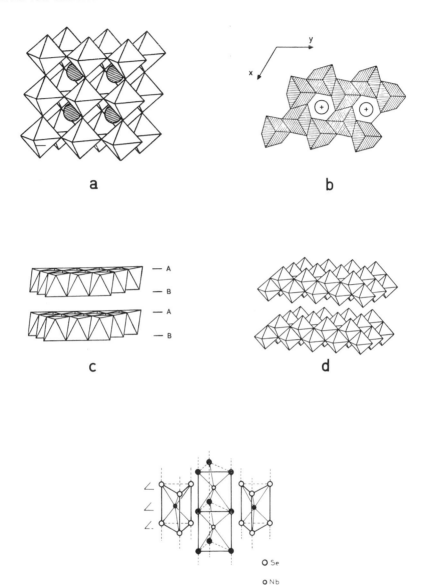

Fig. 1. Scheme of host lattices with different structural dimensionality. Framework structures: (a) WO_3, (b) Nb_3S_4 (spheres are indicating guest ion positions on lattice sites); layered structures: (c) TiS_2, (d) MoO_3; chain structure: (e) $NbSe_3$

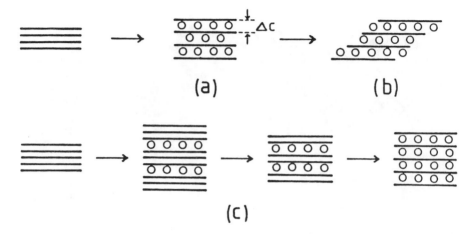

Fig. 2. Layered host lattices; major geometrical changes upon intercalation. (a) change in interlayer spacing, (b) change in stacking
modification, (c) intermediate phases at low guest concentrations
(sequence: empty host lattice, third stage phase, second stage phase,
first stage phase

available van der Waals gaps would result in an unfavourable arrangement and the system may relax energetically via forming intermediate
phases with ordered sequences of filled and unfilled van der Waals
gaps, leading to higher concentrations of guest species in restricted
spatial areas of the solid. A more realistic model which is supported
by various experimental data describes staging in terms of domain formation (7). The appearance of staging (and of two phase ranges in the
course of intercalation reactions) has been explained in terms of lattice strain mediated repulsive interactions between inserted guest cations (8,9).

Under equilibrium conditions the initial intercalation in layered
systems is usually a two phase reaction. In geometric terms this process can be described to start at the basal planes of a crystallite,
the reaction boundary zones subsequently progress from both basal planes of the crystal towards the centre (Fig. 3). For kinetic reasons,
however, this reaction is usually more complex (10).

A second principal point with respect to intercalation via elec-
tron/ion transfer concerns the electronic aspect. The simple rigid
band model puts a constraint on the upper limit for the stoichiometry
that is reached, when energetically favourable bands are filled up with
electrons. There is thus a competition between the geometric restriction
(number of available sites) and the electronic control (number of electrons acceptable by the conduction band) which can be described in a
simple equation (11) that does, however, not take into account specific

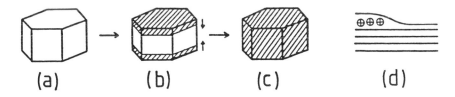

(a)   (b)   (c)   (d)

Fig. 3. Layered host lattices: scheme of phase boundary region progres-
sing perpendicular to basal planes in the initial two phase region
under equilibrium conditions (a-c); scheme of elastic deformation of
layer units upon intercalation (d)

interactions host/guest and guest/guest and their variation with stoichiometry or changes in band structure. The extent of electron transfer in one phase regions is directly correlated with the small changes
in the dimensions of the host lattice matrix. Recently two border cases
have been observed that concern the existence of attractive guest/guest
interaction via partial charge transfer at high guest concentration and
the observation of stoichiometry dependent changes in the oxidation
state of the guest cation in correlation with the absence of charge
transfer to or from the host matrix (12,13).
    A third important point pertains to specific constraints exerted
by the host matrix upon the dynamics of intercalated species and is related to particular anisotropy phenomena. These concern the restricted
dimensionality of translational motion of simple cations as well as
the temperature dependent anisotropy of rotational motion of polyato-
mic guest species. We shall discuss in the following section selected
examples of intercalation systems to illustrate the basic correlations
given in the foregoing text.

REACTIONS OF LAYERED HOST LATTICES

A particularly large variety of mutual interactions host/guest and related geometrical constraints are encountered in the case of host lattices   with two-dimensional structure (2,3,14-18). Layered systems
with electronic conductivity are able to intercalate atomic ions, molecular ions of different geometry and dimensions as well as solvated
ions. Because of their structural simplicity, high electronic conductivity and the possibility to vary both M and X, hexagonal transition
metal dichalcogenides MX_2 (e.g. TiS_2, NbS_2) have been widely studied
under different aspects as host lattices. The reversible intercalation
of atomic ions (alkali metals, Cu, Ag, Tl etc.) leads to ternary phases
A_xMX_2 (eq. 2)

$$xA^+ + xe^- + MX_2 \rightleftharpoons A_x[MX_2]^{x-} \qquad (2)$$

some of which are of interest with respect to their use as electrodes
for secondary batteries (19). At low concentrations of guest species
in most cases staging is observed for intercalation at ambient temperature (20) with a sequence

$$3^d \text{ stage} \rightarrow 2^d \text{ stage} \rightarrow 1^{st} \text{ stage}$$

while for reactions at higher temperature (250 ^oC) 4th stage compounds
have been found (4, 21). Lithium ions are exceptional: because of the
small ionic radius and the corresponding minor change in interlayer
spacing no staging occurs under equilibrium conditions, but concentration dependent intermediate ordering states of the Li sublattice are
observed (22). Pseudo-staging described recently for $Li_x TiS_2$ is due to
kinetic reasons (10). Staging intermediates are also absent, if the
host layers have a higher thickness (e.g. Ta_2S_2C) which has been attributed to energetic reasons (activation energy for elastic deformation
of the host layers (23). Since there is only one octahedral (O) or
trigonal prismatic site (TP) per MX_2 unit, the geometrical restraint
would lead to a limiting composition A_1MX_2, while under electronic
aspects the maximum composition allowed should be A_2MS_2 in the case of
group IVb chalcogenides (filling of the t_{2g} band with two electrons).
Experimentally both compositions can be obtained (24). Recently a structural study has been performed on Li_2TiS_2 which shows that the system
relaxes geometrically by a change in coordination of Li from octahedral
to tetrahedral; since the ratio tetrahedral sites/octahedral sites is
2 : 1 for close packing of the anions, the TiS_2 lattice can take up two
Li and two electrons under these conditions. In the case of Li_xMoS_2
rhombohedral lattice distortion results for x = 1; a similar lattice
distortion has been observed for thermal alkali phases A_xMoS_2, A_xWS_2
and is probably correlated with changes in band structure (24).

Strong constraints concerning reactivity and ionic mobility are
found on deviations $M_{1+y}X_2$ from the ideal stoichiometry which are to be
attributed to the increased activation energy for intercalation, since
the additional fraction y of the transition metal atoms occupy positions in the van der Waals gap (19).

Changes in coordination geometry of A (O, TP) are responsible for
stacking changes in the host layers upon intercalation. In the case of
alkali ions, these transformations depend upon the ionic radius, the
stoichiometry (x) and the ionicity of the M-X-bond (4,21,25). Similar
results were obtained in the case of layered alkali oxometallates (26).

Highly polarizable cations, which show partial covalent bonding to
the host lattice like Cu, Ag, prefer tetrahedral lattice sites in thermal phases Cu_xMX_2 and Ag_xMX_2. The intercalation of these ions into
layered dichalcogenides at room temperature proceeds structurally in a
more complex way. Cu_xTiS_2 exhibits an initial one phase region with
0 < x < 0.5 and copper ions on octahedral sites (3,27). In the range
0.5 < x < 0.7 a two phase region appears which is due to a stacking
transition (s = number of MX_2 slabs per unit cell):

$$1s \text{ (octahedral)} \rightleftharpoons 3s \text{ (tetrahedral)}$$

This is followed by a one phase region $0.7 < x < 1$. The phases with $0 < x < 0.5$ and copper on octahedral positions are metastable; $Cu_{0.5}TiS_2$ transforms at 320 °C to the thiospinel with cubic structure:

$$2 \ Cu_{0.5}TiS_2 \ (hex.) \rightarrow CuTi_2S_4 \ (cubic)$$

Similar reactions are found for related copper phases e.g. Cu_xVS_2 which shows six different intermediate phases; $Cu_{0.5}VS_2$ decomposes to the spinel CuV_2S_4 at elevated temperature (28).

Although a large variety of intercalation compounds is known with molecular ions, few detailed structural studies have been performed on these compounds which could give information on the position of the intercalate species. For the metallocene compounds $(Cp_2M^+)_x[MX_2]^{x-}$ it was concluded from NMR data that the Cp_2M^+ units are located between the MX_2 sheets with the planes of the C_5H_5 aromatic rings perpendicular to the layers (29). For the intercalation of pyridinium ions an initial stage was observed with the molecular ring planes parallel to the MX_2 units, while at the maximum composition $(pyrH^+)_{0.25}(pyr)_{0.25}[MX_2]^{0.25-}$ the ring planes are perpendicular to the MX_2 sheets (2,32).

The intercalation of hydrated alkali, alkaline earth and transition metal ions in dichalcogenide host lattices (eq. 3) provides an interes-

$$xA^+ + yH_2O + xe^- + MX_2 \rightleftharpoons A_x^+(H_2O)_y[MX_2]^{x-} \qquad (3)$$

ting model for the study of constrained two-dimensional electrolyte phases (30). The formation of monolayers or bilayers of water between the MX_2 sheets is dependent upon the charge/radius ratio (e/r) of the intercalated cation. For hydronium cations an extended series of water structures is observed with decreasing concentration of H^+ in the electrolyte:

$$\text{monolayer} \rightarrow \text{bilayer} \rightarrow \text{triple layer} \rightarrow \text{colloid state}$$

While mono- and bilayer phases are highly ordered, one finds total one-dimensional disorder for the triple layer phases; at very low acid concentrations due to weak cohesion resulting from the large interlayer spacing the system desintegrates and transforms reversibly to colloid solutions (3,5,6). Other examples have been reported recently for layer and chain type compounds (e.g. V_2O_5 (31), $RuBr_3$ (32)) that are able to form colloid systems upon intercalation. Hydrated intercalation compounds with chalcophilic cations may dehydrate spontaneously (irreversible step) at high cation concentration, e.g.

$$(Ni^{2+})_x(H_2O)_y[NbS_2]^{-2x} \rightarrow Ni_x[NbS_2] + y \ H_2O \qquad (4)$$

due to the strong intercalation of the host lattice chalcogenide anions with the guest cations (32). All of the hydrated phases of dichalcogenides show a transition series 3^d stage/2^d stage/1st stage (30) at lower guest ions concentrations, usually correlated with stacking transitions. The latter are also found on ion exchange reactions which proceed rapidly at room temperature. A simple example is the exchange

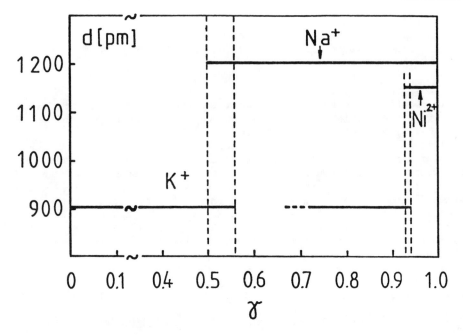

Fig. 4. Ion exchange equilibria K^+/Na^+ and K^+/Ni^{2+} for hydrated layered phases $A^+_{1/3}(H_2O)_y$ $[NbS_2]^{1/3-}$: dependance of interlayer spacing d upon the mole fraction $\gamma(Na)$ and $\gamma(Ni)$, respectively, in the equilibrium electrolyte (30, 32).

equilibrium between two monovalent ions:

$$Na^+_x(H_2O)_y[TiS_2]^{x-} + K^+ \rightleftharpoons K^+_x (H_2O)_y [TiS_2]^{x-} \qquad (5)$$

1s(O)   3s (TP)

These reactions are correlated with a structure induced ion exchange selectivity in the case of cations that form different hydrate structures (Fig. 4) (30). In the titanium sulfide system the K^+ form is a 3s monolayer phase, while in the 1s Na^+ form the cations are located in octahedral positions between bilayers of water. The phase transition, which involves simultaneously (i) a change in interlayer spacing, (ii) a change in the coordination sphere of the guest cation and (iii) a change in the layer stacking sequence, occurs at a critical mole fraction of the ions in a small two phase region and excludes mixed single phases (Na/K) for structural reasons. The mole fraction at which the transition sets in depends upon the valency of the exchangeable ions (Fig. 4).

The reactions of the layered ruthenium(III)chloride $RuCl_3$ are quite similar to those of the transition metal dichalcogenides (33), but in the case of the hydrated phases an interesting electronic re-

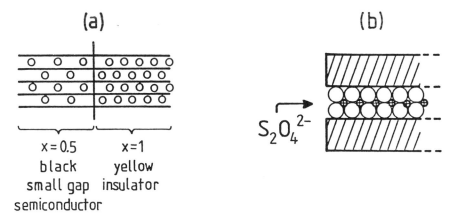

Fig. 5. Formation of $Na_x^+(H_2O)_y[RuCl_3]^{x-}$ from $RuCl_3$: (a) interface semi-conductor (x = 0.5) and insulator (x = 1); (b) scheme of chemical reaction via cointercalation of reducing species into the hydrate layers

striction is observed (Fig. 5). On electrochemical reduction of $RuCl_3$ in aqueous alkali ion electrolytes (eq. 6) the upper limit for y is equal to 0.5. These phases are black mixed valence semiconductors. The

$$xA^+ + xe^- + y\ H_2O + RuCl_3 \rightleftharpoons A_x^+(H_2O)_y[RuCl_3]^{x-} \qquad (6)$$

chemical reduction of $RuCl_3$ in aqueous sodium dithionite $Na_2S_2O_4$ results, however, in the formation of $Na_1^+(H_2O)_y[RuCl_3]^{1-}$, which is a yellow transparent wide gap semiconductor with integral valency of ruthenium. This difference in reactivity can be explained in electronic terms. The electrochemical reaction must terminate at x = 0.5, since the compound with x = 1 is an insulator phase and electron transport across the phase boundary semiconductor/insulator is not possible. The situation is different for $Na_2S_2O_4$: here the electron donating $S_2O_4^{2-}$ ions (or $SO_2^{\cdot-}$ radical anions) can enter in a low concentration (Donnan potential) the interlayer space, since the sodium compound forms a bilayer hydrate with sufficiently large interlayer spacing. The electrons required for formation of $Na_1(H_2O)_y[RuCl_3]$ are now transported via thermal diffusion of the electron donor molecules in the interlayer space, i.e. the low electronic conductivity of the resulting product is not a limiting factor under these conditions.

The observation that neutral molecular species with Lewis base properties (e.g. ammonia or pyridine) and high ionization energies can be intercalated in layered dichalcogenides remained a puzzling problem for some time with respect to the nature of the interaction host lattice/guest species. It could be shown, however, that the limiting condition for intercalation in these hosts, i.e. reaction via electron/ion transfer and formation of ionic structures applies also to these

systems. The reactions proceed via partial irreversible oxidation of
the guest species and formation of protonated molecular ions solvated
by neutral molecules (2,3,12). A different redox mechanism is observed
in the reaction of aqueous Lewis base solutions (e.g. NH_3/H_2O) and
aqueous alkali hydroxides with dichalcogenides. In this case the electrons required are provided by partial irreversible solvolysis of the
host layer matrix leading to redox active free chalcogenide ions. (e.
g. $S^{2-} \rightarrow S_{2/2}^- + e^-$).

For small polar solvent molecules in solvated phases
$A_x^+ (solv)_y [MX_2]^{x-}$ strong and unusual restrictions in the mobility between the layers have been observed. The most simple species that can be
cointercalated along with atomic cations is water. At ambient temperature strong anisotropic mobility was found by NMR studies for H_2O molecules in monolayer hydrates (Fig. 6) (34). The H_2O units can undergo
translational motions to vacant lattice sites, but the proton/proton
vectors remain parallel to the hexagonal c-axis and reorientation of
the C2 axis is thus restricted to the midplane between the adjoining
MX_2 layers. For the series of alkali ions there is a linear correlation
between the enthalpy of hydration of the cations and the temperature
that governs the transition from the rigid lattice state to the region
of ordered activated mobility. Lithium is exceptional and undergoes a
change in coordination geometry. The two spin splitting of the ^{1}H–NMR
signal collapses at low pH values (exchange of alkali cations for H_3O^+)
and at high pH values (cointercalation of K^+OH^- ion pairs) due to rapid

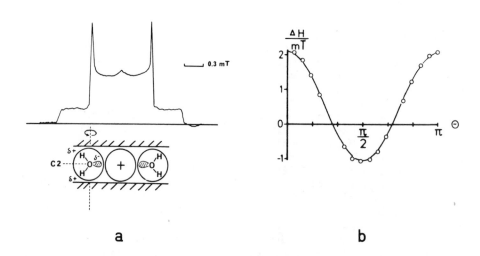

a   b

Fig. 6. Anisotropy of the dynamics of water molecules in hydrated
layered chalcogenides $A_x^+ (H_2O)_y [MX_2]^{x-}$ (monolayer hydrate at 300 K).
(a) ^{1}H–NMR spectrum and structural model; (b) line pair separation
measured as a function of the angle between H_0 and the hexagonal
c-axis of single crystals (34)

proton exchange. In monolayer hydrates the translational motion is two-dimensional in character, while significant contribution from three-dimensional motion appear in bilayer hydrates. Hydrogen disulfide in solvated phases $A_x^+(H_2S)_y[MX_2]^{x-}$ behaves quite similar to water in hydrated dichalcogenides (34).

The corresponding ammonia compounds $(A^+)_x(NH_3)_y[MX_2]^{x-}$ also show strong anisotropy for the motion of NH_3 (35-38). The reorientation of the C3 axis is restricted to the midplane between the layers, but rotation around the C3 axis is activated at ambient temperature. One-dimensional rotation of $-CH_3$ and $-NH_3$ groups has also been observed for methylamines as solvate molecules (38).

FRAMEWORK HOSTS

In contrast with layered systems strong geometric constraints are governing the uptake of guest species in three-dimensional host lattices. While open framework structures are found in the case of insulator host lattices (e.g. zeolites) that allow the insertion of molecular species, all compounds with three-dimensional structures that undergo intercalation via electron/ion transfer provide systems of empty sites with relatively small critical diameters. The intercalation is thus restricted to atomic guest species, i.e. metal ions and hydrogen. An exception is hexagonal tungsten trioxide with parallel lattice channels which is able to intercalate NH_4^+ and NH_3 at higher temperatures. The structure of monoclinic WO_3 provides large cubooctahedral empty sites; the diffusion must proceed, however, via small square windows of oxygen atoms so that only Li^+ and protons can be intercalated at room temperature.

Irreversible major structural transitions have been described in some cases upon intercalation of lithium into framework structures, e.g. the transformation of β-MnO_2 with rutile structure to the tetragonally distorted spinel $Li_x Mn_2O_4$ (39) (Fig. 7). Lithium insertion into the anatase polymorph of TiO_2 yields $Li_{0.5}TiO_2$ which is converted to the spinel $LiTi_2O_4$ at 500 °C (40). Inversely, copper can be removed from the thiospinel $CuTi_2S_4$ by anodic oxidation; the resulting cubic binary phase Ti_2S_4 is a metastable compound that is converted to hexagonal TiS_2 at 500 °C (41). The Cu/Ti/S compounds provide an interesting example for a system in which transformations between layer and framework structures can be achieved via a combination of chemical and thermal reactions in a cyclic process (Fig. 8).

The corresponding removal of Li from $LiTi_2O_4$ does not lead to the binary phase "Ti_2O_4" and ion exchange with H^+ results already at room temperature in the formation of an amorphous phase (40). This is most likely due to differences between Ti/O and Ti/S in the extent of covalent bonding that is governing the kinetics of transformation of these metastable phases to the thermodynamically stable structures.

Most flexible three-dimensional host lattice systems with respect to the variety of guest cations that can be inserted are the molybdenum cluster chalcogenides Mo_6S_8 and Mo_6Se_8 which provide a system of vacant intersecting lattice channels along the rhombohedral axes of

the structure. Cations which can be intercalated by electron/ion trans-
fer into Mo_6S_8 at ambient temperature are Li^+, Na^+, Mn^{2+}, Fe^{2+}, Co^{2+},
Ni^{2+}, Cu^+, Zn^{2+} and Cd^{2+} (2,3). Since Ag^+ cannot be inserted, the critical radius r_c (in terms of hard sphere ionic radii) is $r_{Cd} < r_c < r_{Ag}$.
Mo_6Se_8 which has somewhat larger lattice parameters may in addition
intercalate Ag^+ ions so that $r_{Ag^+} < r_c < r_{K^+}$ applies.

Since the number of available vacant lattice sites for small ions
is considerably larger in these phases than the maximum stoichiometries actually observed, it can be concluded that the electronic factor
governs the limit of uptake of guest ions. From ionic bonding models
as well as from band structure calculations it was derived that the
maximum number of electrons that the Mo_6S_8 system can accept under the
assumption of a rigid band model is equivalent to four, i.e. $[Mo_6S_8]^{4-}$
(42). The experimental data show in fact that the upper limit of intercalation corresponds formally to this value, e.g.
$(Cu^+)_4[Mo_6S_8]^{4-}$, $(Ni^{2+})_2[Mo_6S_8]^{4-}$. These formulae are based, however,
on the assumption of a quantitative transfer of the electrons to the
host lattice according to

$$4A^+ + 4e^- + Mo_6S_8 \rightarrow (A^+)_4[Mo_6S_8]^{4-} \qquad (7)$$

$$2A^{2+} + 4e^- + Mo_6S_8 \rightarrow (A^{2+})_2[Mo_6S_8]^{4-} \qquad (8)$$

Recent studies on these systems yielded evidence that this assumption
needs not necessarily be true for all guest cations and different concentrations of the latter (12, 43). For lithium intercalation it was
found by ^7Li-NMR studies that the line phase $Li_1Mo_6S_8$ can be interpreted in terms of quantitative charge transfer according to
$(Li^+)_1[Mo_6S_8]^{1-}$; $Li_4Mo_6S_8$, however, shows a very strong Knight shift
indicating partial charge transfer and attractive guest/guest interactions via electron deficient Li/Li bonding. Similarly, the intercalation of Ni^{2+} from aqueous solution results in the formation of
$(Ni_2H_2)[Mo_6S_8]$ with a formal transfer of 6 electrons according to

$$2 Ni^{2+} + 2 H^+ + 6e^- \rightarrow (Ni_2H_2)[Mo_6S_8] \qquad (9)$$

Because of the restriction that the Mo_6S_8 band can accept only a maximum of four electrons, it must be concluded that the extra electrons
are stored in Ni/H and Ni/Ni guest/guest bonds providing again attractive interaction between the intercalated species.

The limiting situation with no charge transfer during the redox
reaction has been found in studies on the deintercalation of $Tl(V_5S_8)$
which consists of a V_5S_8 matrix with isolated parallel lattice channels
occupied by Tl^+ cations. The reaction proceeds according to

$$(Tl^+)[V_5S_8]^- \rightarrow (Tl^{3+})_{1/3}[V_5S_8]^- + 2/3 \ Tl^{3+} + 2e^- \qquad (10)$$

The redox process involves only changes in the oxidation state of the
guest cation, while the negative excess charge of the V_5S_8 matrix remains constant (13, 44).

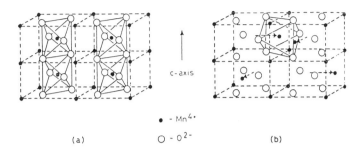

(a)

● - Mn⁴⁺

○ - O²⁻

(b)

Fig. 7. The transformation β-MnO₂ → Li[Mn₂O₄]. Tetragonal rutile structure (a) and cation displacement leading to the cubic [MnO₄] framework (b) (39)

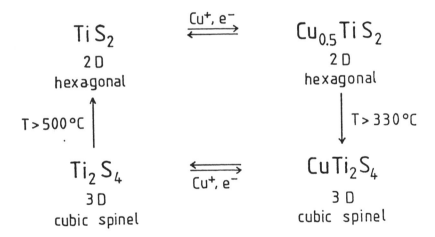

Fig. 8. Transformations between layer and framework structures in the Ti/S system (32, 41)

Monovalent cations in Mo_6S_8 show rather high diffusion coefficients e.g. $\tilde{D} = 1.1 \cdot 10^{-7}cm^2s^{-1}$ for $Li_1Mo_6S_8$ at 300 K. The fact that unsolvated small cations of the transition metals also exhibit relatively high diffusion coefficients at room temperature (e.g. Ni in $Ni_1Mo_6S_8$ $\tilde{D} = 1.7 \cdot 10^{-9}cm^2s^{-1}$, Zn in $Zn_1Mo_6S_8$ $\tilde{D} = 1.5 \cdot 10^{-9}cm^2s^{-1}$ in aprotic electrolytes at 300 K) is not well understood. Several models have been proposed to account for this unusual observation (43).

ELECTRON PROTON TRANSFER

Protons are the smallest guest ions that can be intercalated into host lattices of different dimensionality with suitable band structure (eq. 11). As a consequence of partial covalent bonding to the host lattice

$$xH^+ + xe^- + [Z] \rightarrow H_x[Z] \qquad (11)$$

the mobility of protons in the resulting hydrogen bronzes $H_x[Z]$ is relatively low in most cases.

Although a large series of these compounds is known (3), few systems have been studied in more detail, among these the hydrogen bronzes derived from the layered molybdenum(VI)oxide MoO_3 (45, 46). The latter is an illustrative example for the strong influence of kinetics on the reaction and anisotropic dynamics of hydrogen. Four phases

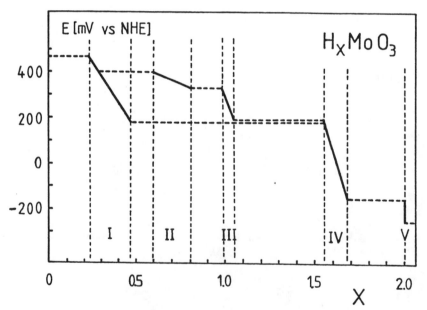

Fig. 9. Phase diagram and hysteresis of the hydrogen bronzes H_xMoO_3 at 300 K (potential E versus stoichiometry) (32)

H_xMoO_3 have been described so far, recently an additional monoclinic phase (II) with $0.6 < x < 0.8$ has been found (47). Figure 9 gives an overview on the phase diagram at 300 K. Electrochemical studies indicate a strong hysteresis: phases II und III can only be obtained on anodic oxidation of compounds with higher hydrogen content and do not appear on cathodic reduction. Figure 10 shows the unit cell volumes of the different phases; it is easy to see that on cathodic reduction via

$$MoO_3 \rightarrow I \rightarrow IV \rightarrow V$$

the transition I → IV is related to a strong volume increase and a corresponding lattice distortion. The activation energy for the formation of IV is obviously lower as compared to the formation of III and II, so that the latter phases can only be obtained on the way back which proceeds via the sequence IV → III → II → I. No kinetic influence is found for the transition IV ⇌ V. Phases above the potential of the normal hydrogen electrode are metastable with respect to slow decomposition into H_2 and hydrogen bronzes H_xMoO_3 with lower x values at room temperature. The initial cathodic reduction in the two phase range $MoO_3 \rightarrow H_{0.2}MoO_3$ is possible, since the insulator MoO_3 transforms to $H_{0.2}MoO_3$ which shows metallic conductivity. Anodic reoxidation in this region is, however, kinetically hindered due to the metal/insulator transition at the phase boundary which inhibits further transport of electrons (43).

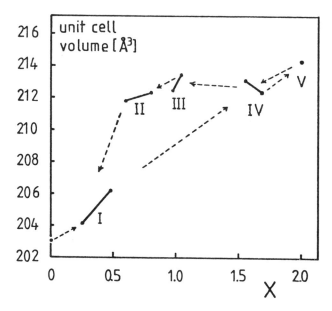

Fig. 10. Unit cell volumes of hydrogen bronzes H_xMoO_3 and reaction hysteresis (32)

Because of its small size and the well-known appearance of tunneling effects one would assume that the translational mobility of hydrogen is approximately isotropic. If not so, one could expect two-dimensional mobility of protons in the layered $H_x MoO_3$ bronzes. H-NMR
studies revealed, however, that $H_{0.3} MoO_3$ is a one-dimensional proton
conductor (48). The hydrogen atoms, which are located in intralayer
split positions, move parallel to the b-axis of MoO_3. A geometric explanation for the phenomenon is based on the fact that the MoO_3 layers
themselves have anisometric structure: chains of edge sharing MoO_6
octahedra are connected via corner sharing of the octahedra. It is
most likely that $H_{0.3} MoO_3$ is also a one-dimensional electronic conductor.

One-dimensional proton conductivity has also been observed for
hexagonal $H_{0.3} WO_3$ which is characterized by large parallel lattice
channels (48). The location of the hydrogen atoms has not yet been
determined for this compound. Neutron diffraction and ^{1}H-NMR studies
on the layered chalcogenide hydrogen bronze $H_x NbS_2$ revealed stoichiometry dependent location and mobility of hydrogen (37). At concentrations up to x = 0.1 the hydrogen atoms occupy positions in the van der
Waals gap, while at higher concentrations (x ~ 0.8) hydrogen is located
inside the NbS_2 layers on trigonal prismatic sites (50); this transition is correlated with a strong increase in the activation energy for
hydrogen motion.

Although the low lattice energies and the very low electronic
conductivities of molecular solids seem to exclude this group of compounds for topotactic redox reactions, it has been demonstrated recently that electron/ion transfer reactions may occur also in these
systems (3.49). The limiting conditions are (i) small guest cations
(ii) a mechanism that provides electron transport between the molecular species in at least one lattice direction and (iii) diffusion
pathways of low energy for the guest cations. The studies performed
so far pertain mainly to reversible electron/proton transfer reactions
in partially oxidized planar metal complex compounds with columnar
structure and one-dimensional electronic interaction between the metal
atoms. The proton transfer proceeds via molecular species that can be
reversibly protonated, e.g. HSO_4^- ions in the partially oxidized form
of Magnus Green Salt $[Pt(NH_3)_4 PtCl_4]^{0.5+} (H_{1.5} SO_4)^{0.5-} (H_2O)_y$.

CONCLUSIONS

Intercalation reactions proceeding via electron/ion transfer are rather complex systems and it is easy to understand that our knowledge
on the geometrical and electronic constraints which are related to the
upper stoichiometry limit, the appearance of intermediate phases and
structural transformations, the interactions host/guest and guest/guest
and the dynamics of guest species, is rather modest in many respects.
One major problem is the dominant influence of kinetics due to non-
equilibrium states which is quite naturally a consequence of the fact
that these solid state reactions proceed at low temperatures.

REFERENCES

1) J.L. Atwood, J.E.D. Davies, D.D. Mac Nicol, Inclusion Compounds, Vol. I, Academic Press, New York (1984)
2) R. Schöllhorn, Angew. Chem. 92, 1015 (1980); Angew. Chem. Int. Ed. Engl. 19, 983 (1980)
3) R. Schöllhorn, Inclusion Compounds, J.L. Atwood, J.E. D. Davies, D.D. Mac Nicol, eds., Vol. I., Academic Press, New York, (1984), p. 249
4) J. Rouxel, Intercalated Layered Materials, F. Lévy, ed., D. Reidel, Dordrecht (1979), p. 201
5) A. Lerf, R. Schöllhorn, Inorganic Chem. 16, 2950 (1977)
6) D.W. Murphy, G.W. Hull, J. Chem. Phys. 62, 973 (1975)
7) A. Hérold, Intercalated Layered Materials, F. Lévy, ed., D. Reidel, Dordrecht (1979), p. 321
8) W.R. McKinnon, R.R. Haering, Solid State Ionics 1, 111 (1980); J.R. Dahn, D.C. Dahn, R.R. Haering, Solid State Comm. 42, 179 (1982)
9) S.A. Safran, Phys. Rev. Letters 44, 937 (1980); Synthetic Metals 2, 1 (1980)
10) T. Butz, A. Lerf, J.O. Besenhard, Rev. Chim. Min. 21, 556 (1984)
11) M.B. Armand, Materials for Advanced Batteries, D.W. Murphy, J. Broadhead, B.C.H. Steele, eds., Plenum Press, New York (1980)
12) R. Schöllhorn, Comments in Inorg. Chem. 2, 271 (1983)
13) W. Schramm, R. Schöllhorn, H. Eckert, W. Müller-Warmuth, Mat. Res. Bull. 18, 1283 (1983)
14) Intercalated Layered Materials (Physics and Chemistry of Materials with Layered Structures, Vol. VI), F. Lévy, ed., D. Reidel, Dordrecht/Boston (1979)
15) J. Rouxel, Rev. in Inorg. Chem. 1, 245 (1979)
16) Physica 99B, (Proc. Int. Conf. on Layered Materials and Intercalates Nijmegen, Netherlands), C.F. van Bruggen, C. Haas, H.W. Myron, eds., North Holland, Amsterdam (1979)
17) Intercalation Chemistry, M.S. Whittingham, A.J. Jacobson , eds., Academic Press, New York (1982)
18) Physics of Intercalation Compounds, L. Pietronero, E. Tosatti, eds., Springer, Berlin/New York (1982)
19) M.S. Whittingham, Progr.Solid State Chem. 12, 41 (1978)
20) T. Hibma, J. Solid State Chem. 34, 97 (1980); Physica 99B, 136 (1980)
21) J. Rouxel, Physica 99B, 3 (1980)
22) A.H. Thompson, Phys. Rev. Letters 40, 1511 (1978); J.R. Dahn, R.R. Haering, Solid State Comm. 40, 245 (1981)
23) R. Schöllhorn, W. Schmucker, Z. Naturforsch. 30b, 975 (1975)
24) R. Schöllhorn, M. Kümpers, D. Plorin, J. Less-Common Metals 58, 55 (1978); D.W. Murphy, J.N. Carides, J. Electrochem. Soc. 126, 341 (1979); J.R. Dahn, R.R. Haering, Mat. Res. Bull. 14, 1259 (1979); M. Kümpers, R. Schöllhorn, Z. Naturforsch. 35b, 929 (1980); R. Schöllhorn, U. Bethel, W. Paulus, Rev. Chim. Min. 21, 545 (1984)
25) J. Rouxel, Ann. Chim. (1976), 199
26) C. Delmas, J.J. Braconnier, C. Fouassier, P. Hagenmuller, Solid State Ionics 3/4, 165 (1981)

27) R. Schöllhorn, Physics of Intercalation Compounds, L. Pietronero, E. Tosatti, eds., Springer Berlin/New York (1981), p. 33
28) P. Rathner, Dissertation, University of Münster (1985); P. Rathner, R. Schöllhorn, in preparation
29) M.B. Dines, Science 188, 1210 (1975)
30) R. Schöllhorn, Intercalation Chemistry, M.S. Whittingham, A.J. Jacobson, eds., Academic Press, New York (1982)
31) P. Aldebert, N. Baffier, J.J. Legendre, J. Livage, Rev. Chim. Min. 19, 485 (1982)
32) R. Schöllhorn, P. Buller, R. Steffen, N. Janzen, to be published
33) R. Schöllhorn, R. Steffen, K. Wagner, Angew. Chem. 95, 559 (1983); Angew. Chem. Int. Ed. Engl. 22, 555 (1983)
34) U. Röder, W. Müller-Warmuth, R. Schöllhorn, J. Chem. Phys. 70, 2864 (1979); J. Chem. Phys. 75, 412 (1981); U. Röder, W. Müller-Warmuth, H.W. Spiess, R. Schöllhorn, J. Chem. Phys. 77, 4627 (1982)
35) B.G. Silbernagel, M.B. Dines, F.R. Gamble, L.A. Gebhard, M.S. Whittingham, J. Chem. Phys. 65, 1906 (1976); B.G. Silbernagel, F.R. Gamble, J. Chem. Phys. 65, 1914 (1976)
36) R. Schöllhorn, H.D. Zagefka, Angew. Chem. 89, 193 (1977); Angew. Chem. Int. Ed. Engl. 16, 199 (1977)
37) W. Müller-Warmuth, E. Wein, R. Schöllhorn, to be published
38) W. Molitor, W. Müller-Warmuth, H.W. Spiess, R. Schöllhorn, Z. Naturforsch. 38a, 237 (1983)
39) J.B. Goodenough, M.M. Thackeray, W.I.F. David, P.G. Bruce, Rev. Chim. Minérale 21, 435 (1984)
40) D.W. Murphy, M. Greenblatt, S.M. Zahurak, R.J. Cava, J.V. Waszak, G.W. Hull, R.S. Hutton, Rev. Chim. Minérale 19, 441 (1982)
41) R. Schöllhorn, A. Payer, Angew. Chem. 97, 57 (1985); Angew. Chem. Int. Ed. Engl. 24, 67 (1985)
42) K. Yvon, Current Topics in Materials Science, E. Kaldis, ed., North Holland, Amsterdam, Vol. 3 (1979), p. 83
43) R. Schöllhorn, Pure & Appl. Chem., 56, 1739 (1984)
44) H. Eckert, W. Müller-Warmuth, W. Schramm, R. Schöllhorn, Solid State Ionics 13, 1 (1984)
45) O. Glemser, G. Lutz, Z. Anorg. Allg. Chem. 264, 17 (1951); O. Glemser, U. Hauschild, G. Lutz, Z. Anorg. Allg. Chem. 269, 93 (1952); O. Glemser, G. Lutz, H. Meyer, Z. Anorg. Allg. Chem. 285, 173 (1956);P.G. Dickens, J.H. Moore, D.J. Neild, J. Solid State Chem. 7, 24 (1973); P.J. Wiseman, P.G. Dickens, J. Solid State Chem. 6, 374 (1973)
46) R. Schöllhorn, R. Kuhlmann, J.O. Besenhard, Mat. Res. Bull. 11, 83 (1976)
47) R. Schöllhorn, F. Rüschendorf, to be published
48) Cl. Ritter, W. Müller-Warmuth, H.W. Spiess, Ber. Bunsenges. Phys. Chem. 86, 1101 (1982)
49) R. Schöllhorn, K. Wagner, H. Jonke, Angew. Chem. 93, 122 (1981); Angew. Chem. Int. Ed. Engl. 20, 109 (1981)
50) C. Riekel, H.G. Reznik, R. Schöllhorn, C.J. Wright, J. Chem. Phys. 70, 5203 (1979); A.R. Kleinberg, A.J. Jacobson, B.G. Silbernagel, T.R. Halbert, Mat. Res. Bull. 15, 1541 (1980)

TRANSITION METAL BRONZES: Properties and Reactivity

Paul A. Sermon
Department of Chemistry,
Brunel University,
Uxbridge, Middx,
UB8 3PH, UK

ABSTRACT. The properties and reactivities of non-stoichiometric bronzes of transition metal oxides, sulphides and fluorides are reviewed with special attention to their transient and sustainable catalytic and electrochemical activity.

1. INTRODUCTION

The properties of tungsten bronzes are used to illustrate those of bronzes more generally.

1.1. Properties

Non-stoichiometric bronzes of tungsten X_xWO_3 were prepared [1] well over a hundred years ago by insertion of X cations into the interstitial channels of the host trioxide and electrons into the oxide conduction band CB. The product is a nonstoichiometric semi- or metallic conductor whose colour, conductivity, electrical, spectral and crystallographic properties vary with the value of x (see Table 1), and also the nature of X and that of the host oxide lattice used.

Table 1: Properties of Some Oxides and Bronzes

WO_3: yellow, monoclinic (290–593K), diamagnetic, semiconductor, stable in acid, soluble in alkali, metastable hexagonal phase [2]

MoO_3: white, orthorhombic, weakly paramagnetic, layer structure with octahedra edge sharing on two levels

Na_xWO_3: yellow-green-blue-red-brown colour depending upon x, metallic conductivity at x> 0.25, resistivity decreases as x increases, monoclinic-tetragonal-cubic lattice (depending upon x) which contains triagonal (56pm), square (96pm) and hexagonal (163pm) interstitial channels which may be occupied by Na cations while electrons are donated to the oxide CB

R. Setton (ed.), Chemical Reactions in Organic and Inorganic Constrained Systems, 341–360.

The trioxide WO_3 is normally an antiferroelectric semiconductor with a
high dielectric constant at room temperature and a monoclinic structure
being formed from an arrangement of WO_6 octahedra (see Figure 1). WO_3 is
among the materials with a framework host lattice of a distorted
ReO_3-type. This can be (a) reduced to a lower oxide WO_z ($0<z<3$) with
greater edge-sharing of octahedra and the formation of crystallographic
shear CS planes, (b) have W^{6+} ions replaced by other ions in equivalent
lattice positions (examples of such crystallographic shear phases are
$Mo_xW_{1-x}O_3$ and $(Mo_xW_{1-x})_nO_{3n-1}$ [3]), and (c) converted to 'metallic'
bronzes X_xWO_z where X sits in $\triangledown$ sites in hexagonal (H), trigonal (T)
or square (S) interstitial channels of the oxide lattice in a positive
oxidation state having donated at least a fraction of its valence
electrons to the oxide CB. Many examples of the latter bronzes are now
known [4] where the maximum value of x is determined by temperature, the
ionic size and charge of M, but at constant temperature as x increases:
(a) the resistivity of the trioxide decreases [5], (b) the monoclinic
structure converts to tetragonal to pseudo-cubic and expands (where
lattice expansions are only linear with small increments of x [6]), (c)
the concentration of shear planes increases and (d) the Fermi level Ef
rises.

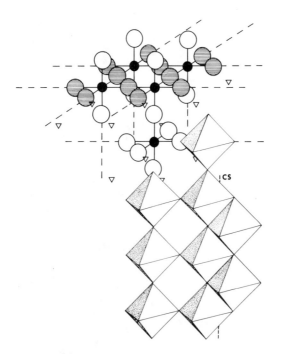

Figure 1: WO_6 octahedra (W^{6+}:●; O^{2-} in the surface:⊖; O^{2-} above or
below the surface:○) corner-shared except at CS plane. $\triangledown$ denotes sites
which may be occupied by cations inserted during bronze formation.

However, the properties of such bronzes is by no means as unequivocal as this statement suggests. Samples of sodium tungsten bronzes Na_xWO_3 can be prepared as single crystals and powders. They are semiconducting below x=0.25 and metallic at higher x. Mott [7] suggested that the semiconductor-metal transition was related to the degree of localisation of electrons introduced by Na. It was thought at an early stage [8] that the CB was formed by overlap of Na p orbitals, but an alternative model [9] considered this was formed by overlap of O sp hybrid orbitals and W 5d orbitals; the inserted Na atoms are then ionised with the Ef falling within the partly filled t_{2g} band. The results of experimental investigations of the electronic structure of Na_xWO_3 have been conflicting. Hall measurements show [10] that the number of conduction electrons rises with x, while NMR and theory confirm the involvement of O_{2p} and W_{5d} states in the CB formation. Some work [11] suggests that the rigid-band model is a reasonable description of the electronic structure of Na_xWO_3, but others have reached a contradictory conclusion [12]. Photoelectron spectroscopy has shown satellites associated with the W 4f doublet [13] and one explanation has been that the samples contain W in different valence states. Although, much of this evidence is contradictory, recent suggestions [14] are that the Na 3s electron goes into the W 5d state (either localised with a particular W^{5+} at low x or delocalised in a band at higher x).

There have been significant advances [2] in the development of new tungsten trioxide hydrates $WO_3.1/3H_2O$ obtained by hydrothermal treatment of a tungstic acid gel which on total dehydration at about 573K yields a metastable hexagonal tungsten oxide with a zeolitic-type structure containing interconnected hexagonal tunnels approaching 0.26-0.32nm width (see Figure 2). This has WO_6 in six-membered rings in the (001)

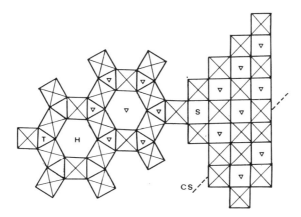

Figure 2: Modes of corner-sharing of WO_6 octahedra to produce hexagonal (H), trigonal (T) and square (S) channels into which cations can be inserted at sites ∇ . Reduction with loss of lattice oxygens is accommodated with edge-sharing at a CS plane.

plane and such layers stacked along the hexagonal axis. This too can accept Li, Na, K and H ions and electrons simultaneously, but there is a rather low upper limit of temperature, above which the new phase reverts to the traditional structure. This limits the stable reactivity of such bronzes. A bronze with a pentagonal tunnel structure [15] has also been reported.

Hydrogen bronzes H_xWO_3 ($0 < x < 0.5$) involve hydrogen which is not fully protonic [16] but transiently bound to different lattice oxygens [17] and is therefore able to move rapidly through the trioxide lattice [18] with a diffusion coefficient which may depend upon sample crystallinity. The hydrogen bronzes of Mo and V [19,20] show layer structures and substantially higher levels of hydrogen incorporation than that of W. There is ESR [21] and XRD evidence of structural rearrangements during MoO_3 transformation and reduction. In H_xMoO_3 and H_xWO_3 most of the lattice cations were in the +4 and +6 states but the surface in the case of $H_xV_2O_5$ is judged [22] to have been composed of $52\%V^{4+}$, $36\%V^{3+}$ and $12\%V^{5+}$ states. Many other transition metal oxides form bronze structures; although hygroscopic ReO_3 may be unique in its ability to form the hydrogen bronze by reaction with water or alcohol [23]:

$$(1+x)ReO_3 + xH_2O = H_xReO_3 + xHReO_4$$

Enthalpies of hydrogen bronze formation relate to OH bond strengths [24].

Intergrowth tungsten bronzes can also be prepared at high temperature (>1073K) and low values of x ($0.06< x <0.10$) with K, Rb, Cs and Tl as X. Those of Bi have been reported more recently [25] where $x<0.02$. In addition phosphate tungsten bronzes (e.g. $Rb_{0.4}P_2W_8O_{28}$) have also been reported [26] with ReO_3-type structural blocks linked by P_2O_7 groups. Bronzes are also formed by transition metal selenides, sulphides (e.g. $K_xV_5S_8$ where $0.5<x<0.7$ and $K_xTi_3S_4$ where $x=0.3$ [27]) and fluorides (e.g. $K_{0.3}NbF_3$ [28]) . The formation of such non-stoichiometric intercalation compounds by the simultaneous uptake of guest ions and electrons by host lattices, rather than the exchange processes in zeolites, has been well reviewed by Schollhorn [4]. Figure 3 suggests how such bronze phases might be immobilised in a highly dispersed state on an inorganic surface. However, almost all work has been on lower surface area unsupported bronzes of transition metals [29].

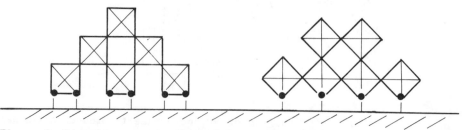

Figure 3: Possible ways in which WO_6 octahedra could be anchored to inorganic surfaces in highly dispersed states [29]. ● denotes oxygen atoms at the support surface.

1.2 Preparation

Generally unsupported bronzes may be prepared by thermal and
electrochemical methods. Wohler first prepared [30] tungsten bronzes by
reducing a mixture of sodium tungstate and tungsten trioxide in gaseous
hydrogen; this was followed by a preparation [31] with potassium
tungstate. Later [32] reducing agents such as Sn, Zn, and Fe were used.
Subsequently, Brunner [33] prepared K_xWO_3 by heating a mixture of K_2WO_4,
WO_3 and WO_2 in vacuo, while others [34] used W as the reducing agent on
heating at 1173K

$$3xX_2WO_4 + (6-4x)WO_3 + xW = 6X_xWO_3$$

where X is K or Na. Others [35] have used the metal chloride rather than
the tungstate reactant. The azide-WO_3 reaction has also been used [6].

Electrolytic reduction is generally used to obtain single bronze
crystals and was applied at an early stage [36] using a molten mixture
of metal tungstate and tungstic acid. The preparation of Na_xWO_3 by this
route has been refined [37].

Tl_xWO_3 single bronze crystals have also been prepared [38] by vapour
phase reaction between metallic Tl and WO_3 at 1173K; the method is
suitable if the inserted metal is sufficiently volatile and has been
used more recently for Bi insertion [25].

Hydrogen bronzes may also be prepared by spillover from phases active in
hydrogen dissociation [39]. Thus the formation of $H_{0.5}WO_3$, $H_{1.7}MoO_3$ and
$H_{3.3}V_3O_8$ has been reported [19,20] and also $H_{1.4}ReO_3$ [40]. Preparation
of the supported bronzes shown in Figure 3 (other than those of
hydrogen) is more difficult because of the uncertainty in the location
of the host and inserted phases and there is the possibility of the
formation of tungstate phases such as $Al_2(WO_4)_3$.

The reactivity of these non-stoichiometric transition metal bronzes and
how these can be controlled must now be considered.

2.   REACTIVITY

Figure 4 shows the surface states of the bronze phase in Figure 1 as
this reacts with hydrogen, producing first OH groups and then removing
some terminal oxygens on dehydration. The potential energy well in which
reactivity at bronze surfaces might be considered to be constrained is
shown in Figure 5. With layer bronzes reactivity would be within a
two-dimensional well. The reactivity of the centre in Figure 5 will be
very different from that seen on the (100) plane of the transition metal
because of the greater metal-metal distances, the effect of electron
withdrawal by O (or S or F) ligands and electron donation by inserted
cations at ∇ sites. Here one has an ionic-polar surface with metallic
conductivity; this unusual combination suggests the potential for
unusual reactivity. The properties of such bronze reaction centres can
be modified by the nature and concentration of inserted ions (and the
associated electrons which change the Ef) but also by changing the
lattice framework cations and anions. In addition it must be remembered

that the reaction 'crucible' in Figure 5 can be modified during reaction by O loss (or replacement) and is also permeable to proton and electron transfer to and from the bulk. This combination makes these constrained centres of bronze reactivity unique.

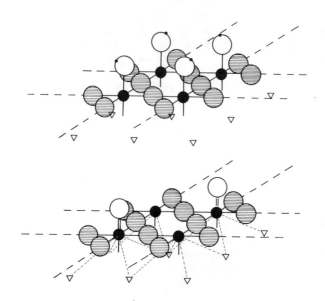

Figure 4: States upon the surface of the oxide-bronze shown in Figure during interaction with hydrogen. First surface OH groups (◯●) are produced, which then dehydrate with loss of a half of the terminal oxygens.

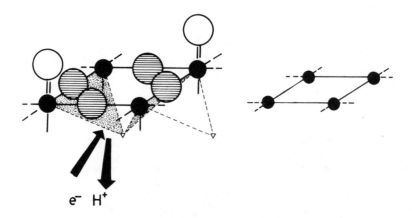

Figure 5: Possible constrained centre at the surface of bronze which is compared to that on the (1Q0) plane of the parent transition metal.

2.1. Hydrogen Acceptance

The rate and extent of acceptance and desorption of hydrogen by
platinised WO_3, MoO_3 and V_2O_5 has been discussed in detail [19,20].
Figure 6 shows that the equilibrium extent of hydrogen sorption by
Pt/WO_3 (which is not greatly dependent upon the Pt loading in the range
0.2-5%). This shows isothermal reversibility and hysteresis (with a
critical temperature at about 490K). In-situ X-ray analysis of 5%Pt/WO_3
at ambient temperature during hydrogen absorption revealed that for
x<0.03 the monoclinic structure of WO_3 was retained. At 0.06<x<0.18 a
tetragonal phase (a=0.523nm and c=0.387nm) was observed which converted
to a tetragonal-pseudocubic structure (a=0.382nm and c=0.375nm) when
x>0.2. It remains uncertain whether the phase changes cause the observed
hysteresis. Figure 6 also shows the hydrogen absorption isotherms of
Pt/Na_xWO_3 as x is increased; clearly as x increases the extent of
sorption y decreases, but (x+y) increases. First it must be noted that
non-platinised samples neither absorbed nor desorbed hydrogen at a
significant rate. Second, although the total amounts of hydrogen
desorbed decrease as x increases, approximately the same high percentage
(as much as 78%) of the absorbed hydrogen can be later desorbed. Third,
hydrogen bronzes of Mo and V oxides sorb much more hydrogen but desorb
only a much smaller fraction of their sorbed hydrogen; some [41] have
found that only H_2O is desorbed from H_xMoO_3 at 374-883K in vacuo.
Nevertheless, it remains surprising that partially protonic H in such

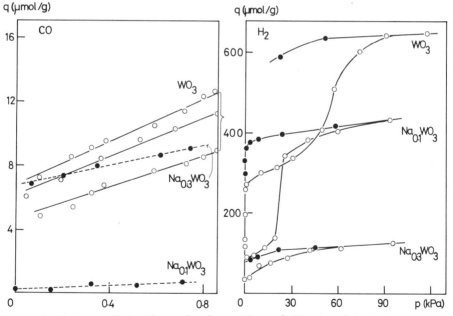

Figure 6: Extent of isothermal adsorption of CO at ambient temperature
on 3%Pt/WO_3 samples (O) of different surface area and 3%Pt/Na_xWO_3 (●)
and the extents of isothermal sorption of H_2 at 423K on samples (O and ●
denote absorption and desorption points) [19,20].

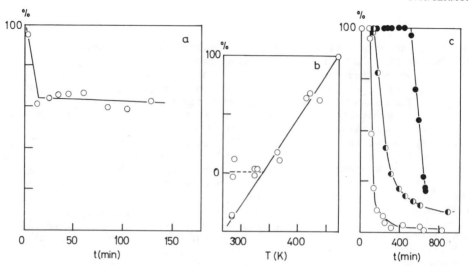

Figure 7: (a) % dehydrogenation of cyclohexene as a function of time over 1%Pt/WO$_3$ at 422K occurring for far longer than would be required to convert the support to the fully saturated hydrogen bronze. (b) Variation of % dehydrogenation of cyclohexene in the absence of gaseous hydrogen over Pt/WO$_3$ with decreasing temperature; the use of the bronze as a hydrogen reservoir was seen with some samples allowing hydrogenation at low temperature. (c) % hydrogenation of pent-1-ene at 373K (O), 273K (◑) and 323K (●) over 1%Pt/H$_x$WO$_3$ [19,20].

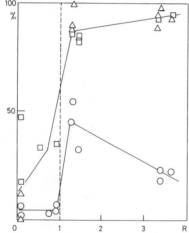

Figure 8: % removal of NO (O), CO (△) and ethene (□) over 2% Pt/WO$_3$ at 541K and a space velocity of 15000 per h with a varying net oxidant/reductant ratio R defined by inlet reactant ratios (2000ppm NO, 3%CO, 100ppm C$_2$H$_4$ and 0-5.5%O$_2$). Activity is a maximum at close to R=1, but is also sustained far above this stoichiometric composition. Bronzes have been found to be active in NO decomposition [42].

close proximity to O^{2-} ions can be desorbed from some activated bronzes rather than proceeding to water formation and later reduction. Kinetic profiles clearly illustrate induction, acceleratory and then deceleratory periods. This suggests that hydrogen absorption sites are generated on the surface of the oxide during the observed profile; there is some controversy as to whether in the intermediate region of the profile there can be direct absorption of hydrogen from the gas phase onto the oxide surface (once this has been activated by Pt). The ability to recover sorbed hydrogen is potentially important in defining the reactivity of these non-stoichiometric solids. The extents of uptake of CO and hydrogen can also be compared in Figure 6; these bronzes clearly absorb 100 times more hydrogen than CO. Such solids might therefore selectively remove hydrogen from process streams leaving CO behind, and the hydrogen may then be desorbed on heating to produce a hydrogen-rich stream. The selectivity of sorption must depend upon the local properties of the surface (and bulk) bronze centres shown in Figure 5. Any adsorptive sieving action must be at very small adsorbate sizes, but then the bronze channels shown in Figure 2 are smaller than those normally used in molecular sieves. Removal of some liganding lattice oxygens or replacement by S all allow such properties to be controlled.

2.2. Transient Catalysis

In 1980 Sermon and Bond reported [20] that pent-1-ene hydrogenation occurred for far greater periods of time over 1%Pt supported upon $H_xWO_3^-$ and H_xMoO_3 (see Figure 7c) than when supported upon silica; the fraction of retrieved hydrogen from the support was greater for H_xWO_3 reaching a maximum of 46.8% at 300K when thermodynamic and kinetic effects were balanced. The rate and extent of ethene hydrogenation over Pt/H_xMoO_3 was then reported to increase with Pt loading at 433K. With $Pt/H_xV_2O_5$ CO_2 and H_2O production in ethene hydrogenation [22] reduced the yield of ethane and suggested the loss of lattice oxygen; as expected the ease of dehydration meant that further additions of hydrogen and oxygen did not and did regenerate this activity. Hence the need to compromise between reactivity and stability. In each case described the hydrogen was introduced as dihydrogen. However, Sermon and Bond [20] also showed that hydrogen accepted by Pt/H_xWO_3 from cyclohexene at high temperature during catalysed dehydrogenation, was then sometimes available at low temperature to hydrogenate the cyclohexene (see Figure 7b); therefore this bronze is capable of acting as a hydrogen reservoir. Figure 7a shows the ability of Pt/WO_3 to dehydrogenate cyclohexene over such long periods of time at 422K that the hydrogen liberated must have been partly lost to the gas phase; thereby illustrating sustained catalysis but independent evidence of this has yet to be published. Returning to the ability of Mo or W bronze to act as hydrogen reservoirs relates to their use in car exhaust catalysis [42] where they would be subject to fluctuations between oxygen- and fuel-rich conditions; the latter condition in the presence of platinum group metals could facilitate extraction of hydrogen from hydrocarbon pollutants which they could make available to reduce NO under oxygen-rich conditions. This is illustrated in Figure 8 where NO removal is seen over Pt/WO_3 at 541K under net oxidising conditions. This 'transient catalysis' is likely to be of value in similar situations where fluctuating reaction conditions prevail. It is in this context that hydrogen bronzes may find use as micro hydrogen storage reservoirs, rather than in the formal gas-cylinder sense [43]. It is also possible that transient catalysis upon bronzes might be useful in gas purification (i.e. the removal of oxygen from argon as in USSR patent 842359 using the hydrogen storage capacity of bronzes).

2.3. Oxidative Catalysis

$X_xV_2O_5$ (where x=0.35 and X is Ag or Cu) has been used [44] to catalyse CO oxidation. Activity of $Ag_xV_2O_5$ may reside on the Ag cations in the vanadia matrix since activity increases with x, although increasing the extent of reduction of that matrix increases the oxidation activity. At 548K the orders with respect to CO and O_2 are 0.23 and 0.01, but the reaction only occurred at a rather high temperature (773K) compared to platinum-group metals (ca. 523K). ESR of Cu(II) ions in $Cu_xV_2O_5$ suggests greater distortion as a result of CO oxidation; this may also suggest a role for Cu(II) in the reaction. Therefore it is consistent that pseudo-first order pre-exponential factors for CO oxidation on these bronzes were 90-110 times greater than for the pentoxide alone.

$Mo_yW_{1-y}O_3$ crystallographic shear structures show [45] lower selective oxidation of propene than $(Mo_yW_{1-y})_nO_{3n-1}$, but the selectivity also increases as y increases and as the density of CS planes increased [3]. The importance of CS structures in the selective insertion of O into hydrocarbons has been amply demonstrated by Haber and coworkers [46] and shown in Figure 9 ; it appears that the metal-oxygen bond strength defines catalytic anisotropy [46]. MoO_3^- is a good catalyst for selective oxidation of methanol to formaldehyde [47] and it suggests 2 adsorbed species: methoxy -OMe and weakly chemisorbed methanol hydrogen bonded to lattice oxygen. Catalytic activity of Mo catalysts in oxidative dehydrogenation of n-butane has been considered, although the activity in binary transition metal oxides appears to depend on the metal- oxygen bond strength with Mo showing higher activity and selectivity than W [48]. Bismuth molybdate catalysts are common for selective oxidation [49]; however, the formation of Bi_xWO_3 bronzes [25] may not be irrelevant. Nevertheless, at present the role of hydrogen bronzes formed insitu is unclear.

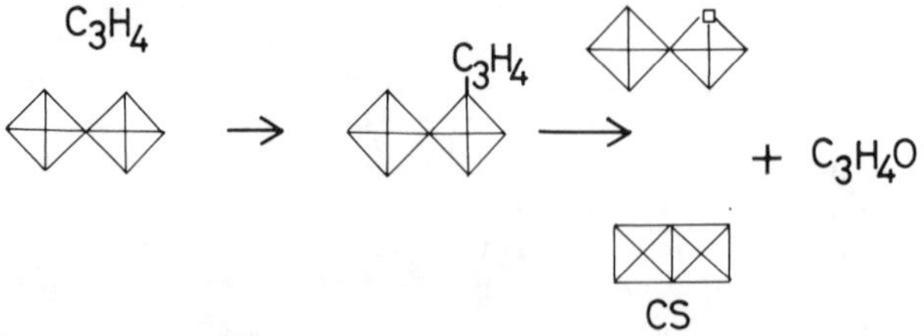

Figure 9 : Mechanism proposed [46] for selective oxidation on a transition metal oxide surface.

2.4. Hydrogenative Catalysis

Pt/MoO$_2$ has been reported to show structure sensitivity in benzene
hydrogenation [50] although there is great difficulty in determining the
dispersion of the Pt without the support chemisorbing and also obscuring
the Pt particles to TEM and XRD. The support may have been converted to
the hydrogen bronze during the reaction. In the work of Marcq etal [20]
it is not clear whether the bronze support in Pt/H$_x$MoO$_3$ is itself
directly active in continuous ethene hydrogenation in the presence of
hydrogen at 433K. This is also true for the continuous dehydrogenation
of cyclohexene followed by Sermon and Bond over Pt/WO$_3$ and Pt/MoO$_3$
[19,20].

Bearing in mind the ease of dehydration of H$_x$V$_2$O$_5$ [19,20,22] it may not
be surprising that the activity-selectivity of Pd/V$_2$O$_5$ in CO
hydrogenation [51] is very low.

In this catalytic context it is relevant that Hall has extensively
studied the ease of reduction and hydrogen acceptance of supported MoO$_z$
phases and their subsequent activity in alkene hydrogenation [52] and
also sulphided phases [52].

Hydrodesulphurisation activity over alumina-supported MoS$_2$ increases
with the extent of hydrogen sorption by the catalyst [53] and it has
been postulated that the active phase is H$_x$MoS$_2$ [4]. Certainly MoS$_2$
absorbs hydrogen in two different sites to a value of x of 0.3 .
INNS has been used to probe this sorbed hydrogen [53]. Radiochemical
methods suggest [54] irreversible adsorption of H$_2$S on MoS$_2$, it may be
that this adsorption is partly dissociative and that the product is a
partial hydrogen bronze. Certainly group VIII metals promote the
hydrotreating performance of MoS$_2$ and WS$_2$ [55]. Carbon-supported Mo
catalysts might under reforming reactions [56] convert to the sulphides
and hence the hydrogen sulphide bronzes.

In each of these cases the role of any hydrogen bronze formed insitu is
unclear. To determine this it is necessary to measure turnover numbers
for hydrogenation-dehydrogenation reactions and compare these for the
activating phase alone. This assumes that the dispersion of the
activating phase can be independently determined.

2.5. Other catalysis

Selective alcohol dehydration to higher hydrocarbons over MoO$_3$/H$_x$MoO$_3$
[57] is as chemically intriguing as methanol conversion over zeolitic
solids [58] the more so since XRD has confirmed [59] that ethanol
or methanol conversion over MoO$_3$ (even in the absence of Pt activation)
produces the bronze H$_x$MoO$_z$ (2.58<z<3.00; x=0.31). Yields of (CH$_3$)O(CH$_3$)
and (CH$_3$O)$_2$CH$_2$ from methanol depended upon the prevailing partial
pressure of oxygen and also whether the oxide is hexagonal or
orthorhombic [60]. On WO$_z$ three types of surface site exist:two which
adsorb ethanol strongly with an interaction between W-O and alcohol O-H

group producing a polar cyclic structure and inducing dehydration/dimerisation and the third which adsorbs precursor butene species [61] and it appears that there is high surface mobility of protons on such surfaces.

Isomerisation and metathesis of butenes over alumina-supported molybdena is sensitive to the Mo oxidation state, it is uncertain what role hydrogen insertion may have [62].

A special type of reactivity is observed in the case of hydrogen bronzes with a layer structure which may act as a solid Bronsted acid with a neutral or ionic Lewis base [63]:

$$HxZ + L = (LH)^{x+}.Zx^-$$

Thus ammonia and pyridinium species can be held within the layers of TaS_2 and MoO_3 [63] (see Figure 10).

2.6. Electrochemical Reactivity

It is interesting to note that Pt/TiO_2 is useful in detection of hydrogen and that Ag/MnO_2 sorbs significant volumes of gaseous hydrogen [64].

The band gap between valence and conduction bands in WO_3 (2.8–3.5 eV) allows its catalysis of photoassisted electrochemical oxidations (e.g. liberation of chlorine from aqueous NaCl solutions [65]). There is much momentum behind the need for new anode materials for photoelectrolysis [66] and it is possible that these materials with their variable Ef could have a role to play; even WO_3 may be useful as the semiconducting anode for water photoelectrolysis [67].

Bronzes such as $Li_xMo_6S_8$ may also be involved in the chemistry of batteries [68] where their fast electronic and X^+ ionic conduction may be particularly useful [69]. Thus protons and electrons can be simultaneously inserted into the PbO_2 cathode in the lead-acid secondary battery:

$$xH^+ + xe^- + PbO_2 = H_xPbO_2$$

and since the host structure does not change and proton mobility is high, a high-power discharge can be obtained. One approach has been to design solid-solution cathodes capable of a reversible, topotactic cation insertion. The Cd-Ni commercial battery uses an aqueous electrolyte and a hydrogen insertion cathode:

$$NiOOH + xH_2O + xe^- = NiO_{1-x}(OH)_{1+x} + xOH^-$$

Fuel cell electrocatalysts using bronzes have also been postulated which use the dissociative power of Pt and ionicity and conductivity of $H_yNa_xWO_3$ [70] in conjunction with acid electrolytes:

$$y/2\ H_2 + yPt = yPt-H$$
$$yPt-H + Na_xWO_3 = Pt + H_yNa_xWO_3$$
$$H_yNa_xWO_3 = Na_xWO_3 + yH^+ + ye^-$$

the use of alkaline electrolytes is prevented by electrode solubility (i.e. $4Na_xWO_3 + 4OH^- + O_2 = xWO_4^= + 2H_2O + 4xNa^+$). The anodic behaviour of Na_xWO_3 has been studied [70] and even in acid solutions preferential dissolution of Na may occur leaving a depleted surface layer. Hibbert and Churchill [71] support suggestions from Fripiat and coworkers that hydrogen can be directly absorbed on the H_xWO_3 bronze surface after this has been activated.

Electrochromic displays using WO_3 [72] are available which involve simultaneous proton and electron insertion into thin WO_3 films to form the blue bronze (application of 1.2V) reversibly:

$$xH^+ + xe^- + WO_3 \text{ (transparent)} = H_xWO_3 \text{ (blue)}$$

Here adsorbed water may be important.

3. FUTURE CONSTRAINED REACTIVITY IN BRONZES

Methanation of CO has been shown to be catalysed by supported Ni, Fe, Pt, Ru, and Rh at temperatures (870-970K [73]) at which most bronzes would be reduced-dehydrated. Nevertheless, it has been shown [74] that the hydrated Rh oxide $Rh_2O_3.5H_2O$ was stable under CO hydrogenation conditions (unlike the anhydrous material which reduced to the zero-valent metal) and produced a high concentration of oxygenated hydrocarbons (ie acetaldehyde) in addition to CH_4 and alkenes, while the unsupported metal was mediocre even in methanation. Selective dehydrogenation of alkanes over bronzes (suggested as a possibility in Figure 7a) would be promising for alkane activation; but there would remain a problem with the reduction of many bronzes.

The use of bronzes in selective hydrogenation may be rather more hopeful since these exothermic reactions occur at much lower temperatures where the bronzes retain their structural integrity. Thus butadiene is required with purities of >99.0% (ASTM-D1717) and close boiling point butenes and vinyl acetylenes might be selectively removed by hydrogenation [75] or dehydrogenation and selective hydrogenation of ethyne in the presence of ethene [76] is also catalytically important. The potential of bronze catalysts in these areas has yet to be realised and Figure 7 only indicates potential directions.

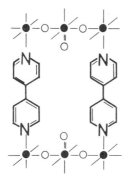

Figure 10 : Mode of interaction of bipyridine with MoO_3 [63] where ● denotes Mo atoms at the surface of oxide layers.

It would clearly be useful if a bronze catalyst exhibited greater activity and selectivity than the group VIII metal it might replace. Figure 8 shows at least the possibility of bronze activity in catalysing the reduction (and possibly decomposition) of NO; it is not surprising that activity is a maximum at the stoichiometric point, but more that NO removal activity is as high as it is at such high values of R. It should be remembered that WO_3 is only active in its own right at substantially higher temperatures. Lewis base molecules adsorb more strongly upon bronzes than the parent oxides. It is therefore interesting to note that Na_xWO_3 may be more active in decomposing NO than WO_3. Thus bronze catalysts may have value in certain environmental control reactions. It is interesting that S addition might not impair the production and reactivity of such bronzes, since in many cases sulphide bronzes exist.

Consideration of the energy levels of the zero-valent atomic states suggest that N_2 (and isoelectronic CO) should bond better to W and Mo than elements further to the right in the periodic table, since their d levels have a relatively closer energy compared to the 2s-2p states of C, N and O and also the ls state of H. However, it must be remembered that the lattice cations in the bronzes described will have rather few d-electrons and a low crystal field stabilisation energy favouring the octahedral or tetrahedral states. This may mean that the energies of the surface transformations during adsorption (and also the precursor steps shown in Figure 4) are very small. It is known [77] that converting WO_3 to $W_{20}O_{58}$ only requires 20.9kJ/mol and hence that oxygen vacancies at the surface can be readily generated. This suggests that such non-stoichiometric solids may have activity in ammonia synthesis and also oxidative dehydrogenation.

In catalysis of oxidation reactions many transition metal bronzes are able to use a combination of hydrogen removal with selective addition of lattice oxygen.

In electrochemical conversions bronzes have the ability to exhibit metallic conductivity with ionic surfaces. This is likely to be developed quickly.

4.   CONCLUSIONS

First, just as it is necessary to justify the logic of first hydrogenating molecular dinitrogen to ammonia only to reoxidise this to $NO-N_2O-NO_2$, here it is necessary to consider what advantages flow from converting a metal to a semiconductor oxide (or sulphide) and then reconverting this to a metallic solid by ion-electron insertion; can the properties of the metallic non-stoichiometric bronze be an improvement upon those of the metal (or its oxides)? First, we have used O (or S) ligands to modify the surface properties of W (or Mo, etc.). Clearly, although the Ef of the solid may be similar to that in the metal the d-electron density at the W or Mo lattice cations is far lower than in the metal. This modifies the adsorptive and catalytic properties. Thus it now has the ability to heterolytically adsorb hydrocarbons like an ionic solid, but also to delocalise electrons like a metal. It is in this context that the bronzes show unique chemical reactivity at constrained reaction centres with diffusion of reactants to and therefrom.

Consider then the adsorption of water (and alcohols) upon W and Na_xWO_3 at temperatures which do not induce volatilisation of the solid. The O-H stretching and bending frequencies are much higher on the bronze surface than on transition metals as a result of the more specific interaction between the water dipoles and the charge centres on the bronze [79]. Methanol adsorbs dissociatively on a clean W(100) surface at 310K to produce hydrogen and CO [80]; it seems likely that the hydrogen extracting capacity of the bronze would enhance this dissociative adsorption.

It appears that the reactivity of non-stoichiometric bronze phases may already find use in catalysis and electrocatalysis without their presence and effect being fully determined in that they may be produced during use; although their properties such as surface ordering may also be temperature-dependent [81]. This emphasises the need for refinement of insitu chemical and structural analyses [82] in preference to traditional post-mortems analyses. However, the full potential of the chemical reactivity of these constrained non-stoichiometric solids is only likely to be realised when their chemistry (i.e. surface and bulk properties) can be understood and controlled. Reversibility of hydrogen sorption, changes in oxygen coordination, replacement of O by S or F and changes in X and x are all parameters which may be used to control uniquely and precisely the chemical properties of these interesting constrained solids. There is every reason to suppose that research on their reactivity will achieve intriguing and useful goals.

REFERENCES

1   F.Wohler Ann.Phys. $\underline{2}$,350,(1824)

2   B.Gerand, G.Nowogrocki and M.Figlarz J.Solid.State.Chem. $\underline{38}$,312,(1981); M.Figlarz, B.Gerand, B.Beaudoin, B.Dumont and J.Desseine 10th Int.Symp.React.Solids.Dijon (1984) p279; B.Gerand, G.Nowogrocki, J.Guenot and M.Figlarz J.Solid.State.Chem. $\underline{29}$,429,(1979); B.Gerand, G.Nowogrocki and M.Figlarz J.Solid.State.Chem. $\underline{38}$,312,(1981); B.Gerand, J.Desseine, P.Ndata and M.Figlarz Stud. Inorg. Chem. $\underline{3}$,457,(1983); B.Gerand and M.Figlarz p.275 'Spillover of Adsorbed Species' ed. G.M.Pajonk, S.J.Teichner and J.E.Germain, Elsevier (1983); M.Figlarz and B.Gerand Mater.Sci.Monog. $\underline{10}$,887,(1982)

3   S.Barber, J.Booth, D.R.Pyke, R.Reid and R.J.D.Tilley J.Catal. $\underline{77}$,180,(1982); E.Salje and G.Hoppmann Philo.Mag. $\underline{43B}$,105,(1981)

4 P.G.Dickens and M.S.Whittingham Quart.Rev. 22,30,(1968);
P.Hagenmuller Progr.Solid State Chem. 5,71,(1971); D.J.M.Bevan and
P.Hagenmuller 'Non-Stoichiometric Materials' Pergamon (1973);
R.Schollhorn Angew Chem 19,983,(1980); L.Kihlborg Stud.Inorg.Chemistry
3,143,(1983)

5 B.W.Brown Phys.Rev. 84,609,(1951)

6 B.L.Chamberland Inorg.Chem. 8,1183,(1969)

7 N.F.Mott Philo.Mag. 35,111,(1977)

8 A.R.Mackintosh J.Chem.Phys. 38,1991,(1963); R.Fuchs J.Chem.Phys.
42,3781,(1965)

9 M.J.Sienko J.Amer.Chem.Soc. 81,5556,(1959); J.B.Goodenough
Bull.Soc.Chim.France 1200,(1965); R.G.Egdell, H.Innes and M.D.Hill
Sur.Sci. 149,33,(1985); D.W.Bullett Solid State Comm. 46,575,(1983)

10 P.A.Lightsey, D.A.Lilienfield and D.F.Holcomb Phys.Rev. 14B,
4730,(1976); D.P.Tunstall Phys.Rev. 14B,4735,(1976); L.Kopp,
B.N.Harman and S.H.Liu Solid State Comm. 72,677,(1977)

11 D.W.Bullett J.Phys. 16C,2197,(1983); J.N.Chazalviel, M.Campagna,
G.K.Wertheim and H.R.Shanks Phys.Rev. 16B,697,(1977); G.K.Wertheim and
J.N.Chazalviel Solid.State.Commun. 40,931,(1981); G.Hollinger,
F.J.Himpsel, B.Reihl, P.Pertosa and F.P.Doumerc Solid State Comm.
44,1221,(1982)

12 H.Hochst Solid State Comm 37,41,(1980); H.Hochst, R.D.Bringans and
H.R.Shanks Phys.Rev. 26B,1702,(1982)

13 M.Campagna, G.K.Wertheim, H.R.Shanks, F.Zumsteg and E.Banks
Phys.Rev.Lett. 34,738,(1975); R.G.Egdell, H.Innes and M.D.Hill
Sur.Sci. 149,33,(1985)

14 M.D.Hill and R.G.Egdell J.Phys. 16C,6205,(1983); R.G.Egdell,
H.Innes and M.D.Hill Sur.Sci. 149,33,(1985); R.G.Egdell and M.D.Hill
Chem.Phys.Lett. 88,503,(1982); R.G.Egdell and M.D.Hill Chem.Phys.Lett.
85,140,(1982)

15 C.Michel J.Solid.State.Chem. 52, 281,(1984)

16 A.Fujimori, H.Nozaki, N.Kimizuka, N.Tsuda, K.Tahara and H.Nagasawa
Phys,Lett. 80A,188,(1980)

17 E.Salje and G.Hoppmann Philo.Mag. 43B,105,(1981)

18 M.A.Vannice, M.Boudart and J.J.Fripiat J.Catal. 17,359,(1970);
M.L.Hitchman Thin Solid Films 61,341,(1979)

19   P.A.Sermon and G.C.Bond J.Chem.Soc.Faraday.Trans. I,72,730,(1976);
G.C.Bond,P.A.Sermon and C.J.Wright Mater.Res.Bull. 19,701,(1984)

20   P.A.Sermon and A.R.Berzins 'Metal Hydrogen Systems' ed
T.N.Verziroglu p.451,(1982) Pergammon; D.Tinet, H.E.Szwarckopf and
J.J.Fripiat 'Metal Hydrogen Systems' ed T.N.Verziroglu p.459,(1982)
Pergammon; R.Erre, H.Van Damme and J.J.Fripiat Sur.Sci. 127,48,(1983);
A.R.Berzins and P.A.Sermon Nat. 303,506,(1983); A.R.Berzins and
P.A.Sermon 10th.Int.Symp.React.Solids. Dijon (1984); P.A.Sermon and
G.C.Bond JCSFT I 76,889,(1980); G.C.Bond, P.A.Sermon and C.J.Wright
Mater.Res.Bull. 19,701,(1984); J.P.Marcq, X.Wispenninckx,
G.Poncelet,D.Keravis and J.J.Fripiat J.Catal. 73,309,(1982);
C.Marinos, S.Plesko, J.Jonas, D.Tinet and J.J.Fripiat Chem.Phys.Lett.
96,357,(1983)

21   E.Serwicka J.Solid.State.Chem. 51,300,(1984)

22   J.P.Marcq, G.Poncelet and J.J.Fripiat J.Cat. 87,339,(1984)

23   N.Kimizuka,T.Akahane,S.Matsumoto, K.Yukino Inorg.Chem.
15,3178,(1976)

24   M.T.Weller and P.G.Dickens Solid State Ionics 9-10,1081,(1983)

25   A.Ramanan, J.Gopalakrishnan, M.K.Uppal, D.A.Jefferson and
C.N.R.Rao Proc.Roy.Soc. 395A,127,(1984); D.A.Jefferson, M.K.Uppal,
D.J.Smith, J.Gopalakrishnan, A.Ramanan and C.N.R.Rao Mater.Res.Bull.
19,535,(1984)

26   J.P.Giroult, M.Goreaud, Ph.Labbe and B.Raveau Acta.Cryst.
36B,2570,(1980)

27   K.D.Bronsema Mater.Res.Bull. 19,555,(1984); R.Schollhorn,
W.Schramm and D.Fenske Angew.Chem. 19,492,(1980)

28   R.Masse, S.Alecward, M.T.Averbuch- Pouchot J.Solid.State.Chem. 53,
                                                           136,(1984)
29   L.Salvati, L.E.Makovsky, J.M.Stencel , F.R.Brown and D.M.Hercules
J.Phys.Chem. 85,3700,(1981); A.J.V.Roosmalen, D.Koster and J.C.Mol
J.Phys.Chem 84,3075,(1980); J.B.Peri J.Phys.Chem. 86,1615,(1982);
A.Iannibello, S.Marengo, P.Tittarelli, G.Morelli and A.Zecchina
J.Chem.Soc.Far.Trans. I, 80,2209,(1984)

30   F.Wohler Ann.Chim.Phys. 43,29,(1823)

31   A.Laurent Ann.Chim.Phys. 67,215,(1838)

32   E.Banks and A.Wold Prepn.Inorg. Reactions 4 (1968), Interscience

33   O.Brunner Diss. Zurich (1903)

34 H.Straumanis J.Amer.Chem.Soc. 71,679,(1949); E.O.Brimm,
J.C.Brantley, J.H.Lorenz and M.H.Jellink J.Amer.Chem.Soc.
73,5427,(1951)

35 L.E.Conroy and T.Yokakawa Inorg.Chem. 4,994,(1965); L.E.Conroy and
G.Poddsky Inorg.Chem. 7 614,(1968)

36 E.Zettnow Pogg.Ann. 130,240,(1967)

37 G.Hagg Z.Phys.Chem. 29B,129,(1935); L.D.Ellerbeck, H.R.Shanks,
P.H.Sidles and G.C.Danielson J.Chem.Phys. 35,298,(1961); W.Kunnmann
and A.Ferretti Rev.Sci.Instrum. 35,465,(1944)

38 M.J.Sienko J.Amer.Chem.Soc. 81,5556,(1959)

39 P.A.Sermon and G.C.Bond Catal.Rev. 8,211,(1973)

40 P.G.Dickens and M.T.Weller J.Solid.State.Chem. 48,407,(1983);
M.T.Weller and P.G.Dickens Solid State Ionics 9-10,1081,(1983);
R.Schollhorn, F.K.Reesink and R.Reimold J.Chem.Soc.Chem.Comm.
398,(1979); S.Horiuchi, N.Kimizuka and A.Yamamoto Nature
279,226,(1979)

41 N.Sotani, M.Kunitomo and M.Hasegawa Chem.Lett. 647,(1983);
N.Sotani, Y.Kawamoto and M.Inui Mater.Res.Bull. 18,797,(1983)

42 German patent 2509204 (1975)

43 H.C.Angus Chem.Ind. 68,(1984)

44 J.Van den Berg, J.H.L.M.Brans-Brabant, A.J.Van Dillen, J.C.Flach
and J.W.Geus Ber.Bunsen. 86,43,(1982); J.Van den Berg,
J.H.L.M.Brans-Brabant, A.J.Van Dillen, J.W.Geus and M.J.J.Lammers
Ber.Bunsen. 87,1204,(1983)

45 J.C.Volta and J.M.Tatibouet J.Catal. 93,467,(1985)

46 J.Haber, J.Janas, M.Schiavello and R.J.D.Tilley J.Catal.
82,395,(1983); J.Ziolkowski J.Catal. 80,263,(1983)

47 R.P.Groff J.Catal. 86,215,(1984); J.N.Allison and W.A.Goddard
J.Catal. 92,127,(1985)

48 G.A.Stepanov, A.L.Tsailingol'd, V.A.Levin and S.F.Pilipenko
Stud.Sur.Sci.Catal 7,1293,(1981)

49 F.Theobald, A.Laarif and M.Tachez J.Catal. 73,357,(1982)

50 M.Astier, A.Bertrand and S.J.Teichner Bull.Chim.Soc.Fr. I-218,(1980)

51 E.K.Poels, E.H.Van Broekhoven, A.A.V.Barneveld and V.Ponec
React.Kin.Catal.Lett. 18,223,(1981)

52 W.K.Hall, and W.S.Millman Stud.Sur.Sci.Cata. 7,1304,(1981);
J.Valyon, R.L.Schneider and W.K.Hall J.Catal. 85,277,(1984)

53 D.Fraser, R.B.Moyes, P.B.Wells, C.J.Wright, and C.F.Sampson
Stud.Sur.Sci.Catal. 7,1424,(1981); J.Laine ACS Div.Pet.Chem.
25,438,(1980); Bull.Soc.Belg. 90,1215,(1981); S.Vasudevan, J.M.Thomas,
C.J.Wright and C.Sampson J.Chem.Soc.Chem.Comm. 418,(1982)

54 M.L.Mirza, K.C.Campbell, S.J.Thompson and G.Webb
Radiochem.Radioanal.Lett. 54,249,(1982)

55 D.S.Thakur and B.Delmon J.Catal. 91,308,(1985)

56 A.J.Bridgewater, R.Burch and P.C.H.Mitchell J.Chem.Soc.Far.Trans.
I,76,1811,(1980)

57 P.Vergnon and J.M.Tatibouet Bull.Soc.Chim.Fr. 455,(1980);
B.H.Davis J.Catal. 79,58,(1983)

58 M.M.Wu and W.W.Kaeding J.Catal. 88,478,(1984);
E.Santacessaria,D.Gelosa, E.Giorgi and S.Carra J.Catal. 90,1,(1984)

59 J.Guidot and J.E.Germain React.Kin.Catal.Lett. 15,389,(1980)

60 J.M.Tatibouet and J.E.Germain Compt.Rend. 290,321,(1980)

61 F.Gobal J.Chem.Res.(S) 322,(1980) & 248,(1981)

62 T.Okuhara and K.Tanaka J.Catal. 42,474,(1976); Y.Iwasawa,
H.Ichinose, S.Ogasawara and M.Soma J.Chem.Soc.Far.Trans.I 77,
1763,(1981)

63 J.W.Johnson, A.J.Jacobson, S.M.Rich and J.F.Brody J.Amer.Chem.Soc.
103, 5246,(1981)

64 L.Harris J.Electrochem.Soc. 127,2657,(1980); A.Kozawa and
K.V.Kordesch Electrochim.Acta. 26,1489,(1981)

65 J.Kiwi and M.Gratzel J.Chem.Soc.Far.Trans.II 78,931,(1982)

66 G.B.Daar, M.P.D.Edwards, J.B.Goodenough and A.Hamnett
J.Chem.Soc.Far.Trans.I 79,1199,(1983)

67 M.A.Butler, R.D.Nasby and R.K.Quinn J.Solid.State.Chem.
19,1011,(1976)

68   P.J.Mulhern and R.R.Haering Can.J.Phys. $\underline{62}$,527,(1984)

69   J.B.Goodenough Proc.Roy.Soc. $\underline{393A}$,215,(1984)

70   B.S.Hobbs and A.C.C.Tseung Nature $\underline{222}$,556,(1969); B.Broyde J.Catal. $\underline{10}$,13,(1968); T.Kishi, Y.Muranushi and T.Nagai Surf.Tech. $\underline{21}$,351,(1984)

71   D.Hibbert and C.R.Churchill J.Chem.Soc.Far.Trans.I $\underline{80}$,1977,(1984)

72   S.S.Shun and P.H.Holloway J.Vac.Sci.Tech. $\underline{1A}$,529,(1983); M.Fujii, T.Kawai, H.Nakamatsu and S.Kawai J.Chem.Soc.Chem.Comm. 1428,(1983); N.Yoshiike and S.Kondo J.Electrochem Soc. $\underline{130}$,2283,(1983); F.G.K.Baucke and J.A.Duffy Chem.Britn. 643,(1985)

73   H.J.Jung, P.L.Walker and M.A.Vannice J.Catal. $\underline{75}$,416,(1982); J.Happel, H.Y.Cheh, M.Otarod, S.Ozawa, A.J.Severdid, T.Yoshida and V.Fthenakis J.Catal. $\underline{75}$,314,(1982); F.Fajula R.G.Anthony and J.H.Lunsford J.Catal. $\underline{73}$,237,(1982); N.W.Cant and A.T.Bell J.Catal. $\underline{73}$,257,(1982)

74   P.R.Watson. and G.A.Somorjai J.Catal. $\underline{72}$,347,(1981)

75   Y.Okaamoto, K.Fukino, T.Imanaka and S.Teranishi J.Catal. $\underline{74}$,173,(1982); J.Grant, R.B.Moyes and P.B.Wells J.Catal. $\underline{51}$,355,(1978); R.G.Oliver and P.B.Wells J.Catal. $\underline{47}$,364,(1977)

76   W.T.McGown, C.Kemball, D.A.Whan and M.S.Scurrell J.Chem.Soc.Far.Trans.I, $\underline{73}$,632,(1977)

77   M.W.M.Hishman and S.W.Benson J.Phys.Chem. $\underline{89}$,1905,(1985)

78   W.Palczewska Adv.Catal. $\underline{24}$,245,(1975)

79   D.G.Aitken, P.A.Cox, R.G. Egdell,M.D.Hill and I.Sach Vacuum $\underline{33}$,753,(1983); M.J.Cardillo and Y.Look Sur.Sci. $\underline{66}$,272,(1977)

80   R.P.H.Gasser, G.V.Jackson and F.E.Rolling Sur.Sci. $\underline{61}$,443,(1976); W.F.Egelhoff, J.W.Linnett and D.L.Perry Far.Disc.Chem.Soc. $\underline{60}$,127,(1975)

81   C.J.Schramm, M.A.Langell and S.L.Bernasek Sur.Sci. $\underline{110}$,217,(1981)

82   J.M.Thomas and R.M.Lambert 'Characterisation of Catalysts' Wiley (1980)

CRYSTALLINE SILICIC ACIDS

G. Lagaly,
H.-M. Riekert, and
H.H. Kruse
Institut für anorganische Chemie
der Christian-Albrechts-Universität
Olshausenstraße 40
D-2300   K i e l
Germany

ABSTRACT. Different types of crystalline silicic acids
are obtained from layer silicates by proton exchange.
Hydrated alkali layer silicates formed from SiO_2 dispersions in alkaline solutions at 100-150°C are precursors
of several acids which are thus available in large amounts.
    The crystalline acids intercalate many organic com-
pounds. The reactivity differs widely. The rate of intercalation is high and, in each case, quantitative reaction is achieved.
    The catalytic activity is related to the high sur-
face acidity, the particle morphology and distinct pore
shapes.

1.   INTRODUCTION

Lamellar compounds with intracrystalline reactivity provide peculiar types of interface reactions. Interac-
ting molecules have to penetrate between the layers,
and sterical constrains are more dominant than on external surfaces.
    Adsorption of molecules or ions on the internal
surfaces requires the expansion of the interlamellar
space or the displacement of interlayer molecules.
Whether, in the latter case, the spacing increases or
decreases, depends on the size and packing of the guest
compounds and the displaced interlayer molecules.
    The internal surfaces are not accessible to all
types of adsorptives. Unpolar compounds can not penetrate between the layers; many bases, in particular amines, are adsorbed strongly. The chemical nature of the
host compound is only one of the factors determining the

361

R. Setton (ed.), Chemical Reactions in Organic and Inorganic Constrained Systems, 361–379.
© *1986 by D. Reidel Publishing Company.*

reactivity. The type of interlayer bonds and the stacking
of the layers are of influence and the elasticity of
the layers plays an important part during the penetration of the guest compound between the layers. Lamellar
crystalline silicic acids reveal these dependences very
clearly. Many different types of these acids are known
and are easily prepared. They are of the same chemical
nature as expressed by the general formula $SiO_2 \cdot xH_2O$. The
layers are connected by hydrogen bonds which form either
directly between silanol groups or between surface silanol
groups and interlayer water molecules. Thus, differences
in the intracrystalline reactivity of the distinct crystalline acids are not the result of different types of
interlayer bonds but are related to the density of the
hydrogen bonds and the competition between formation of
 energetically favoured hydrogen bonds and dense-packing
of the layers.

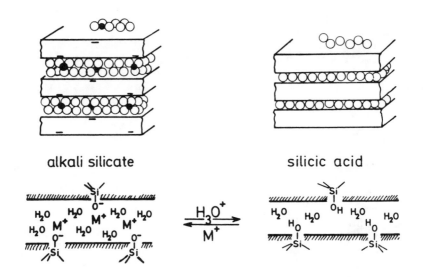

Figure 1. Formation of crystalline silicic acids from
alkali or alkaline-earth layer silicates.

2.   PREPARATION

Most of the crystalline acids are prepared by replacing
the alkali or alkaline-earth interlayer cations of the
layer silicates by protons (Fig. 1) (Iler, 1964; Wodtcke
and Liebau, 1965; Wey and Kalt, 1967b; Lagaly et al., 1975;

Frondel, 1979; Kalt et al., 1979; Lagaly, 1979; Wolf and
Schwieger, 1979). Most of the parent lamellar silicates
are synthetic products; only a few acids are obtained
from minerals that cannot be synthezised so far.

In Table 1 are listed the acids and their parent
compounds that can be easily prepared in larger amounts.
(The complete list reported earlier (Lagaly, 1979) has to
be supplemented by three further crystalline acids: two
modifications of $H_2Si_3O_7$ from $Na_2Cu(Si_3O_7)_2 \cdot 5H_2O$ and
$Na_2Si_3O_7$ (Guth et al., 1977) and $H_2Si_2O_5 \cdot 0.7H_2O$ from
synthetic $Na_2CuSi_4O_{10}$ or litidionite $NaKCuSi_4O_{10}$ (Guth
et al., 1978).)

TABLE I: Crystalline silicic acids from synthetic alkali
silicates

Silicic acid	Parent Compound	Synthesis[1]
$H_2Si_2O_5$-I, II	α-$Na_2Si_2O_5$	a: Wodtcke a. Liebau, 1965
$H_2Si_2O_5$-III	$KHSi_2O_5$	b: Wey a. Kalt, 1967a
$H_2Si_2O_5$-I, II, III	$NaHSi_2O_5 \cdot 3H_2O$	Beneke a. Lagaly, 1977
	(kanemite)	
$H_2Si_4O_9 \cdot xH_2O$	$Na_2Si_4O_9 \cdot xH_2O$	b: Beneke a. Lagaly, 1983
	(makatite)	
$H_2Si_8O_{17} \cdot xH_2O$	$Na_2Si_8O_{17} \cdot xH_2O$	b: Iler, 1964
$H_2Si_{14}O_{29} \cdot xH_2O$	$Na_2Si_{14}O_{29} \cdot xH_2O$	b: Lagaly et al., 1975
	(magadiite)	
$H_2Si_{20}O_{41} \cdot xH_2O$[2]	$K_2Si_{20}O_{41} \cdot xH_2O$[2]	b: Beneke et al., 1984
	$Na_2Si_{20}O_{41} \cdot xH_2O$	b: Beneke a. Lagaly, 1983
	(kenyaite)	

[1]
  a: from Na_2CO_3, SiO_2-melts

  b: from SiO_2, NaOH(KOH), H_2O-dispersions (100-150°C)

[2]
  Previously, this salt was formulated as $K_2Si_8O_{17} \cdot xH_2O$.
  It retains an excess of KOH that cannot be removed
  without replacing structural K ions. Redetermination
  of the formula led to $K_2Si_{20}O_{41} \cdot xH_2O$ (or more likely

$K_2H_2Si_{20}O_{41} \cdot xH_2O)$ and to $H_2Si_{20}O_{41} \cdot xH_2O$ for the corresponding acid.

The reaction of silica with aqueous NaOH or KOH solutions at 100-150°C is the simplest way for preparing the parent alkali silicates. The kind of product is mainly determined by the molar ratio $SiO_2/NaOH(KOH)/H_2O$ and the reaction time (Iler, 1964; Beneke and Lagaly, 1983; Beneke et al., 1984). The Cs^+, Ag^+, Ca^{2+}, Sr^{2+} and Ba^{2+} forms of $K_2Si_{20}O_{41} \cdot xH_2O$ are obtained by the same procedure. The reaction of SiO_2 with KOH at 100°C (for instance: 3 moles SiO_2, 2 moles KOH, 18 moles H_2O) is also recommanded for the preparation of $KHSi_2O_5$.

3. STRUCTURE, MORPHOLOGY AND SURFACE PROPERTIES

3.1. Structure

Liebau (1964) and Le Bihan et al. (1971) determined the crystal structure of three modifications of $H_2Si_2O_5$ and their parent compounds. The structure of the layers is maintained during the proton exchange with slight changes of the position and orientation of the $[SiO_4]$-tetrahedra. Such changes are generally observable by some changes of the lattice dimensions (cf. Lagaly, 1979). The silicate layers, even of the highly-condensed acids, retain a flexibility that makes possible some adaption onto the interlayer cations. The acids derived from $Na_2Si_{20}O_{41} \cdot xH_2O$ (kenyaite) and $K_2Si_{20}O_{41} \cdot xH_2O$ show identical powder diagrams but differ slightly by their intracrystalline reactivity. Both acids can be converted into each other by treatments with KOH and NaOH, resp., and hydrochloric acid. The structural principle of the layers must be the same in both silicates but the structure may adapt to the interlayer cations by slight changes of the tetrahedral network (Beneke and Lagaly, 1983).
    Nuclear magnetic resonance spectroscopy provides an useful method for characterising the silanol groups (Voitländer et al., 1975; Lagaly, 1979) and the interlayer hydrogen bonding scheme (Rojo et al., 1983).
    Specification of the silicic acids by chemical formulae involves several problems (Lagaly, 1979; Beneke et al., 1984). The notation in Table 1 indicates the number of silanol protons that are replaceable by alkali ions. The ratio: total number of silanol groups/ the number of silicon ions is often higher (for instance: 4/20 for $H_2Si_{20}O_{41} \cdot xH_2O$). This ratio is difficult to measure. Formulae derived from water desorption data and weight loss curves are often incorrect. The complete desorption of interlayer water requires such high temperatures that the

dehydration of the silanol groups is initiated. Furthermore, water molecules enclosed between collapsing
layers are tenaciously retained, and isolated silanol
groups survive even at high temperatures (Lagaly et al.,
1975, 1979).

3.2. Morphology

The highly-condensed acids ($H_2Si_{14}O_{29} \cdot xH_2O$, $H_2Si_{20}O_{41} \cdot xH_2O$)
are distinguished by an unusual particle morphology.
Plate-like crystals are intergrown to spherical nodules,
and wedge-shaped pores form between the silicate lamellae
(Fig. 10 in: Lagaly, 1979; Fig. 2 in: Beneke and Lagaly,
1983). The silicic acid $H_2Si_8O_{17} \cdot xH_2O$ crystallizes in
isolated or intergrown retangular plates. The acids from
makatite and $KHSi_2O_5$ form lath-shaped crystals.

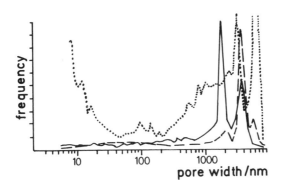

Figure 2. Pore size distribution of $H_2Si_{14}O_{29} \cdot xH_2O$ (S =
41 m^2g^{-1},—compacted at about $2 \cdot 10^8$ Pa, -- uncompacted)
and of "Kieselgel 60" (...), Hg-penetration method, samples outgased at 150°C.

3.3. Porosity

The porosity of the crystalline acids exhibits some features that were not expected initially. The pore distribution of a sample of $H_2Si_{14}O_{29} \cdot xH_2O$ (Fig. 2) is obtained by mercury penetration. A sharp maximum occurs
for pores of 3-4 μm. The pore width is reduced to about
2 μm by compacting the sample. No pores are detectable
below 1 μm. The quite different pore distribution of
porous "Kieselgel" is inserted in Fig. 2 for comparison.
     The pore distribution below 10 nm is derived from
N_2 adsorption (Fig. 3). The small hysteresis between adsorption and desorption isotherms and the t-plot indicate

a very low proportion of meso- and micropores. Analysis
of the desorption branch by the Kelvin equation confirms
the absence of these pores. The porosity is thus mainly
determined by interparticle pores larger than 1 μm.

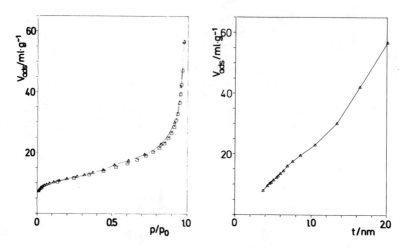

Figure 3. Nitrogen adsorption isotherm and t-plot of
$H_2Si_{14}O_{29} \cdot xH_2O$ (as in Fig. 2)

3.4. Surface acidity

The crystalline silicic acids are strong solid acids
with surface acidities ranging from $H_o < -3$ ($H_2Si_{14}O_{29} \cdot xH_2O$)
to $H_o \sim 2.3$ ($H_2Si_2O_5$) (Werner et al., 1980). The acidity
measured in aqueous dispersions with Hammett indicators
does not differ from the values obtained in dry benzene.
(It should be mentioned that, for measurements in dry
benzene, water cannot be removed completely from the
surface without condensation of neighbouring silanol
groups.) The surface acidity is decreased after dehydrating the acids at higher temperatures. For instance,
$H_2Si_{14}O_{29} \cdot xH_2O$ has $H_o \sim -3$ when heated to 300°C and 2.3-3.3
when heated to 300-900°C. After tranformation into cristo-
balite at $\geqslant$ 900°C (Lagaly et al., 1979), H_o becomes
4.9-6.8.
    Orthosilicic acid is very weak ($pK_s \sim 10$). The acidi-
ty increases to $H_o = 6$-7 by condensation to oligomeric
forms (Strazkesko et al., 1974; Iler, 1979). The cause
is seen in the formation of $d\pi$-$p\pi$ Si-O interactions (cf.
Werner et al., 1980). The acidity of amorphous silica
($Ho = 4.9$-6.8) is thus not very different from the acidity
of soluble oligomeric forms. The further increase of
acidity associated with formation of crystalline modifica-

tions is related to the hydrogen bonding system. On the
surface of the crystalline acids the network of hydrogen
bonds between silanol groups and surface water molecules
will be continuous and much regular than on the surfaces
of amorphous products. It is assumed that this regular
arrangement enhances the surface acidity. A strong increase
of acidity by formation of intramolecular hydrogen bonds
was postulated by Middleton and Lindsey (1964).

Titration of the acidic sites with butylamine according to Benesi (Werner et al., 1980) reveals two types
of acidic centres; sites with H_o < 3.3 and sites with H_o >
4.9. With regard to the above model this result can be
an artifact. If the enhanced acidity is really caused by
the continuous network of hydrogen bonds, than small amounts
of butylamine adsorbed on the surface will destroy this
continuity and the acidity of the remaining sites is decreased.

4.   INTERACTION WITH ORGANIC COMPOUNDS

4.1. Adsorption of alkanes

Alkanes cannot penetrate between the layers and are only
adsorbed on the external surfaces. Adsorption enthalpies
of different alkanes on crystalline acids were derived
from gas-chromatographic data.

The fine-powdered silicic acids were compacted to
pellets (optimal pressure: 2000 kp·cm^{-2} = 2 x 10^8 Pa), the
pellets broken to pieces and the fraction 0.25-0.5 mm
(60-35 mesh) separated by sieving. This material was
packed into gas-chromatographic columns. Adsorption isotherms were obtained from analysis of the peak profiles
(Riekert, 1982).

The adsorption enthalpy is obtained (1) from the
isotherms at different temperatures (100-120°C) and (2)
from the retention volumes V_s by plotting ln V_s/T versus
1/T (Riekert, 1982). At the very low surface coverages
(about 0.1-2 μ moles/g) in these experiments the isosteric adsorption enthalpies are independent on the
amounts adsorbed. The data derived from the adsorption
isotherms and the retention volumes are in best agreement
(Table 2).

Table 2 collects the adsorption enthalpies of different alkanes, ethene and ethyne on three different acids
and "Kieselgel 60" at 90-120°C. The enthalpies of the
crystalline samples are very similar and considerably
larger than for Kieselgel 60. They exceed even those of
porous silica with the narrowest pores (Table 3) and
approximate values on zeolites (Steinberg et al., 1976).

Table II: Differential adsorption enthalpies of hydrocarbons on silicic acids and "Kieselgel 60" (Merck,
Germany), in kJ/mole, from $\lg(V_s/T)$ vs. $1/T$, 90-120°C;
acids activated at 140°C

Adsorptive	Kiesel-gel 60	Silicic acids		
		A	B	C
Ethane	22.2	27.7	28.1	27.8
n-Propane	27.1	34.0	33.7	36.0
n-Butane	35.2	42.1	41.6	42.7
n-Pentane	39.5	48.7	50.7	50.7
n-Hexane	45.0	56.8 [1) 56.2]	57.3 [1) 57.3]	61.4 [1) 60.6]
Ethene	27.7	33.2	31.9	31.7
Ethyne	30.3	35.5	36.2	34.3
Benzene	55.1	65.7	64.6	65.4

A: $H_2Si_2O_5$-I, B: $H_2Si_8O_{17} \cdot xH_2O$, C: $H_2Si_{14}O_{29} \cdot xH_2O$

1)
   isosteric enthalpies from adsorption isotherms (100-120°C)

Table III: Differential adsorption enthalpies of hydrocarbons on porous silica ("Kieselgel", Kiselev et al.,
1964), and $H_2Si_{14}O_{29} \cdot xH_2O$ in kJ/mole

Kieselgel	Ethane	Propane	Butane	Ethene
3.2 [1)]	24.9	33.1	40.6	28.9
4.6	23.4	31.4	39.3	27.2
7	22.6	30.1	38.3	26.8
10	22.0	28.9	36.0	26.0
30	19.7	24.3		
$H_2Si_{14}O_{29}$	27.8	36.0	42.7	31.7

1)
   pore width (nm) of Kieselgel

The high adsorption enthalpies of the crystalline samples
are especially noteworthy because micro- and mesopores
are absent. Rather, wedge-shaped pores are typical. They
are formed by expansion of the interlayer spaces towards

the crystal edges or between intergrowing plates. A simple model in Fig. 4 illustrates that the adsorption enthalpy increases in the order: planar surface < large cylindrical pores < wedge-shaped pores.

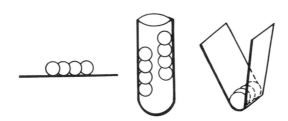

Figure 4. Adsorption of alkanes on planar surfaces, in cylindrical and wedge-shaped pores.

Due to the strong interaction with hydrocarbons the crystalline acids appear to be useful adsorbents for gas-chromatographic separations. Fig. 5 demonstrates the separation of different hydrocarbons by a column packed with $H_2Si_2O_5$-I. (The HETP of this acid at 51°C is 3.2 mm for CO_2, carrier gas (He) with flow rates > 5 cms^{-1}, column length 118 cm, internal diameter 0.45 cm, Riekert, 1982).

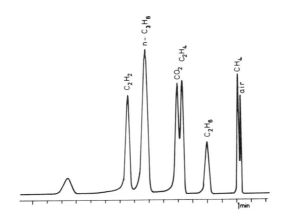

Figure 5. Gas-chromatographic separation of different hydrocarbons and CO_2 by $H_2Si_2O_5$-I (uncompacted, column length 118 cm, diameter 0.45 cm, carrier gas flow 0.943 ml·s^{-1}, column temperature 60°C)

4.2. Reaction with surface active agents

Surface active agents are adsorbed by $H_2Si_{14}O_{29} \cdot xH_2O$
($x = 4$) onto the external surfaces only (Dörfler et al.,
1984). Cetylpyridinium chloride, sodium undecylsulfonate
and octaethyleneglycol dodecyl ether are found in levels
of $2-3 \times 10^{-4}$ moles g^{-1}. On the basis of the BET-surface
area (83 m^2g^{-1}) the area per surfactant is 0.42, 0.28
and 0.77 nm^2. Addition of Na_2SO_4 enhances the adsorption
of the non-ionic agent.

4.3. Intercalation

Intercalation requires guest molecules with polar
groups that can accept or donate hydrogen bonds. Even
highly polar molecules, e.g. organochlorine compounds and
nitrils, are not intercalated, if such groups are absent.
The importance of hydrogen bonds is clearly evidenced by
infrared studies (NMF and DMFA on $H_2Si_{14}O_{29} \cdot xH_2O$ and
$H_2Si_{20}O_{41} \cdot xH_2O$, Rojo and Ruiz-Hitzky, 1984). Sterical aspects are more dominant than for the adsorption on external surfaces. The ability of the guest molecules to arran-
ge in close contacts to the surface atoms is also important.

When intercalation occurs, complete degrees of
reaction are always achieved, and the rates of intercalation are high. In several acids, the guest compounds
have to displace the interlayer water molecules. An increasing amphiphilic character complicates the displacement and renders more difficult the direct intercalation
of, for instance, long chain alkylamines. It is recommanded to intercalate a short chain alkylamine and to displace it by the longer amine.

The selection of reacting guest molecules is facilitated by the results of earlier studies on kaolinites.
Possible compounds are: urea and urea derivatives, short
chain acid amides and their methyl and ethyl derivatives,
dimethyl sulfoxide, amine-N-oxides, pyridine-N-oxides etc..
In contrast to kaolinite, a large variety of amines are
intercalated.

Basal spacings after intercalation of various compounds are discussed in detail (Lagaly, 1979). Generally,
the interlamellar orientation of homologous molecules
(e.g. fatty acid amides and derivatives) changes with increasing molecular weight because of a competition between
formation of hydrogen bonds and optimal space filling.
Thus, the layer separation does not increase continuously
with the molecular volume (Fig. 13 in Lagaly, 1979).

In most cases, the interlamellar orientation of the
guest molecules can be deduced approximately from the observed layer separations. Principle new informations are

not gained by complete crystal structure analysis which,
in most cases, is difficult or impossible to be carried
out.

4.4. Intercalation of amines

     Various types of amines are intercalated by crystal-
line silicic acids. Unbranched primary alkylamines are
arranged in bilayers with the chain axes tilted towards
the layer normal. The chain tilt decreases with increasing
chain length, so that very long chains are standing per-
pendicularly to the layers (Lagaly et al., 1974; Lagaly,
1979). Shorter chain alkylamines may also be intercalated from aqueous solutions. Kalt et al. (1979) reported
the presence of water to be mandatory for intercalating
alkylamines (propylamine to decylamine) into $H_2Si_2O_5$-III.
The chains are then oriented normal to the layers, but
as indicated by the absolute values of the spacings, some
C-C bonds are in gauche-conformation or the alkyl chains
interpenetrate to some extent.

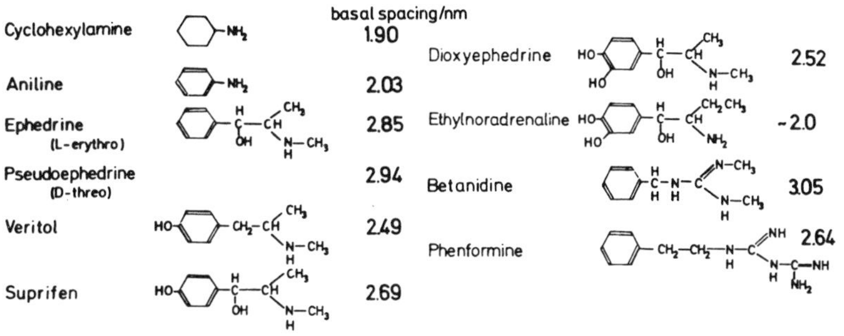

Figure 6. Layer separation d_L after intercalation of
ephedrine and derivatives into $H_2Si_{20}O_{41} \cdot xH_2O$.

     In most cases, alkylation of the amine group im-
pedes the bimolecular arrangements and promotes monolayers
with the amines lying flatly between the layers (spacings 2.0-2.15 nm). For instance, butylamine increases
the spacing to 2.94 nm, but alkylated derivatives such as
N-methyl butylamine (2.12 nm), N-ethyl butylamine (2.13 nm),
dibutylamine (2.15 nm) form monolayers. In contrast, the
N,N-dimethyl derivative assumes a tilted orientation with
a spacing of 2.80 nm.
     The organic-chemical drugs ephedrine and derivatives
are preferentially intercalated (Fig. 6). The adsorption
isotherm for ephedrine from ethanolic solution approximates

a saturation value of 90 mmoles/mole SiO_2 (= 1.8 moles/mole
$H_2Si_{20}O_{41}$) at concentrations $\geqslant 0.3$ moles/l. The maximal
spacing of 2.90 nm already is observed at concentrations
of 0.15 moles/l with a coverage of 70 mmoles/mole SiO_2
(Beneke et al., 1984).

4.5. Comparison between different acids

Distinct differences in the reactivity of the silicic
acids are observable. The reactivity increases in the
order $H_2Si_2O_5$ (modifications III-V), $H_2Si_4O_9 \cdot xH_2O$ (from
makatite) $< H_2Si_2O_5$-I, $H_2Si_2O_5$-VI $< H_2Si_4O_9 < H_2Si_8O_{17} \cdot$
$xH_2O < H_2Si_{14}O_{29} \cdot xH_2O < H_2Si_{20}O_{41} \cdot xH_2O$.
        In general, the reactivity increases with the de-
gree of condensation. The last three acids contain interlayer water molecules, and intercalation proceeds by
displacement. The reactivity is maintained when this
water is removed from the interlayer space under conditions that interlamellar condensation of silanol groups
is impeded (drying at low humidity or below 120°C). The
above classification of the acids is deduced from the
behavior towards typical guest molecules which themselves are reactive in the order: urea and derivatives,
acid amides $<$ N-oxides, DMSO $<$ alkylamines. Thus, several
silicic acids do not react with urea and acid amides,
but all acids react with alkylamines.
        The reactivity can depend on the conditions of
preparation. When mineral acids of too high concentra-
tion are applied in preparing $H_2Si_{14}O_{29} \cdot xH_2O$, dehydrated
forms with reduced reactivity are obtained (Lagaly et al.,
1975). Long reaction periods during the synthesis of the
parent alkali silicates can reduce the reactivity of
$H_2Si_{20}O_{41} \cdot xH_2O$ (Beneke et al., 1984). One of the causes
is the condensation of silanol groups. Disilicic acids
become less reactive when stored at room temperature for
prolonged periods (Lagaly et al., 1974). The reactivity
of the higher-condensed acids remains unaltered, even
after storage for years.

4.6. Comparison with other layer compounds

The crystalline silicic acids evidence that the chemical
composition is not solely decisive for different intracrystalline reactivities. This conclusion is also drawn
from the variety of oxidic lamellar compounds listed in
Table 4. The highly-condensed silicic acids and zirconium
phosphate are the most reactive host materials. Intriguing
is the contrasting behaviour of kaolinite and the other
layer compounds. Only kaolinite does not intercalate alkylamines directly. The cause may be found in different
initiation processes. According to a model of Weiss,

the interaction of the dipole moment of the guest molecule with the kaolinite layer induces reorientations of the OH groups in the structure which initiate an elastic deformation of layers and the opening of the interlayer spaces (Weiss et al., 1981).

TABLE IV: Reactivity of lamellar compounds towards three groups of guest molecules

| | Direct Intercalation of | | |
	Urea	DMSO	Alkylamines
Kaolinite	+	+	–
Silicic acids:			
$H_2Si_{14}O_{29}\cdot xH_2O$	+	+	+
$H_2Si_{20}O_{41}\cdot xH_2O$	+	+	+
Zirconium phosphate			
$H_2Zr(PO_4)_2\cdot H_2O$	+	+	+
Vanadyl phosphate			
$VOPO_4\cdot 2H_2O$		+	+
FeOCl	–	+	+
Silicic acids:			
$H_2Si_2O_5$-I	–	+	+
$H_2Si_2O_5$-III	–	–	+
Niobyl phosphate			
$NbOPO_4\cdot 3H_2O$	–	–	+
Krautite			
$MnHAsO_4\cdot H_2O$	–	–	+
Molybdic acid			
$MoO_3(OH_2)\cdot H_2O$	–	–	+
Uranium mica			
$HUO_2PO_4\cdot 4H_2O$	–	–	+
$Ag_6Mo_{10}O_{33}$	–	–	+

In contrast to kaolinite, dipole-dipole interactions may not govern the initiation process of other host compounds. The intercalation is not initiated by the ben-

ding up of the layer under the influence of adsorbed molecules. Rather, the initiation starts when the guest molecules squeeze into the interlayer space with subsequent
elastic deformation of the layers. This process requires
strong interactions of the guest molecules with interlayer ions or internal surface groups. Thus, the intercalation of amines may be initiated by protonation of
the bases by protons from acidic surface OH groups or
interlayer H_3O^+ ions (as present in uranium mica). Even
in case of stoichiometric FeOCl, the reaction may be
initiated by protons from interstitial sites which compensate the $Fe^{2+}_{Fe}3+$ defects (Weiss and Choy, 1984).
Intercalation of alkylamines into $Ag_6Mo_{10}O_{33}$ occurs by
complexing the silver interlayer ions (Rösner and Lagaly,
1984).

4.7. Interlamellar grafting

Grafting is a highly effective method for making silicate
surfaces hydrophobic. For instance, reaction between surface silanol groups and chlorosilanes leads to covalent-
ly bound surface silane groups. When vinyl chlorosilanes
are applied, the vinyl groups remain accessible to subsequent reactions and a complete coverage of the surface
by polymeric material may be attained (Ruiz-Hitzky and
Fripiat, 1976).
     Direct reaction of silylating agents with crystal-
line acids is restricted to the external silanol groups.
The interlamellar silanol groups become accessible when
the interlayer space is expanded by intercalation of
certain organic molecules. A considerable part of the
internal surface silanol groups (30-50 percent for
$H_2Si_{14}O_{29} \cdot xH_2O$) can be silylated by reacting the acid
with trimethyl chlorosilane after intercalation of DMSO
(Ruiz-Hitzky and Rojo, 1980; Ruiz-Hitzky et al., 1985).

5.   DEHYDRATION OF ALKOHOLS

Because of their high surface acidity several crystalline silicic acids are suitable catalysts for acid catalyzed reactions. The formation of tetraphenyl porphyrine
from benzaldehyde and pyrrole is catalyzed by $H_2Si_8O_{17} \cdot xH_2O$, $H_2Si_{14}O_{29} \cdot xH_2O$ and $H_2Si_{20}O_{41} \cdot xH_2O$ (Werner et al.,
1980). In quest of practical applications, studies on
dehydration or isomerization reactions are in consideration. As an example, Fig. 7 shows the ability of
different acids to dehydrate tert. butanol. The highly
active $H_2Si_2O_5$-I and $H_2Si_{20}O_{41} \cdot xH_2O$ (sample 43 T) dehydrate tert. butanol to isobutylene below 150°C. The other
acids require temperatures up to 240°C for complete conver-

sion. $H_2Si_2O_5$-III is inactive.

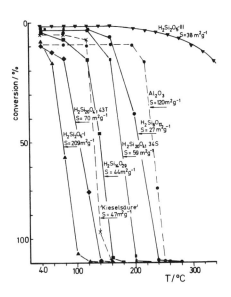

Figure 7. Dehydration of tert. butanol by different crystalline acids, "Kieselsäure, gefällt" and γ-Al_2O_3.

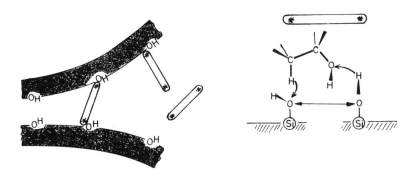

Figure 8. Two-centre mechanism for alkanol-alkene conversion (schematically) and the advantage of substrates with
wedge-shaped pores.

The activity is not related to the surface area. An
increase of the total surface area seems to be not as

decisive as the related increase of the edge surface area.
    Dehydration of alkohol is initiated by proton
transfer from acidic silanol groups to the alkanol groups
and decomposition of water. The carbenium ion then do-
nates back a proton to a basic site on the surface. The
alkohol-alkene conversion is strongly promoted when the
proton transfer from and to the surface proceeds in a
concerted reaction. This requires distinct distances between the reacting surface sites. On the surfaces of amorphous products the butanol molecules may find reacting
sites in adequate positions. On the basal planes of the
crystalline acids the silanol groups are located too di-
stantly from each other. Fig. 8 illustrates that in wedge-
shaped pores the molecules can easily find two reacting
sites in appropriate distance. It is thus the very structure of the crystal edges with the openings of the interlayer spaces or the pores between intergrowing plates which
determines the catalytic activity.

References

Beneke, K. and Lagaly, G. (1977) 'Kanemite - innercrystal-
    line reactivity and relations to other sodium sili-
    cates." Am.Min. 62, 763-771

Beneke, K. and Lagaly, G. (1983) 'Kenyaite - synthesis and
    properties." Am.Min. 68, 818-826

Beneke, K., Kruse, H.-H., and Lagaly, G. (1984) 'Eine kri-
    stalline Kieselsäure mit hoher Einlagerungsfähig-
    keit.' Z. anorg. Allgem. Chem. 518, 65-76

Le Bihan, H.-Th., Kalt, A., and Wey, R. (1971) 'Étude
    structurale de $KHSi_2O_5$ et $H_2Si_2O_5$.' Bull. Soc. fr.
    Minéral. Cristallogr. 94, 15-23

Dörfler, H. D., Bergk, K.-H., Müller, K., and Müller, E.
    (1984) 'Adsorption von kationischen, anionischen und
    nichtionischen Tensiden an Na-, Ca-, und H-Magadiit.'
    Tenside Detergents, 21, 226-234

Frondel, C. (1979) 'Crystalline silica hydrates from
    leached silicates.' Am. Min. 64, 799-804

Guth, J.-L., Hubert, Y., Jordan, D., Kalt, A., Perati, B.,
    and Wey, R. (1977) 'Un nouveau type de silice hydra-
    tée cristallisée de formule $H_2Si_3O_7$.' C. R. Acad. Sc.
    Paris, t. 285, 1367-1370

Guth, J.-L., Hubert, Y., Kalt, A., Perati, B., and Wey, R. (1978) 'Un nouveau type de silice hydratée cristal-lisée.' C. R. Acad. Sc. Paris, t. 286, 5-8

Iler, R. K. (1964) 'Ion exchange properties of a crystal-line hydrated silica.' J. Colloid Sci. 19, 648-657

Iler, R. K. (1979) 'The chemistry of silica.' J. Wiley and Sons, New York

Kalt, A., Perati, B., and Wey, R. (1979) 'Intercalation compounds of $KHSi_2O_5$ and $H_2Si_2O_5$ with alkylammo-nium ions and alkylamines.' Proc. Intern. Clay Conf. Oxford 1978, M. M. Mortland and V. C. Farmer (eds.), Elsevier, Amsterdam, 509-516

Kiselev, A. V., Nikitin, Yu, S., Petrova, R. S., Shcherba-kova, K. D., and Yashin, Ya.I. (1964) 'Effect of pore size of silica gels on the separation of hydro-carbons.' Anal. Chem. 36, 1526-1533

Lagaly, G., Beneke, K., Dietz, P., and Weiss, A. (1974) 'Intracrystalline reactivity of phyllodisilicic acid $(H_2Si_2O_5)_\alpha$.', Angew. Chem. internat. Edit. 13, 819-820

Lagaly, G., Beneke, K., and Weiss, A. (1975) 'Magadiite and H-Magadiite.' Am. Min. 60, 650-658

Lagaly, G. (1979)'Crystalline silicic acids and their interface reactions.'Adv. Colloid Interf. Sci. 11, 105-108

Lagaly, G., Beneke, K., and Kammermeier, H. (1979) 'Neue Modifikationen des Siliciumdioxids.' Z. Naturforsch. 34 b, 666-674

Liebau, F. (1964) 'Über Kristallstrukturen zweier Phyllo-dikieselsäuren $H_2Si_2O_5$.' Z. Kristallographie 120, 427-449

Middleton, W. J., and Lindsey, R. V. (1964) 'Hydrogen bonding in fluoro alcohols.' J. Am. Chem. Soc. 86, 4948-4952

Riekert, H.-M. (1982) 'Gaschromatographische und kataly-tische Untersuchungen an kristallinen Kieselsäuren.' Thesis, Universität Kiel, Germany

Rösner, Ch. and Lagaly, G. (1984) 'Interlayer reactions
    of the silver molybdate $Ag_6Mo_{10}O_{33}$.' J. Solid State
    Chem. 53, 92-100

Rojo, J. M., Ruiz-Hitzky, E., Sanz, J., and Serratosa, J.M.
    (1983) 'Characterization of surface SiOH-groups in
    layer silicic acids by IR and NMR spectroscopies.'
    Rev. Chim. min. 20, 807-816

Rojo, J. M. and Ruiz-Hitzky, E. (1984) 'Interaction des
    amides N-substituées avec des acides siliques lamel-
    laires.' J. Chim. Phys. 81, 625-628

Ruiz-Hitzky, E., and Fripiat, J. J. (1976) 'Organomineral
    derivatives obtained by reacting organochlorosilanes
    with the surface of silicates in organic solvents.'
    Clays Clay Min. 24, 25-30

Ruiz-Hitzky, E., and Rojo, E. (1980) 'Intracrystalline
    grafting on layer silicic acids.' Nature 287, 28-29

Ruiz-Hitzky, E., Rojo, J. M., and Lagaly, G. (1985)
    'Interlamellar grafting of layer silicic acids.' Col-
    loid Polym. Sci. in press.

Steinberg, K.-H. (1976), Bremer, H., and Falke, P. (1976)
    'Integrale Adsorptionswärmen basischer Verbindungen
    an ionenausgetauschten Y-Zeolithen.' Z. Phys. Chem.
    257, 151-160

Strazkesko, N.D., Strelko, V. B., Belyakov, V. N., and
    Rubanik, S. C. (1974) 'Mechanism of cation exchange
    on silica gels.' J. Chromatogr. 102, 191-195

Voitländer, J., Wittich, E. K. H., and Lagaly, G. (1975)
    'Kernmagnetische Relaxationszeit-Messungen an OH-
    Protonen einer kristallinen Kieselsäure.' Z. Natur-
    forsch. 30 a, 1330-1331

Weiss, A., and Choy, J. H. (1984) 'Phase changes in Fe(III)
    OCl-RNH_2-Intercalation complexes.' Z. Naturforsch.
    39 b, 1193-1198

Weiss, A., Choy, J. H., Meyer, H., and Becker, H. O. (1981)
    'Hydrogen reorientation, a primary step of interca-
    lation reactions into kaolinite.' Proc. internat.
    Clay Conf. Bologna, Pavia 1981, Abstracts, 331

Werner, H.-J., Beneke, K., and Lagaly, G. (1980) 'Die
    Acidität kristalliner Kieselsäuren,' Z. anorg. allgem.
    Chem. 470, 118-130

Wey, R., and Kalt, A. (1967 a) 'Contribution à l'étude
    de l'hydrogénodisilicate de potassium. Obtention et
    synthèse.' C. R. Acad. Sci. Paris, t. 265, 1-3

Wey, R., and Kalt, A. (1967 b) 'Synthèse d'une silice hy-
    dratée cristallisée. C. R. Acad. Sc. Paris, t. 265,
    1437-1440

Wodtcke, F. and Liebau, F. (1965) 'Über die Darstellung
    zweier Modifikationen der Phyllodikieselsäure $H_2Si_2O_5$.'
    Z. anorg. allgem. Chem. 335, 178-188

Wolf, F., and Schwieger, W. (1979) 'Zum Ionenaustausch
    einwertiger Kationen an synthetischen Natriumpoly-
    silicaten mit Schichtstruktur.' Z. anorg. allgem.
    Chem. 457, 224-228

THE BEHAVIOUR OF SOME ALKALI METAL GRAPHITE COMPOUNDS TOWARDS
SUNDRY REAGENTS

Ralph Setton[1] and Charles Mazières[2]
1 C.N.R.S., Organisation Cristalline,
  1B, rue de la Férollerie
  45071 Orléans Cedex 2
  France;
2 Laboratoire de Physicochimie Minérale,
  Université de Paris Sud
  91405 Orsay Cedex
  France.

ABSTRACT. The conditions of formation of some alkali metal-molecule-graphite compounds are re-examined. A structure is proposed for the compound $KH_{0.67}C_8$ formed when gaseous H_2 reacts with KC_8; the structure fits quite well with data recently obtained by X-ray diffraction, is consistent with the value of the electronic charge residing on the C atoms and explains the low reactivity of RbC_8 and the lack of reactivity of CsC_8 towards H_2. Similarly, a critical study of the compounds formed by the binary intercalation compounds with furan, tetrahydrofuran, benzene or ammonia shows that charge transfer is of paramount importance in determining the overall composition of the ternary compounds formed since the molecules are competing with the graphitic C atoms for the valence electron of the alkali metal which affects the apparent van der Waals bulk of the third component.

1. INTRODUCTION

Graphite shares with clays and a few other species the ability to form intercalation compounds in which foreign atoms or molecules are taken up in the interlaminar space between the sheets of "graphene", the plane layers of aromatic carbon atoms. This causes the distance between the planes of graphene to change from the original 3.35 Å to a value which depends on the size of the intercalated species in a still somewhat obscure way.

Broadly speaking, two possibilities are met: in the first, the intercalate forms a two-dimensional (2D) lattice characterised by a and b parameters which are commensurable with the 2D lattice of the C atoms in the graphene layers while, in the second, the layers of intercalate and the hexagons are not commensurable.

As can be expected, commensurability is often accompanied by "nesting", especially – but not exclusively – when the guest species is

381

R. Setton (ed.), Chemical Reactions in Organic and Inorganic Constrained Systems, 381–400.
© *1986 by D. Reidel Publishing Company.*

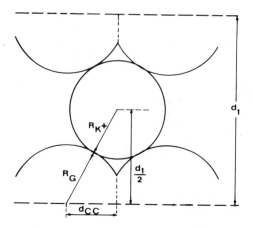

Figure 1. Hard-sphere model for the nested binary intercalation compounds of graphite.

constituted by metal atoms or ions with spherical symmetry. In the graphite intercalation compounds (GIC) formed with metals, the spheres lie over and under the centers of two hexagons in two successive graphene layers separated by a distance d_1 which can be calculated [1] with good accuracy by

$$d_1 = 2[(R_{M^+} + R_G)^2 - d_{CC}^2]^{1/2} \qquad\qquad (1)$$

where R_{M^+} is the standard radius of the ion of the intercalated metal, R_G, the van der Waals radius of the C atoms in the graphene layers, is 1.675 Å and the in-plane nearest neighbour C-C distance is $d_{CC} \simeq 1.421$ Å.
    The rationale behind this expression is obvious from Figure 1: it implies that the charge transfer to the graphene layers does not affect too markedly the π orbitals which constitute their van der Waals boundary and that, even if the charge transfer from M is not complete, only the core electrons of M are responsible for its steric characteristics. This apparently oversimplified argument has recently been vindicated by the demonstration [2] that the assumption of full charge transfer in the computation of the structural energy of the MC_8 compounds by a Thomas-Fermi pseudo-potential development leads to the correct equilibrium values of d_1 (within 2.5%) for the three MC_8.
    Table 1 shows that (1) gives, on the whole, good agreement with experiment for all the known metal compounds*, with no particular bias towards too small or too large values for d_1, for values ranging from 3.76 Å for LiC_6 to 6.06 Å for CsC_8. The values thus obtained are not sensitive to small changes in R_{M^+} since $\Delta d_1/\Delta R_{M^+} \simeq 0.4$ to 0.6; in fact,

---

*Equally good agreement is obtained with compounds containing acceptor species such as Br_2, ICl, IBr, etc.

TABLE 1 – Characteristics of some binary metal-graphite compounds

M_1	I_2	Compound$_3$	x_4	R_{Mx+} $_5$	d_1 $_6$	d_1 (obs.) $_7$
Yb	6.22	YbC$_6$	3	.86	4.20	
			2	(1.03)	4.61	4.57
Ca	6.11	CaC$_6$	2	.99	4.51	4.52
			1	1.18	4.95	
Sr	5.69	SrC$_6$	2	1.12	4.82	4.94
			1	(1.30)	5.23	
Eu	5.64	EuC$_6$	3	.98	4.49	4.87
			2	1.29	5.20	
Sm	5.60	SmC$_6$	3	1.00	4.53	4.58
			2	(1.17)	4.93	
Li	5.40	LiC$_6$	1	.68	3.76	3.63–3.74
Ba	5.21	BaC$_6$ (a)	2	1.34	5.32	5.25–5.28
			1	1.53	5.75	
Na	5.14	(NaC$_8$)	1	.97	4.46	4.48–4.52
K	4.34	KC$_8$	1	1.33	5.30	5.32–5.41
Rb	4.17	RbC$_8$	1	1.47	5.61	5.61–5.68
Cs	3.89	CsC$_8$	1	1.67	6.06	5.93–5.94

(a) A composition close to BaC$_8$ has also been reported [6].
Column 2 gives the ionisation potential of the metal in eV. Col. 4 is the valence of the ion whose radius is given in col. 5 [3], [4]. The values in columns 5 to 7 are in Å. Values in brackets () have been estimated using data in [5]. When two values are given in col. 7, they correspond to the lowest and highest values reported.

one could compute R_{M+} from d_1 for any specific boundary adopted for the orbitals on the graphene layers [7] and this, in turn, could give an estimate of the charge transfer if the relation between R_{M+} and the statistical or fractional degree of ionisation were known with sufficient accuracy, which unfortunately is not yet the case. For want of a more explicit relation, we shall retain (1) and similar expressions derived below as a means of probing the details of various reactions.

2. THE REACTION BETWEEN HYDROGEN AND THE ALKALI METAL BINARIES

The reactivity of molecular hydrogen towards the MC$_8$ binaries seems to be opposite to that of H$_2$ towards the free alkali metals: there is no reaction with CsC$_8$, although the gas immediately forms CsH with Cs; RbH$_{2/3}$C$_8$ is only formed under pressure, under conditions fairly similar to those needed to prepare RbH; KC$_8$ readily absorbs H$_2$ to yield KH$_{2/3}$C$_8$ {8} although pressure must be used to prepare KH. The ternary K-hydrogen and Rb-hydrogen compounds are both second-stage and their structure implies the formation, in the interlaminar space, of triple-decker layers, after the pattern:

$$...\overset{-}{C}\overset{+}{K}\overset{-}{C}\overset{+}{K}\overset{-}{C}\overset{+}{K}\overset{-}{C}\overset{+}{K}... + H_2 \rightarrow ...\overset{-}{C}\overset{-}{C}\overset{+}{K}\overset{-}{H}\overset{+}{K}\overset{-}{C}\overset{-}{C}\overset{+}{K}\overset{-}{H}\overset{+}{K}...$$

More recently [9], a similar compound $KH_{0.8}C_8$ was obtained by the direct action of KH on graphite, as well as a first-stage $KH_{0.8}C_4$. Although the H/K ratio is larger by 20 % in these new compounds than in $KH_{2/3}C_8$, the structure of the triple layer has been stated [9] to be the same. The values of d_1 are of course much larger than in the binaries (cf. Table 2) owing to the formation of the intercalated triple layer.

A number of questions come to mind: what is the actual structure of these ternary compounds? Why is the H/M ratio not equal to 1? Why are there apparently two possible values for the H/M ratio, namely 0.67 and 0.8? What prevents the formation of the Cs ternary?

An attempt has already been made [12] to derive a structure for $KH_{2/3}C_8$ based on results from neutron diffraction [13], EPR and NMR, but as will be seen below, since the neutron diffraction measurements had given an erroneous value for the distance between the two planes of K^+ ions, the structure was perforce incorrect.

Consider now the arrangement of species shown in Figure 2, drawn with potassium as an example and assuming that the composition of the ternary compound is KHC_8 rather than $KH_{2/3}C_8$ or $KH_{0.8}C_8$. Each H^- ion can be viewed as being both in a rhombus of four K^+ ions (placed exactly as in KC_8) and at the center of a slightly flattened tetrahedron of four K^+ ions, themselves in registry with the two graphene layers in the AB (staggered) configuration [9] rather than in the AA stacking usually found in the first-stage binaries. Simple geometric considerations yield

$$d_1(MHC_4) = d_1(MC_8) + h + d_{MH}, \tag{2}$$

$$h = [d_{MH}^2 - (m.d_{CC})^2]^{1/2}. \tag{3}$$

TABLE 2 – Characteristics of some ternary alkali metal-hydrogen-graphite derivatives

Compound 1	d_1 2	$d_1(MC_8)$ 3	d_{MH} 4	R_{M^+} 5	R_{H^-} 6	h 7	d(MH) 8	z 9
$KH_{2/3}C_8$	8.53	5.37	2.87	1.363	1.51	0.29	2.86	3.16
$RbH_{2/3}C_8$	9.03	5.65	2.89	1.485	1.41	0.49	3.02	3.38
$(CsHC_4)$	10.23	5.94	3.08	(1.67)	(1.41)	1.31	3.19	4.39

All values in Å. For $CsHC_4$ (not yet synthesised), values in brackets were assumed, those in *italics* were calculated. Column 2: d_1 observed, from [9] and [10]. Col. 3: mean observed values. Col. 4: center-to-center distance between M^+ and H^- calculated by (4). Col. 5: R_{M^+} deduced from $d_1(MC_8)$ by reversing eq. (1). Col. 6: $R_{H^-} = d_{MH}$ (col. 4) – R_{M^+} (col. 5). Col. 7: h is calculated using (3). Col. 8: center-to-center distance between M^+ and H^- in the salt-like hydride MH [11]. Col. 9 gives the difference in level between the two M^+ layers.

where h is the difference between the level of the H^- ions and that of the K^+ ions, d_{MH} is their center-to-center distance and $m.d_{CC}$ is the projection of d_{MH} on the graphene plane, here with m = 2. From (2) and (3), we obtain

$$d_{MH} = \frac{1}{2}\left\{d_1(MHC_4) - d_1(MC_8) + (m.d_{CC})^2/[d_1(MHC_4)-d_1(MC_8)]\right\}, \quad (4)$$

giving the values shown in column 4 of Table 2. In the same Table, column 8 gives the M-H distance in the corresponding salt-like hydride; comparing the values in columns 4 and 8, we note that the K-H distance in KHC_4 is only about 1 % larger than in KH, a negligible difference which implies that the charge density on the H atoms in the two compounds must be practically identical and is therefore probably 1 e^-/H atom in the ternary compound. We may have here the answer to the second question raised above: the ratio $H/K \simeq 0.67$ or 0.8 entails some empty tetrahedral cavities; this leaves about 0.33 or 0.2 electron available to ensure the presence of a charge on the graphene layers. The values are in acceptable agreement with the ^{13}C NMR results which place about 0.1 e^- per 8 C atoms on the graphene layers [12]. In fact, the calculated charge density on the C atoms due to the valence electrons which have left the K atom and are not on the H^- ion is nearly the same in both cases, namely 0.042 and 0.05 e^-/C atom respectively in $KH_{2/3}C_8$ and $KH_{0.8}C_4$.

A check on this hypothetical value of the charge density on the C atoms is provided by the calculation of d_{CC}, which is known to increase when extra electrons reside on the graphene layers; this increase is given by [14]:

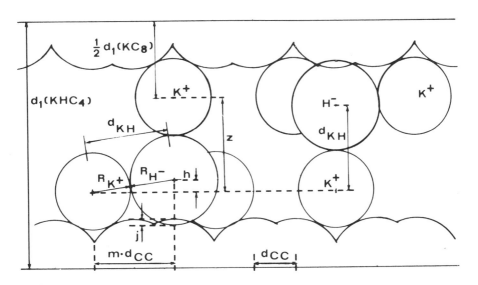

Figure 2. Presumed structure for the ternary hydrogen-alkali metal graphite intercalation compounds.

$$\Delta d_{CC} = 0.157 f_c + 0.146 |f_c|^{3/2} + 0.236 f_c^2 \tag{5}$$

where $\Delta d_{CC} = d_{CC} - d_0 = d_{CC} - 1.421$ Å and f_c = average extra charge
carried by each C atom. For $f_c = 0.042$ or 0.05, we obtain $d_{CC} = 1.429$ or
1.431 Å respectively for $KH_{2/3}C_8$ or $KH_{0.8}C_4$, in good agreement with the
value $d_{CC} = 1.429$ Å which we have calculated from very complete X-ray
diffraction results recently given [9].

A further check on the structure proposed in Figure 2 is provided
by the distance between the two layers of K^+ ions. The initial studies
by neutron diffraction [13] had suggested this distance to be 2.45 Å
but later work [15] by X-ray diffraction on samples prepared from "Highly Oriented Pyrolytic Graphite" gave the distance as 3.18 Å. From Figure
2, this distance z is found to be given by:

$$z = h + R_{M^+} + R_{H^-} \tag{6}$$

and is given in Column 9, Table 2. The value computed for KHC_4, namely
z = 3.16 Å. is in excellent agreement with the experimental value.

A last confirmation could apparently be found in the composition of
the compound $KH_{0.36}C_{24}$ formed when H_2 reacts on the second-stage KC_{24}
[16] since the quantity of hydrogen retained would leave enough charge
from the K atoms to give a concentration $(1 - 0.36)/24 = 0.027$ e$^-$/C atom,
quite comparable to the values found above. It must however be admitted
at this point that one can see no simple means of reconciling the structure shown with a formula such as $KH_{0.36}C_{24}$ which implies that only
one cavity out of three can contain an H^- ion.

Another difficulty becomes apparent when we turn to RbC_8 and CsC_8:
the binary rubidium compound, with $d_{CC} = 1.434$ Å [17], must have a larger electron transfer to the graphene layers than KC_8 with $d_{CC} = 1.432$ Å,
as can be expected from the fact that the ionisation potential of rubi-
ium, I(Rb), is smaller than that of potassium (cf. column 2, Table 1).
If the reason for the lower affinity of RbC_8 for hydrogen is the smaller
number of electrons available to form the hydride ion, one can expect,
by extension, that no ternary Cs-H GIC will be formed since I(Cs) is
smaller than I(Rb), as indeed is the case. Yet, extremely careful measurements have yielded $d_{CC} = 1.427$ Å [18] to 1.428 Å [17] for the in-
plane C-C distance in CsC_8, corresponding to a *smaller* amount of extra
charge on the graphene layers in the Cs binary than in the Rb or the K
binaries. Hence, either (5) above does not apply to the Cs compounds or,
as seems more likely, factors other than mere electron transfer in the
binary MC_8 must govern the reactivity towards molecular hydrogen.

It has been suggested [19] that the lack of reactivity of CsC_8 may
be due to the impossibility for the H_2 molecules to circulate between
the graphene layers owing to the steric hindrance offered by the bulky
Cs^+ ions, but one must confront this argument with the fact that CsC_{24}
physisorbs large quantities of gaseous H_2 at low temperatures [20], that
this gas has been proved to reside between the planes of graphene [21]
and that the process is totally reversible upon raising the temperature.
Apparently, the steric hindrance to a process of diffusion, which at
best would only explain a serious decrease in the kinetics of the reaction but not the total lack of reactivity, cannot be the answer.

Let us then turn again to Table 2. where we note that both the value of the center-to-center distance d_{RbH} and that of the radius of the
H^- ion are smaller by sizable amounts than the corresponding values in
the hydride RbH, but that R_{Rb^+} is not very different from the value
given in Table 1. Assuming now that R_{Cs^+} in the as-yet-unknown $CsHC_4$
would have its usual value as given in Table 1, we can compute the characteristic distances of the ternary cesium-hydrogen-graphite compound
(cf. Table 2); their only remarkable feature is the extremely large
value of h which increases steadily as the size of the M^+ ion increases.
We are thus led to wonder if the key to the ternary compounds could not
be found in an overlap between the 1s orbital of the H^- ion and the π
orbitals of the graphene layers. The hard-sphere model which was
presented   leads to the conclusion that this overlap will be possible
whenever (Figure 2)

$$j = R_{H^-} + R_G - h + d_1(MC_8)/2 > 0$$

and that this overlap exists in the K-H GIC in which $j = 0.2$ Å, could
become possible under pressure in $RbH_{2/3}C_8$ for which $j = -0.2$ Å, but is
likely to be impossible in any cesium compound with this structure
which leads to $j = -1.2$ Å.

3. THE REACTION BETWEEN SOME MORE COMPLEX MOLECULES AND THE ALKALI
   METAL BINARIES

More complex molecules, such as furan (C_4H_4O, here noted Fu), benzene
(C_6H_6, Bz), tetrahydrofuran (C_4H_8O, THF), hexamethylphosphorotriamide
[$O=P(NMe_2)_3$, with Me = CH_3, HMPT] etc. also react with the alkali GIC's.
In many cases [22], the ternary compound can be obtained by letting the
third component react on the binary GIC while, in others, the ternary
is formed directly by co-intercalation when graphite reacts on solutions
of the alkali metal in the appropriate solvent (NH_3 [23], HMPT [24]).
Fundamental to the possibility of intercalating the third component is
the existence of some bond between it and the alkali metal. This bond
may be strong enough to give rise to irreversible reactions between the
intercalated species [22][25], so that, in most cases, the intercalation
of the third component is *not a fully reversible process* and one must be
careful to distinguish between the first intercalation of the third component and subsequent ones since attempts at de-intercalation often lead
to chemical reactions. Understanding the course of any of these subsequent reactions is only possible if the process involved in the first
intercalation is well understood and we shall therefore limit ourselves
to its study.

Painstaking measurements of the exact quantity of third component
taken up during the "ternarisation" of a binary first- or second-stage
compound MC_n revealed at first that whole-number stoichiometries are
purely fortuitous so that, once the ternarisation has started, it will
continue until all the available space is full rather than until the
ratio x/n for the compound MX_xC_n has reached an integral value. In fact,
the value of x can often be obtained within a few percent from a know-

ledge of d_1, of the stage s' of the ternary, and of n, using

$$x = [n'(2.62)(d_1 - 3.35) - V_M]/V_X \qquad (7)$$

in which n' = n/s', 2.62 is the area in $\mathring{A}^2$ attributed to a single C atom
in a graphene layer, V_M is the volume of the metal atom or ion and V_X is
the volume attributed to the molecule, both in $\mathring{A}^3$ [36][27]. The choice
of V_X presents somewhat of a problem and, initially, the molecular volume was obtained from the molecular weight and the density at room temperature. The reason for this choice was the belief that the steric requirements were principally a function of the molecular movements and,
therefore, of the temperature, so that, to a good approximation, the intercalated molecule needed about as much room as the molecule in the
free state. As will be seen later, this assumption turned out to be
wrong, although it does yield substantially correct results in many cases. Consider, for instance, the ternarisation of the second-stage KC_{24}
by benzene at room temperature [27], giving the first-stage $KBz_{2.50}C_{24}$,
with d_1 = 9.23 $\mathring{A}$. Using the values V_{K+} = 10 $\mathring{A}^3$ and V_{Bz} = 148 $\mathring{A}^3$, we
find, from equation (7), x = 2.43, quite close to the experimentally de-
termined value x = 2.50. Note that the molecular volume used in the com-
putation was obtained from the density of liquid benzene at room temperature. Other compounds give equally satisfactory results, as shown in
Table 3 which also contains, for comparison, results obtained with some
acceptor GIC's. Extensive investigations on the applicability of this
"rule" however soon turned up cases which seemed to show either an erra-
tic behaviour or, more interestingly, a systematically deviant behaviour.
The ternarisation of the K-graphite binaries by the heteroaromatic
furan molecule is a case in point which we shall now examine.

3.1. The potassium-furan ternary compounds

Furan is one of the few molecules which will ternarise the first-stage
KC_8 as well as the higher-stage compounds [27][28]. Ternarisation was
carried out by allowing furan vapour to react on the pure first- to
sixth-stage binaries prepared from 50 μm graphite powder, under strictly
controlled conditions of temperature, for the solid phase, and pressure,
for the vapour phase. By working on known masses of the binary GIC and
monitoring the quantity of furan which is absorbed, the overall composition of the ternary formed is easily attained.
    Table 4 [28] gives the stoichiometry of the ternary GIC's thus
obtained, as well as some relevant data.
    The first striking fact related to this ternarisation process is
that, as from the third stage on, a binary of stage s always yields a
ternary of stage s' = s-1. X-ray diffractograms show that the ternaries
are pure single phases and only in the ternarisation of KC_{24} does one
obtain a mixture, and this only under very specific conditions. Since
the stage s' and d_1 are known, we can compare the observed values of x
(column 2, Table 5) to the ones calculated using (7), with V_{Fu} = 120 $\mathring{A}^3$,
the molecular volume at room temperature. Except for $KFu_{1.30}C_{24}$, the
agreement is quite poor and gets systematically worse as the stage increases: the observed values of x slowly increase with the stage while

TABLE 3 - Characteristics of some ternary GIC's and test of the applicability of equation (7)

Compound 1	s' 2	d_1 (Å) 3	x (obs.) 4	x (calc.) 5	
Donor GIC's					
$Na(HMPT)_x C_{27}$	1	7.63	1.05-1.19	1.02	[24]
$Li5HMPT)_x C_{32}$	1	7.48	1.16-1.77	1.16	[24]
$K(Fu)_x C_8$	1	8.77	0.75-0.78	0.88	[28]
$K(THF)_x C_{24}$	1	8.90	2.55	2.54	[29]
$K(Bz)_x C_{24}$	1	9.23	2.50	2.43	[27]
Acceptor GIC's					
$C_{23}SbF_6(NM)_x$	1	8.05	1.70	1.77	[30]
$C_{48}PF_6(NM)_x$	2	8.05	2.10	2.18	[30]
(NM = nitromethane)					

TABLE 4 - Characteristics of the potassium-furan-graphite ternary compounds

Binary KC_n	Stage s	I_c (Å)	Ternary $KFu_x C_n$	Stage s'	I_c (Å)
KC_8	1	5.40	$KFu_{0.75}C_8$	1	8.78
KC_{24} (a)	2	8.73	$\begin{cases} KFu_y C_{24} \\ KFu_{1.30}C_{24} \end{cases}$	1 / 2	8.77 / 12.12
KC_{36}	3	12.12	$KFu_{2.20}C_{36}$	2	12.16
KC_{48}	4	15.26	$KFu_{2.30}C_{48}$	3	15.48
KC_{60}	5	18.75	$KFu_{2.30}C_{60}$	4	18.85
KC_{72}	6	22.15	$KFu_{2.35}C_{72}$	5	22.30

(a) Depending on the initial conditions, the ternarisation of KC_{24} will give either the pure second-stage ternary $KFu_{1.30}C_{24}$ or a mixture of this compound with a first-stage $KFu_y C_{24}$ which eventually is transformed into the second-stage GIC by loss of part of its furan.

the calculated values decrease.

The experimental results can yield a parameter which is somewhat easier to grasp than a "calculated value of x" computed on the basis of an arbitrarily chosen value of V_{Fu}. Using the experimental values of d_1 and of x, let us calculate V_A, the *volume apparently occupied by one molecule* of furan in the ternaries of stage 1 to 5 (column 4, Table 5). We find that V_A decreases from 138 to 85 $\mathring{A}^3$, and Figure 3 shows that the relation between V_A and s' is apparently hyperbolic. A least-squares fit yields:

$$V_A = 67.7 + 85.6/s'. \tag{8}$$

Thus, as s' increases, V_A tends towards the limiting value 67.7 $\mathring{A}^3$, which is over 40 % less than the value 120 $\mathring{A}^3$, the room-temperature molecular volume initially believed to be always applicable. This limiting value is quite close to that of the van der Waals volume which was defined by Bondi [31] as the volume "impenetrable for other molecules with thermal energies at ordinary temperature" and which is an additive property of the constituents of the molecule. In the present case, the van der Waals volume of furan, computed with the atomic and group contributions given by Bondi, is 62.4 $\mathring{A}^3$/molecule.

It is also important to note that for s' = 1, V_A is even larger than the room-temperature molecular volume of furan. Why then do the molecules apparently repel one another, especially in the lower-stage ternaries, and why should this repulsion decrease when the ratio of graphitic C atoms to the number of furan molecules increases? We believe this effect to be related to a *varying amount of charge carried by the molecules or affecting their apparent steric requirements* when, as in THF or ammonia, the electronic structure of the molecule is incompatible with the formation of a stable ion.

It is known that, in some alkali metal ternaries, the organic moity bears a negative charge, as confirmed by the Raman frequencies of the intercalated species [32] or the shifts in ^{13}C NMR spectra [27]. The organic molecules are therefore competing with the graphene layers for the charge donated by the K atom. If we assume, as a first approximation, that any charge not located on the graphene layers is equally shared among all the organic molecules, it should be possible to assess the non-G* electrons by monitoring d_{CC} in the ternaries since this parameter is directly related to the charge on the graphene layers. When the X-ray diffractogram of the ternary can be fully indexed on the basis of a 2D lattice commensurable with the graphene layers, d_{CC} can be obtained from the *a* and *b* cristallographic parameters after adequate statistical treatment and weighting of the parametral *d* values; when full indexing is not possible or when the intercalate forms an incommensurable layer or a 2D liquid, d_{CC} can still be obtained from the two reflexions always found at positions close to 2.13 and 1.23 Å which, in the case of graphite, are indexed as (100) and (110) and are therefore respectively equal to $(3/2)d_{CC}$ and $(\sqrt{3}/2)d_{CC}$.

We have obtained d_{CC} from the diffractograms of all the compounds prepared (column 3, Table 6). In order to compute the amount of non-G electrons, we must first choose between the following hypotheses:

---

*For convenience, we shall refer to these electrons which are not on the graphene layers as "non-G electrons".

TABLE 5 – Comparison of observed and calculated values of x and V_A in the potassium-furan-graphite ternary compounds

Compound KFu_xC_n 1	s' 2	x (calc.) 3	V_A (obs.) 4	V_A (calc.) 5
$KFu_{0.75}C_8$	1	0.87	138	153
$KFu_{1.30}C_{24}$*	2	1.34	123	111
$KFu_{2.20}C_{36}$	2	2.06	113	
$KFu_{2.30}C_{48}$	3	1.81	95	96
$KFu_{2.30}C_{60}$	4	1.70	89	89
$KFu_{2.35}C_{72}$	5	1.66	85	85

*Pure phase

The values of x in column 3 are calculated using eq. (7), with V_{Fu} = 120 $Å^3$ and V_{K^+} = 10 $Å^3$. V_A (column 4) is the volume, in $Å^3$, apparently taken up by a molecule of furan in the interlaminar space, while V_A (column 5) is calculated using eq. (8).

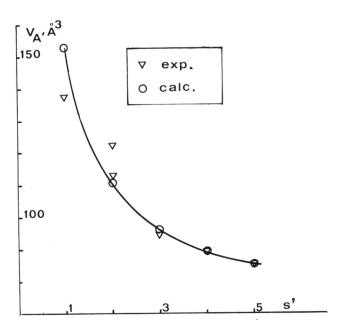

Figure 3. Variation of the apparent volume V_A of furan in the potassium-furan ternary GIC's, as a function of the stage s'. The circles are the values of V_A calculated with the help of equation (8) whose graph is shown by the full line.

    a) only the C atoms in the "bounding" graphene layers (i.e. those immediately in contact with the intercalate) bear a charge;

    b) the charge on the graphene layers is evenly distributed among all the C atoms.

Relation (5) above was given by the authors [14] as being only strictly applicable to second-stage compounds, but as being extendable to higher stages within certain restrictions. In the light of many arguments, first among which being the fact that the diffractograms point to a single value for d_{CC} rather than to a range of values, as already pointed out [33] and confirmed by our own diffractograms, we opt for hypothesis b above and assume that (5) is applicable to all the C atoms of the graphene layers. Column 4 in Table 6 gives f_c, as calculated from d_{CC}, while column 5 gives p, the non-G electrons per molecule, as calculated by dividing out the charge not residing on the graphene layers among the organic molecules present. A plot of V_A vs p (Figure 4) shows a gratifyingly smooth correlation between these parameters. It is interesting that, in the first-stage $KFu_{0.75}C_8$, with the least number of carbon atoms and the highest K/Fu ratio, the organic molecule bears a full charge since $p = (1 - n.f_c)/x = 1$. Thus, it is only because there is less than one molecule of furan per K atom that some charge is left for the graphene layers. The situation is analogous to the one met with the K-H ternaries, which also show a ratio $K/H > 1$ (again to ensure the presence of charge on the graphene layers) but differs in the apparent impossibility to form higher-stage ternaries. In the furan derivatives, the increasing number of C atoms in the graphene layers efficiently trap the charge which has less and less of a chance to jump onto an organic molecule. The value of x, the ratio of furan molecules to K atoms, is mainly fixed by the furan-K coordination or solvation relationship and is therefore approximately constant and independent of the charge on the furan; in the K-H ternaries, the homologous situation cannot arise owing to the necessity of having fully charged H^- ions and to the eventual collapse of the structure shown in Figure 2 if it were to contain an insufficient number of H^- ions to act as spacers.

3.2. Other alkali metal ternary GIC's

We have tried to find confirmation of the above findings in the data on other compounds of similar nature dispersed in the litterature. The main problem lies in the difficulty of obtaining simultaneously the values of the interplanar distance d_1, of the composition x, and of the in-plane C-C distance d_{CC} for a given compound. Only in a very small number of cases were we able to obtain sufficient or complete data.

3.2.1. The potassium-benzene ternaries. Table 7 gives data relevant to three K-benzene ternary compounds [27][34][35], two of which are second stages containing different amounts of benzene. The scarcity of data does not permit to see more than a trend (cf. Figure 5) but this is in the same direction as that of the furan derivatives.

3.2.2. The alkali metal-ammonia ternaries. These compounds, which have been known for over thirty years, have recently been the subject of new

TABLE 6 – Structural and charge characteristics of the potassium-furan-graphite ternary compounds KFu_xC_n

s' 1	Ternary 2	d_{CC} (Å) 3	$10^2 \cdot f_c$ 4	p 5	V_A (Å³) 6
1	$KFu_{0.75}C_8$	1.4272	3.25	1.0	138
2	$KFu_{1.30}C_{24}$	1.2380	1.56	0.48	123
2	$KFu_{2.20}C_{36}$	1.4235	1.41	0.22	113
3	$KFu_{2.30}C_{48}$	1.4234	1.35	0.15	95
4	$KFu_{2.30}C_{60}$	1.4239	1.62	0.013	89
5	$KFu_{2.35}C_{72}$	1.4233	1.30	0.063	85

Column 1 gives s', the stage of the ternary GIC whose formula is given in column 2. Col. 4: f_c is the excess charge carried by each carbon atom in the graphene layers, computed using eq. (5) and the value of d_{CC} in col. 3. Col. 5: non-G electron per furan molecule. Col. 6: apparent volume of each molecule of the third component.

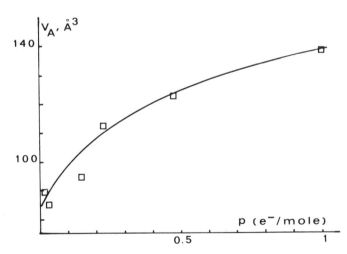

Figure 4. Variation of the apparent volume V_A as a function of the charge borne by each molecule of furan (the line is a guide to the eye).

investigations. The interlaminar distance d_1 is the parameter which is least difficult to obtain but the older references [36][23] give "ideal (i.e. whole number) formulas" rather than actual compositions, while more recent publications [37][38] give the value of x with excellent accuracy but make no mention of d_{CC}. From the few cases where all the necessary information was available [36], we have extracted values of V_A

TABLE 7 - Structural and charge characteristics of some benzene-
potassium-graphite ternary compounds

s'	Ternary	I_c (Å)	d_{CC} (Å)	$10^2 \cdot f_c$	p	V_A (Å³)
1	2	3	4	5	6	7
1	$KBz_{2.33}C_{24}$	9.31	1.4240	1.67	0.26	156
2	$KBz_{2.45}C_{48}$	12.66	1.4213	0.18	0.37	149
2	$KBz_{1.02}C_{24}$	12.45	1.4228	1.10	0.72	168

See the notes in Table 6 for the identification of the items in various
columns.

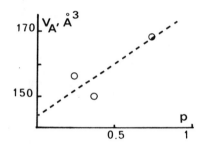

Figure 5. Variation of the apparent volume V_A as a function of the
charge borne by the molecules of benzene in three ternary compounds.

and of p for four compounds, as presented in Table 8 and Figure 6-a.
Assuming that the identity of the alkali metal in the ternary GIC does
not affect V_A, we find a trend of variation vs p which is now opposite
to the ones met previously.
     There is, of course, a major difference between NH_3 and unsatura-
ted molecules such as furan or benzene: the latter have some empty low-
lying antibonding electron states which can accomodate the excess negative charge. Even if the charged entity thus formed now has a markedly
decreased stability (due for instance to the partial loss of resonance
stabilisation) and will therefore undergo chemical changes more readily
[25], it is a perfectly legitimate form, as opposed to NH_3^- which would
not survive its formation. Since the non-G electrons cannot reside on
the molecules, they must merely act as poles towards which the NH_3 dipoles will point their positive end, thus losing some of their freedom
of motion and lessening their steric requirements. One can expect that
this effect should become more important as the concentration of non-G
electrons increases - as in Figure 6-a - but that sufficiently high values of the e^-/molecule ratio could bring about a reversal of the trend
owing to the repulsion of the electrons by each other and to the formation of voids or vacancies, if the present e^-/NH_3 system were finally

TABLE 8 - Structural and charge characteristics of some alkali metal-ammonia-graphite ternary compounds.

s' 1	Ternary 2	I_c (Å) 3	d_{CC} (Å) 4	p 5	V_A (Å³) 6
1	$Li(NH_3)_{2.1}C_{11.6}$	6.62	1.4290	0.25	47
1	$Na(NH_3)_{2.2}C_{16.5}$	6.64	1.4302	0.11	63
2	$Na(NH_3)_{2.3}C_{26.9}$	9.95	1.4264	0.10	48
1	$K(NH_3)_{2.1}C_{12.5}$	6.56	1.4272	0.28	45

See the notes of Table 6 for the identification of the items in the various columns.

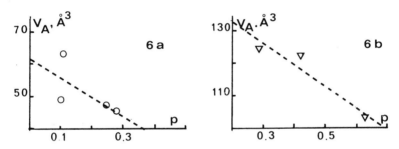

Figures 6-a (left) and 6-b (right). Variation of the apparent volume V_A as a function of the charge per molecule of third component. Figure 6-a: ternary ammonia compounds $M(NH_3)_xC_n$ (cf. Table 8). Figure 6-b: ternary lithium-THF compounds $Li(THF)_xC_n$ (cf. Table 9).

TABLE 9 - Structural and charge characteristics of some lithium-tetrahydrofuran-graphite ternary compounds

s' 1	Ternary 2	I_c (Å) 3	d_{CC} (Å) 4	p 5	non-G 6	V_A (Å³) 7
1	$Li(THF)_{1.4}C_6$	12.53	1.4244	0.63	0.88	102
1	$Li(THF)_{2.3}C_{12}$	12.44	1.4214	0.42	0.97	123
1	$Li(THF)_{3.42}C_{18}$	12.45	1.4216	0.28	0.96	125

Column 6: total number of non-G electrons per Li atom. See the notes of table 6 for the identification of the items in the other columns.

to behave like the solvated electrons found in solutions of alkali
metals in liquid ammonia [39][40] which are, however, much less concentrated, with at most one K atom/5 NH_3 as against one K atom/1 or 2 NH_3
in the present case.

3.2.3. The lithium-tetrahydrofuran ternaries.

Table 9 gives d_1, x [41]
and d_{CC} [42] for three ternaries of Li and THF. They are particularly
interesting since they all are first-stage, have approximately the
same value of d_1 ($\simeq$ 12.5 Å) and of the n/x ratio with about 4 to 5
C atoms/THF molecule, but different quantities of metal. As shown in
Figure 6-b, once again V_A decreases as p increases.

In spite of the likeness to the behaviour of the NH_3 ternaries,
the reason of the variation of V_A is probably quite different: it has
been shown [41] that the Li^+ ion is strongly solvated in the ternaries
and is at the center of a tetrahedron of four THF molecules. Because of
this coordination, the THF molecules lose a large part of their mobility, and this shows up in V_A which changes from 135 $Å^3$ (the value for
liquid THF at room temperature) to 102 $Å^3$ in $Li(THF)_{1.46}C_6$, where the
Li/THF ratio is so high that the chances of finding a non-coordinated
THF molecule must be practically nil. As a result, V_A is now halfway
to the van der Waals value, 77 $Å^3$.

In the other two Li-THF ternaries, V_A is higher, indicating perhaps the presence of some non-coordinated THF, thanks to the fact that
the total quantity of Li has decreased. As seen from the values in
column 6, Table 9, most of the electrons do not reside in the graphene
layers; this is confirmed by the nearly symmetrical and non-Dysonian
shape of the EPR absorption line [41].

With other alkali metals, which do not possess the high polarising
ability of Li, the situation is quite different.

3.2.4. The potassium-tetrahydrofuran ternaries.

Although these ternary
compounds have been studied extensively, once again it proved to be
difficult to obtain coherent sets of values of d_1, x and d_{CC}. Some of
the available data yielded, for instance, values of d_{CC} smaller than
1.421 Å which, if confirmed, would be characteristic of graphene donating rather than receiving electrons; this matter obviously needs
further investigations but, for the time being, we decided to leave
these points out of the group adopted.

There are two series of ternary potassium-THF compounds: one with
$d_1 \simeq$ 7.2 Å, in which the THF molecules are approximately parallel to
the layers, and one with $d_1 \simeq$ 8.9 Å in which the THF stands nearly perpendicular to the layers [26][29][34][43]. The relevant data [34][42]
pertaining to some of these compounds are presented in Table 10 and
Figure 7. Although the data are not as complete as for the furan derivatives and seem more widely dispersed, V_A apparently increases with p,
in spite of the fact that the charge cannot be incorporated in one of
the molecular orbitals of THF. We must then be here in the presence of
the second aspect of the solvation of electrons by polar molecules,
with the excess volume resulting from cavities known to exist in this
type of solution [40].

TABLE 10 – Structural and charge characteristics of some potassium-
        tetrahydrofuran-graphite ternary compounds

s' 1	Ternary 2	I_c (Å) 3	d_{CC} (Å) 4	p 5	V_A (Å^3) 6
1	K(THF)$_{2.55}$C$_{24}$	8.91	1.4216	0.36	133
3	K(THF)$_{3.0}$C$_{72}$	15.65	1.4221	0.18	114
1	K(THF)$_{1.61}$C$_{24}$	7.16	1.4213	0.59	142
2	K(THF)$_{1.76}$C$_{48}$	10.53	1.4216	0.47	131
2	K(THF)$_{2.1}$C$_{60}$	10.46	1.4213	0.42	136
3	K(THF)$_{2.0}$C$_{72}$	13.85	1.4218	0.33	114

See the notes of Table 6 for the identification of the items in the
various columns.

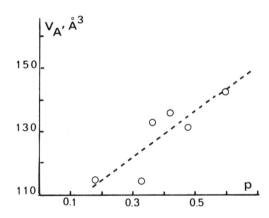

Figure 7. Variation of the apparent volume V_A as a function of p in
six potassium-THF-graphite ternary compounds.

4. CONCLUSION

The process which transforms a binary GIC into a ternary compound is
quite complex and involves specific changes in all three components.
Beyond the obvious modifications – such as an increase in the interlaminar spacing or a change in colour – lie a host of more subtle processes principally connected with changes in the distribution of the
electrons.
    The very fact that the third component has been accepted implies
that some sort of bond, weak or strong, must exist between the inter-

calated metal and the invading molecule. In order to form this bond,
a redistribution of the population of the electronic energy levels
must occur, which often implies, as a first step, an increased strip-
ping of the valence electron from the donor species. The result is not
necessarily an increase in the number of negative charge carriers residing on the graphene layers since the electrons are now offered other
possibilities: forming a loose association with the third component (a
"solvated" electron), or a much stronger association by using one of
the available antibonding orbitals, when they exist. All these are
reversible processes in which the three species are competing for the
valence electron so that, eventually, some kind of equilibrium will be
reached. When the higher stages undergo this "ternarisation" process,
this equilibrium may be very heavily biased in favour of a transfer
to the graphene layers by the sheer weight of number of the graphitic
C atoms present; each C atom acts as a potential well for the electron
but the final concentration of charge carriers in the graphene layers
will be small.

If X, the third component, is unsaturated and therefore has one
or more π orbitals, it can easily accept an extra electron on the first
low-lying antibonding orbital; this leads to a charged ion-molecule X^-
with a diffuse and enlarged outer boundary which will steer away from
other similarly charged ion-molecules, and the average amount of space
needed to accomodate the bulk of X will increase. Counterbalancing this
effect is the attraction of the M^+ ion on the electron-rich π bonds
which tends to form MX_x or $MX_x^{\delta-}$ clusters; this effect must eventually
win out when, in the higher stages, most of the charge is delocalised
in the graphene layers. The decrease in the excess charge and the clus-
tering both contribute to drive the molecular volume of X towards the
van der Waals value, the lowest and irreducible limit.

If X is a saturated molecule, electron-dipole associations are
formed which, at first, tend to restrict the tumbling motion of X and
hence decrease its steric requirements. For some electron-dipole combinations, the increase in the volume apparently occupied by each molecule of X can be due to the formation of cavities or voids already
known to appear when electrons are solvated by such molecules as THF,
ammonia, etc.

The present model, based on the interpretation of the value of x
in MX_xC_n as viewed through V_A, the volume apparently occupied by each
molecule of X, seems consistent with most of the data now on hand. It
will be interesting to check its validity in future research as appropriate measurements become available in this so far little explored
aspect of the ternary graphite intercalation compounds.

ACKOWLEDGEMENTS. Partial financial support by NATO (Research Grant No.
98/80) is gratefully acknowledged.

BIBLIOGRAPHY

1   R. Setton, in *"Les Carbones"*, by the Groupe Français d'Etudes des Carbones, Masson, Paris 1965, T. II, p. 645
2   D.P. DiVincenzo and E.J. Mele, *Phys. Rev. Letters*, 53 (1984), 52
3   D'Ans-Lax, *"Taschenbuch für Chemiker und Physiker"*, Springer-Verlag, Berlin 1970, Band III, p. 258
4   *"Handbook of Chemistry and Physics"*, 44th ed., Chemical Rubber Publishing Company, Cleveland 1963, p. 3507
5   K.W.F. Kohlrausch, *Acta Physica Austriaca*, 3 (1950), 452
6   L.A. Girifalco and T.O. Montalbano, *J. Mat. Science*, 11 (1976), 1036
7   D. Guérard and P. Lagrange, *J. Chim. Physique*, 81 (1984), 853
8   D. Saehr and A. Hérold, *Bull. Soc. Chim. France*, (1965), 3130
9   D. Guérard, C. Takoudjou and F.Rousseaux, *Synthetic Metals*, 7 (1983), 43
10 P. Lagrange, D. Guérard, A. Hérold and D.G. Onn, *Extended Abstracts 15th Biennial Conference on Carbon*, American Carbon Society, Philadelphia (1981), p.383
11 Ref. 3, Band I, p. 196
12 J. Conard, H. Estrade-Swarckopf, P. Lauginie, M. El-Makrini, P. Lagrange and D. Guérard, *Synthetic Metals*, 2 (1980), 261
13 P. Lagrange, Thesis, Nancy 1977.
14 L. Pietronero and S. Strässler, *Phys. Rev. Letters*, 47 (1981), 59
15 D. Guérard, C.Takoudjou and F. Rousseaux, *Abstracts, 6th London International Carbon and Graphite Conference*, Society of Chemical Industry, London 1982, p. 85
16 A. Hérold in Ref. 1, p. 517
17 J. Jegoudez, C. Mazières and R. Setton, unpublished results
18 D. Guérard, P. Lagrange, M. El-Makrini and A. Hérold, *Carbon*, 16 (1978), 285
19 T. Enoki, M. Sano and H. Inokuchi, *J. Chem. Phys.*, 78 (1983), 2017
20 K. Watanabe, T. Kondow; T. Onishi and K.Tamaru, *Chemistry Letters*, (1978), 51
21 A. Hérold, in *"Intercalated Layered Materials"*, F.A. Lévy ed., Reidel, Dordrecht 1979, Vol. 6, P. 394
22 J. Jégoudez, C. Mazières and R. Setton, *Synthetic Metals*, 7 (1983), 85
23 W. Rüdorff and E. Schulze, *Angew. Chemie*, 66 (1954), 305
24 D. Ginderow and R. Setton, *Carbon*, 6 (1986), 81
25 F. Béguin, J. Jégoudez, C. Mazières and R. Setton, *C. R. Acad. Sci. Paris*, 233 II (1981), 969
26 R. Setton, F. Béguin, J. Jégoudez and C. Mazières, *Rev. Chimie Minérale*, 19 (1982), 360
27 F. Béguin, R. Setton and L. Facchini, *Synthetic Metals*, 7 (1983), 263
28 J. Jégoudez, Thesis, Orsay 1985
29 F. Béguin and B. Fernandez, personal communication.
30 D. Billaud, P.J. Flanders, A. Pron and J.E. Fischer, *Mater. Sci. Eng.*, 54 (1982), 31
31 A. Bondi, *J. Phys. Chem.*, 68 (1964), 441
32 S. Matsuzaki and M. Sano, *Chem. Phys. Letters*, 115 (1985), 424

33 D.E.Nixon and G.S. Parry, *J. Phys. C*, $\underline{2}$ (1969), 1732
34 F. Béguin, Thesis, Orléans 1980
35 F. Béguin and R. Setton, unpublished results
36 E. Schulze, Thesis, Tübingen 1954
37 S.K. Hark, B.R. York, S.D. Mahanti and S.A. Solin, *Solid State Common.*, $\underline{50}$ (1984), 595
38 S.K.Hark, B.R. York and S.A. Solin, *Synthetic Metals*, $\underline{7}$ (1983), 257
39 W.C. Johnson and A.W. Meyer, *J. Amer. Chem. Soc.*, $\underline{54}$ (1932), 3621
40 J.L. Dye, *Scientific American*, $\underline{216}$ (1967), 76
41 F. Béguin, H. Estrade-Swarckopf, J. Conard, P. Lauginie, P. Marceau, D. Guérard and L. Facchini, *Synthetic Metals*, $\underline{7}$ (1983), 77
42 F. Béguin, unpublished results
43 F. Béguin, R. Setton, A. Hamwi and P. Touzain, *Mater. Sci. Eng.*, $\underline{40}$ (1979), 167

MECHANISM OF THE CATALYTIC GASIFICATION AND REACTIVITY OF GRAPHITE

Heinz Heinemann and Gabor A. Somorjai
Lawrence Berkeley Laboratory
University of California
Berkeley, CA 94720

ABSTRACT. Graphite is being used as a carbon source for gasification
to insure that hydrogen or hydrogen in hydrocarbons is derived from
water. Relatively low temperatures (500-800K) are used to favor the
equilibrium $C+2H_2O \rightarrow CH_4+CO_2$ which is almost thermally neutral.
In the presence of alkali hydroxide C_2-C_6 hydrocarbons are formed
in addition to H_2, CH_4 and CO_2. Formation of hydrocarbons is a
stoichiometric reaction proceeding on the crystal edges to form a
phenolate and hydrocarbons, e.g. $5C+4KOH \rightarrow 4COK+CH_4$. Surface
spectroscopy has confirmed the presence of phenolate, which can be
decomposed over metal oxide to make the reaction truly catalytic:
$4COK \xrightarrow{MeOx} 2K_2O+2C+2CO$; $2K_2O+2H_2O \rightarrow 4KOH$. In the presence of
both KOH and metal oxide the major products are CO_2 and H_2 with
traces of CH_4 and CO. The 2:1 ratio of H_2/CO_2 appears due to
either or both watergas shift and Boudouart reaction. The mechanisms
will be illustrated by electron microscopy and XPS measurements.

The production of synthesis gas by reaction of various carbonaceous
deposits with steam has been frequently investigated and is a
commercial process. Gasification in the presence of alkali metal
compounds has been reviewed by Wen.[1] The production of low
molecular weight gaseous hydrocarbons directly from carbon or from
carbonaceous deposits provides an intriguing alternative to syngas
production with subsequent methanation or liquefaction. The reaction
of carbon with water to produce methane and carbon dioxide $2C + 2H_2O
\rightarrow CH_4 + CO_2$ is virtually thermoneutral (ΔG_{298K} = 2.89 Kcal/mol)
and thermodynamically feasible at low temperatures. By contrast
studies of coal gasification with water have been always carried out
at high temperature regimes in order to have efficient production of
carbon monoxide and hydrogen, a very endothermic reaction. More
recent studies by Exxon researchers[2] reported the production of
substantial amounts of methane along with CO and hydrogen during coal
gasification using potassium carbonate as a catalyst. These studies
employed relatively high temperatures (900-1100 K).
  We have studied the gasification of graphite in the presence of

401

R. Setton (ed.), Chemical Reactions in Organic and Inorganic Constrained Systems, 401–410.

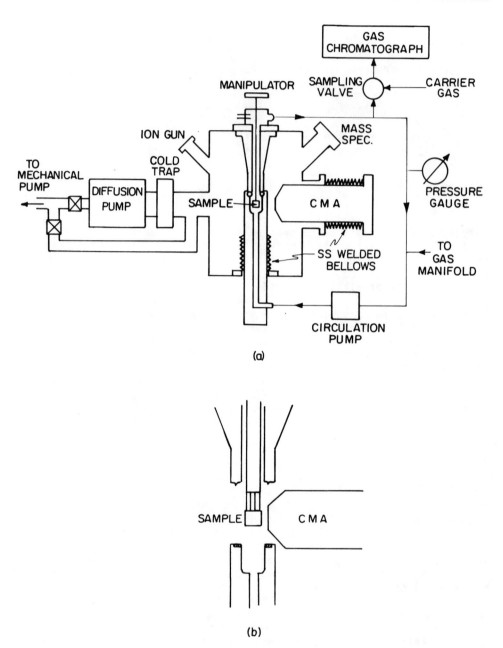

Figure 1.   High pressure-low pressure ultra-high vacuum cell.

alkali hydroxide and in some cases of earth alkali hydroxide at temperatures in the range 500-800 K and atmospheric pressure. We have used graphite as a carbon source because of its lack of hydrocarbonaceous material and in order to be sure that all hydrogen produced as hydrogen or in hydrocarbons was derived from hydrogen in water. In a few experiments we have demonstrated that char derived from an Illinois No. 6 coal gasifies at better rates but with the same product distribution as graphite.

Two types of equipment have been used in our experiments which were designed to study not only gasification but the mechanism of hydrogen and of hydrocarbon production. These are shown in Figures 1 and 2. Figure 1 shows a high pressure-low pressure (ultra high vacuum) cell opperating at a base pressure greater than 10^{-9} Torr. It is equipped to determine surface composition analysis by Auger electron spectroscopy and by XPS. The system is also equipped with a high pressure cell which isolates the sample and permits the performance of chemical reaction studies at high pressure without removing the sample from the UHV chamber. The product distribution is monitored by gas chromatography with a thermal conductivity detector. The second figure shows a flow reactor in which continuous gasification studies were undertaken at atmospheric pressure.

Early work in the high pressure-low pressure cell indicated that methane could be produced at temperatures as low as 475 K in the presence of potassium hydroxide[3] as indicated in Figure 3. Figure 4 shows the methane production with various alkali hydroxides. While the rate of methane production is small as indicated in Table 1 it becomes appreciable if one assumes that not all the geometric surface area of graphite is available for attack and that the attack probably takes place on the edges of the graphite.

We have carried out gasification studies in the environmental cell of a transmission electron microscope similar to the studies that were previously carried out by Baker[4] who gasified graphite crystals in the presence of nickel and hydrogen and showed that gasification occurred by "basal plane penetration in imperfect regions or by altering the rate of reaction at edges or steps". In our studies thin specimens of highly oriented pyrolithic graphite were obtained by cleavage of graphite crystals. Potassium hydroxide was introduced onto the surface of the graphite by dipping the specimen supported on copper or nickel grids into a .38 M solution of potassium hydroxide and then dried. Transmission electron microscopy was carried out in a Hitachi 650 kV microscope.[5] Argon at about 1 atm pressure was bubbled through water at room temperature giving an argon/water ratio of about 40/1 and then introduced into the environmental cell to give a pressure of 50 Torr. At 500°C the potassium hydroxide was dispersed as particles of 0.1 to 0.5 microns in diameter on the surface of the graphite. Figures 5 and 6 show micrographs recorded 11 minutes apart taken from a sequence showing the channel growth at 770 K. Channels are evident in two adjacent graphite crystals emanating from the edges of the crystal each channel with a particle at its head. As the reaction continues the particles move and the length of the channels increase. The channels remain roughly parallel sided indicating that

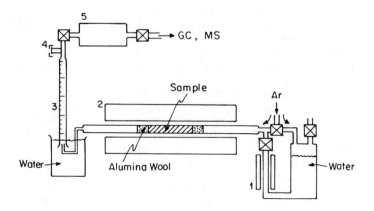

1 Steamer
2 Reactor
3 Burette
4 Septum
5 Vacuum Container

Figure 2. Flow reactor.

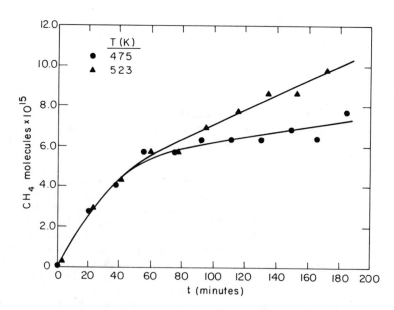

Figure 3. CH_4 production as a function of temperature
in the presence of KOH.

there is little effect of uncatalysed reaction at the channel edges or
of wetting of the channels sides by catalytic material.
Intercallation of potassium into the graphite does not appear to play
a role in the gasification. Samples of potassium intercallated
graphite $(C_{24}K)$ were subjected to reaction with water at 573 K and
exhibited an appreciably lower methane production than graphite
impregnated with potassium hydroxide. Both AES and XPS studies of
potassium intercallated graphite also indicated that potassium
diffuses to the surface under the influence of the electron beam. The
intensity of the potassium peak decreased upon heating.

It was also observed when using calcium hydroxide for the
gasification of graphite that a new photoelectron emission peak
corresponding to a C_{1s} electron binding energy of 290 eV is formed
representing a more reactive form of carbon probably in connection
with oxygen. As we will show later this probably constitutes a
phenolate species also observed by Mims and Pabst.[2]

In experiments with the flow reactor shown in Figure 2 it was
found that two regimens of gas production prevailed as illustrated in
Figure 7. In the first one hydrogen, methane and higher hydrocarbons
were produced with almost no carbon monoxide or carbon dioxide
production.[7] In the second regime hydrogen and CO were produced at a
lower rate. The first regime prevails until exactly one-half mole
of hydrogen as either hydrogen or in hydrocarbon has been produced per
mole of KOH present. This then would indicate that we are dealing
with a stoichiometric reaction in which carbon reacts with potassium
hydroxide to form a phenolate and one-half mole of hydrogen which in
turn may react with carbon to form hydrocarbons. Such a reaction can
be described as: $5C + 4KOH \rightarrow 4COK + CH_4$. The fact that higher

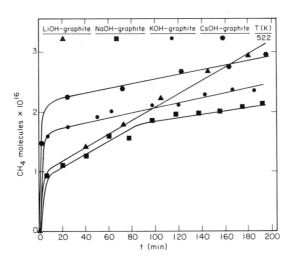

Figure 4. CH_4 production with various alkali hydroxides.

Figure 5. Micrograph of KOH catalyzed reaction.

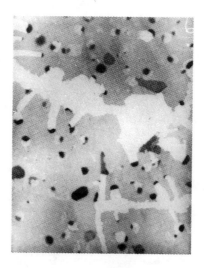

Figure 6. Micrograph of KOH catalyzed reaction ten
           minutes later.

hydrocarbons up to C_6 are also formed is of considerable
interest. Figure 8 shows a gas chromatogram of these hydrocarbons
indicating that many of them are olefinic. The distribution of the
higher hydrocarbons relative to methane and hydrogen is shown in
Figure 9.

The presence of the stoichiometric phenolate compound is confirmed
by XPS as previously shown.

When a sample of potassium hydroxide impregnated graphite which
has been reacted to completion of the stoichiometric reaction is
heated to about 1300 K the phenolate is decomposed with production of
carbon monoxide. Following this, gasification at low temperature can
again begin in line with the regime 1 in Figure 8.

We have found that the overall reaction can be made truly
catalytic by adding a co-catalyst to the potassium hydroxide.
Suitable materials are metal oxides and particularly nickel and iron
oxide as shown in Figure 10. It appears that in the presence of the
metal oxide the phenolate is decomposed as quickly as it is formed and
gasification occurs at 800 K at a steady rate which has been followed
up to 25% conversion of the graphite. This is illustrated in Figure
11. The reaction proceeding can be simulated by the following
equations:

$$4COK \xrightarrow{\text{MeOx}} 2K_2O + 2C + 2CO$$
$$2K_2O + 2H_2O \rightarrow 4KOH.$$

In the gasification with potassium hydroxide plus metal oxide the
products are essentially hydrogen and carbon dioxide in a ratio of
2:1. Only traces of hydrocarbons are found. It was at first
suspected that in these reactions methane might be a primary product
which is then steam reformed to CO and hydrogen. Addition of methane
to the water feed illustrated, however, that only minor amounts if any
of steam-reforming take place. Decomposition of methane over the

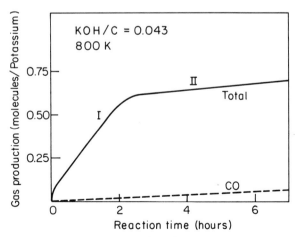

Figure 7. Gas production in the flow reactor.

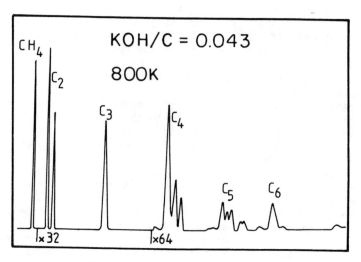

Figure 8.   Gas chromatogram of hydrocarbon produced.

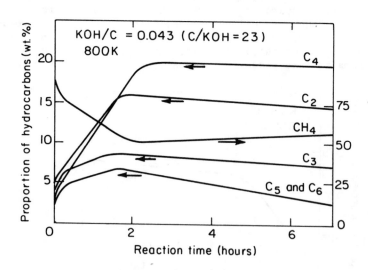

Figure 9.   Distribution of higher hydrocarbons.

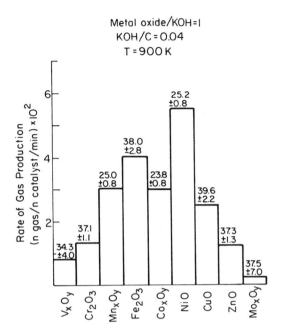

Figure 10. Gas production and activation energy for various metal oxides.

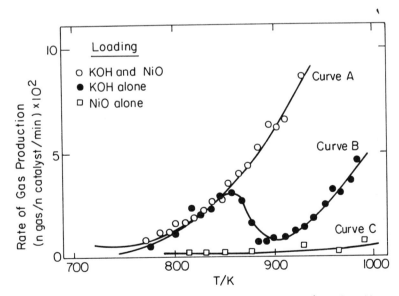

Figure 11. Temperature controlled gas production for three catalysts.

metal oxide may occur to a small extent and this may be the reason why only traces of hydrocarbons are found. The high hydrogen to CO_2 ratio and the absence of carbon monoxide are attributed to a watergas shift reaction proceeding simultaneously with the gasification. We have considered the possibility of a Boudouart reaction with subsequent gasification of the carbon from the disproportionation but consider it less likely than the watergas shift reaction. It is important to note that the production of hydrogen to carbon dioxide in the 2:1 molar ratio allows high hydrogen production directly from the gasification of carbonaceous material without a separate shift reactor and that the reaction proceeds at temperatures appreciably lower than those previously observed. One must also note that the reaction is carried out at essentially atmospheric pressure and that the kinetics should be appreciably improved by operating at higher water partial pressure. Studies in this direction are under way. We have repeated the experiments in the flow reactor with char from Illinois No. 6 coal and have obtained identical product distribution at higher rates than with graphite.

It can be concluded that phenolate type compounds are formed from the reaction of carbon, water and alkali hydroxide as intermediates in the production of hydrogen and/or of hydrocarbons from carbonaceous material and water in the presence of alkali hydroxides.

ACKNOWLEDGEMENT. This work was jointly supported by the Director, Office of Energy Research, Office of Basic Energy Sciences, Chemical Sciences Division and the Assistant Secretary for Fossil Energy, Office of Coal Research, Liquefaction Division and Gasification Division of the U.S. Department of Energy under Contract Number DC-AC03-76SF00098.

REFERENCES.

1. Wen, W.-Y., Catal. Rev. Sci.-Eng. 22, 1, 1, 1980.

2. Mims, C.A. and Pabst, J.K., Fuel 62, 176, 1983.

3. Cabrera, A.L., Heinemann, H. and Somorjai, G.A., J. Catal. 75, 7, 1982.

4. Baker, R.T.K., Catal. Rev., Sci.-Eng. 19, 2, 161, 1979.

5. Coates, D.J., Evans, J.W., Cabrera, A.L., Somorjai, G.A. and Heinemann, H., J. Catal. 80, 215, 1983.

6. Delannay, F., Tysoe, W.T., Heinemann, H. and Somorjai, G.A., Carbon 22, 4/5, 401, 1984.

7. Delannay, F., Tysoe, W.T., Heinemann, H. and Somorjai, G.A., Appl. Catal. 10, 111, 1984.

REACTIVITY OF HYDROCARBONS ON CONTACT WITH THE SOLID SUPERACID :
FIRST STAGE ANTIMONY PENTAFLUORIDE INTERCALATED GRAPHITE

J. Sommer
Université Louis Pasteur de Strasbourg
Institut de Chimie
1, rue Blaise Pascal
67000 Strasbourg

ABSTRACT. Antimony pentafluoride intercalated in graphite is shown to be a very selective catalyst for hydrocarbon reaction at low temperature. Low molecular alkanes, cycloalkanes and bicycloalkanes isomerize at 0°C and below with high conversion. At room temperature and above cracking occurs which causes deactivation of the catalyst but not the destruction of the layered system. The selectivity of the catalyst makes it very suitable to study the initial steps of the reaction mechanism and to show the specificity of the superacid catalysis.

1. INTRODUCTION

Since the first experiments on acid catalyzed isomerization of saturated hydrocarbons by Nenitzescu and Dragan (1) in 1933 a continuous research effect has been devoted to this reaction due to it's large scale industrial applicability (2). In the early sixties, Olah and his group (3) as well as the Dutch group (4) at the Shell laboratories discovered the reactivity of saturated hydrocarbons under superacidic conditions at unusually low temperatures even below 0°C. These results and the difficulties encountered in separating the original liquid superacids from the products stimulated the research effort on solid superacids. The use of SbF_5-intercalated graphite as solid superacid will be discussed here.

2. RESULTS AND DISCUSSION

Antimony pentafluoride vapour inserts readily in graphite at 70° under vacuum and forms the first stage insertion compound : $C_{6,5}SbF_5$. The intercalation of SbF_5 and the advantages of handling the insertion compound were fist described by Lalancette (5) but Herold and Melin (6) have more closely investigated the various stages of insertion.

411

R. Setton (ed.), Chemical Reactions in Organic and Inorganic Constrained Systems, 411–419.
© 1986 by D. Reidel Publishing Company.

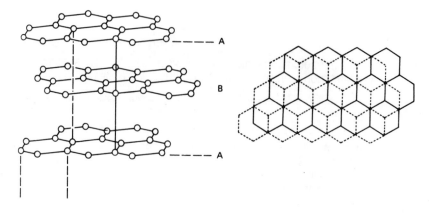

Figure 1.   Graphite lamellar structure

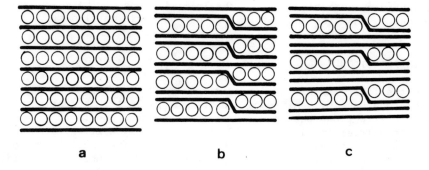

Figure 2.   Graphite intercalation stages : first stage (a) second
stage (b) third stage (c)

The extremely strong Lewis acid SbF_5 is well known to react with
hydrocarbons in the presence or in the absence of a strong Broensted
coacid. Due to the extreme difficulty to conduct the reactions under
absolute proton free conditions the mechanism of the carbenium ion formation from a saturated alkane is still open for discussion. The two
reaction schemes which have been proposed are :

$$RH + HF : SbF_5 \longrightarrow [RH_2]^+ SbF_6^- \longrightarrow R^+ SbF_6^- + H_2 \quad (1)$$

$$RH + 2SbF_5 \longrightarrow R^+ SbF_6^- + SbF_3 + HF \quad (2)$$

Once this fundamental step has been achieved (this step is stoechiometric and consumes acid) the reaction becomes catalytic both for isomerization and for cracking.

In the isomerization process the two following steps form a catalytic cycle which is controlled by two kinetic and two thermodynamic parameters that orient the isomer distribution depending on the reaction conditions.

In the case of hexane isomerization the steps are the following :

step 2 :   isomerization of the carbenium ion

step 3 :   hydride transfer from the alkane to the carbenium ions

Step 2 is kinetically controlled by the rates of hydride and alkide shifts but thermodynamically by the relative stabilities of the secondary and tertiary ions. Step 3 however is kinetically controlled by the rates of hydride transfer from the saturated hydrocarbons and thermodynamically controlled by the relative stability of the various hydrocarbon isomers.

For this reason the outcome of the reaction will be very different depending on the reaction conditions. We have tested the first stage antimony pentafluoride intercalated under various experimental conditions.

The isomerization of alkanes towards the thermodynamic equilibrium between the various isomers can be obtained by contacting one or a mixture of the isomers at temperatures as low as room temperature or 0°C with negligible side reactions.

Starting with a commercial mixture of 61.5% *cis-* and 38.5% *trans-*decalin in the presence of SbF_5-intercalated graphite in the molar ratio (hydrocarbon to catalyst) of ~10:1, we noticed that the isomerization of *cis-* to *trans-*decalin is too fast at room temperature to be conveniently followed by g.l.c. By increasing the molar ratio to 17:1, and operating at 0°C, the reaction can conveniently be followed. After 3 h contact time, the reaction had reached equilibrium. We obtained 98% of *trans-*decalin, which is in agreement with the thermodynamic ratio (99.3%:0.7%) extrapolated from Allinger's data (8). No methylbicyclo-nonanes or dimethylbicyclo-octanes could be detected under these conditions.

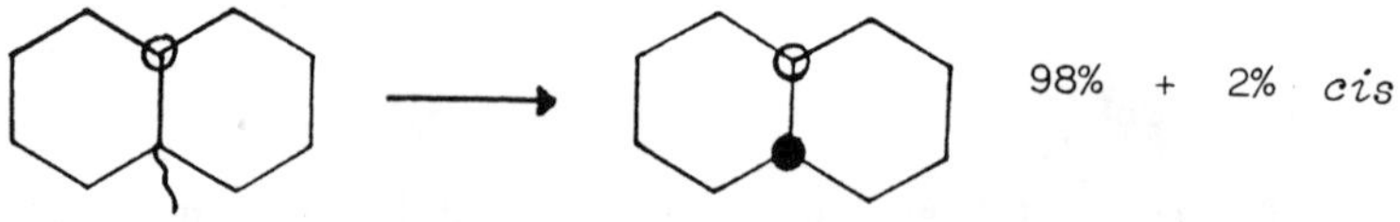

98% + 2% *cis*

The same procedure has been tested with perhydroindan. Starting again with a commercial mixture comprising 8% *trans-* and 92% *cis*-isomer in the presence of catalyst in the molar ratio 8:1 (hydrocarbon to catalyst) we obtained, after 3 h at 0°C, 62% of *trans-* and 38% of *cis*-perhydroindane, a ratio which is close to the thermodynamic ratio (69:31) extrapolated from ref. 9.

When methylcyclopentane is mixed with the catalyst in the ratio 15:1 at room temperature, it yields 91% of cyclohexane in about 4 h (the half-life being 1 h). This reaction, which is mechanistically quite different from the previous examples as it involves a change in branching, can be dramatically accelerated by adding 1% of cyclohexene at the start, in which case the half-life is reduced to 5 min at the same temperature. That the methylcyclopentane to cyclohexane ratio obtained (91%:9%) is indeed the equilibrium ratio at room temperature was shown by the reaction of cyclohexane with the catalyst, which yielded the same mixture after 3 h. The ratio was not affected by addition of olefin or longer contact times.

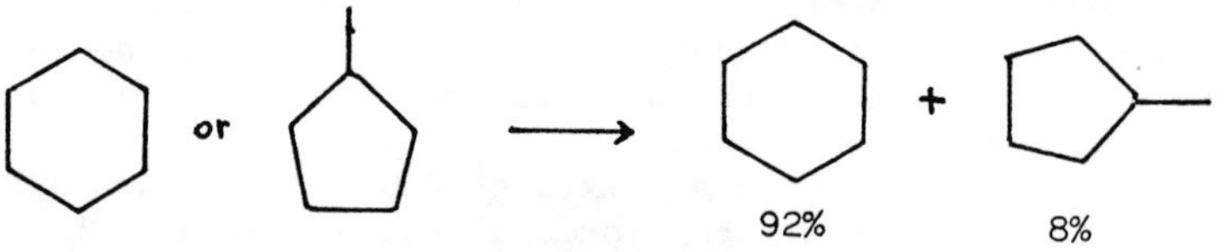

92%   8%

In the last example it is interesting to notice how different the thermodynamic ratio may be if we compare the equilibrium ratio in the neutral hydrocarbon with the ratio of the corresponding ions. The large energy differences (10 Kcal) between the secondary cyclohexyl cation and the tertiary methylcyclopentylion means

that under superacidic conditions in the presence of excess liquid
superacid only the latter can be observed. However when the alkane is
in excess it serves as the solvent, the substrate, and the hydride
donor and despite the extremely low concentration of secondary cations,
the thermodynamic equilibrium between the isomeric alkanes is reached
in a short time.

   The thermodynamic equilibrium between the various isomeric hexanes
can be reached at 0° with negligible cracking side reactions. Starting
from 3-methylpentane rapid isomerization to 2-methylpentane and 2,3-
dimethylbutane is observed (Fig. 3) the formation of n-hexane and
2,2-dimethylbutane is much solwer. The isomerization of alkanes under
these conditions has the advantage that the branched (high octane value)
isomers are thermodynamically favoured at low temperature.

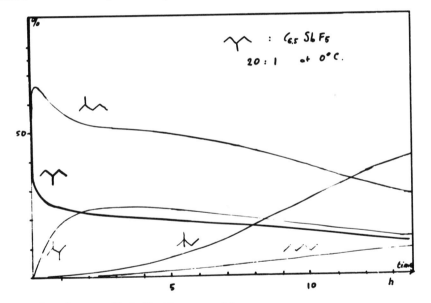

Figure 3.   Isomer distribution vs. time

   However isomerization reactions which have been tested on alkanes
like 1 or 2 and which

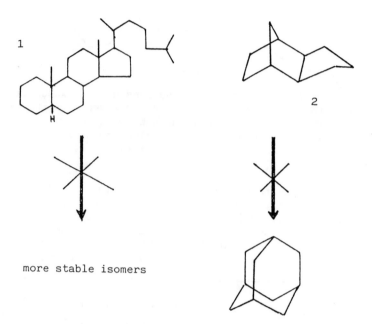

more stable isomers

do occur with other acid catalysts could not be achieved with the graphite intercalated compound. This can be rationalized in terms of steric restrictions for access to the acidic sites.

On the other hand under our experimental conditions the activity of the catalyst decreases with time and for example in the above described reaction the catalyst is deactivated after 50-80 turnovers.

Three causes have been suggested (10) :
- Leaching out of SbF_5
- reduction of SbF_5 to SbF_3
- poisoning of the acidic sites by carbonaceous products.

However the analysis of the deactivated catalyst by X-ray diffraction studies and elemental analysis shows that large quantities of SbF_5 are still present and that large regions of the first stage compounds can be detected (11).

Consequently we consider that the chemistry of the GIC is basically the same as that of SbF_5 with the following differences :
- The reactions are less violent and easier to controll mainly because graphite acts as a solvent which dilutes the superacid. The local overheating problem encountered in liquid superacids seems absent here.
- The product/catalyst separation is very easy and can be done by simple filtration.
- Steric requirements have to be met especially for access to the active sites.

For these reasons SbF_5 intercalated graphite is not a good candidate for industrial development as long as the deactivation process can not be slowed down. However it can be used in the laboratory as a mild and selective isomerization catalyst.

However due to the mildness and efficiency to react with hydrocarbons at room temperature it has been used successfully to study the
initial steps of the isomerization and cracking of hexanes (12).

As a model for cracking of alkanes, the reaction of 2-methylpen-
tane over SbF_5-intercalated graphite has been studied in a flow system,
the hydrocarbon being diluted in a hydrogen stream.   A careful study of
the product distribution vs time on stream showed that propane was the
initial cracking product whereas isobutane and isopentane (as major
cracking products) appear only later

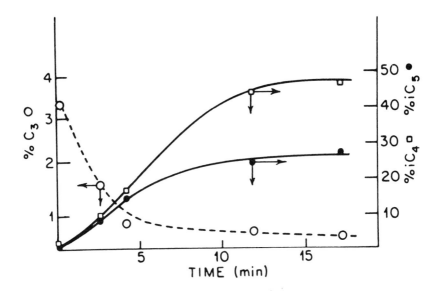

Figure 4.   3-methylpentane cracking products distribution vs. time

This result can only be explained by the β-scission of the trivalent 2-methyl-4-pentenyl ion as the initial step in the cracking process.   Based on this and on the product distribution vs time profile,
a general scheme for the isomerization and cracking process of the
methylpentanes has been proposed.

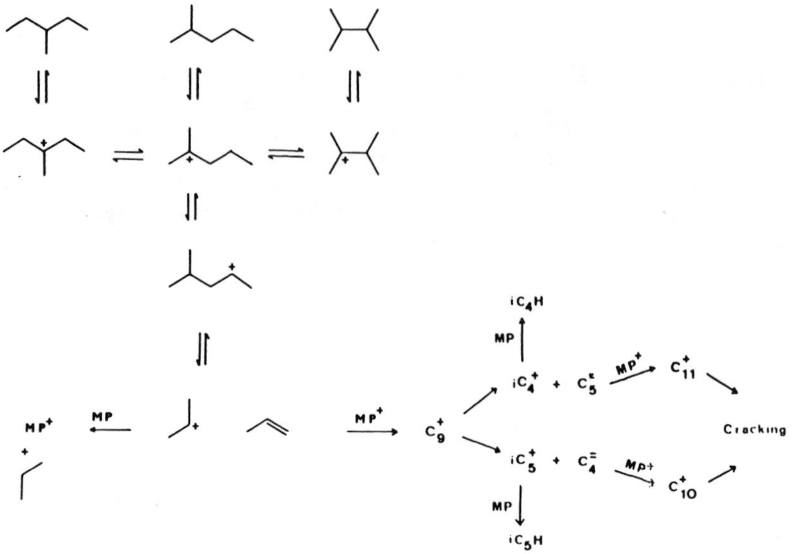

iC$_4$H

MP

iC$_4^+$ + C$_5^=$ MP$^+$ C$_{11}^+$

Cracking

MP$^+$ C$_9^+$

iC$_5^+$ + C$_4^=$ MP$+$ C$_{10}^+$

MP

iC$_5$H

MP$^+$ MP $\searrow$ MP$^+$ C$_9^+$

The propene which is formed in the β-scission step never appears as a reaction product because it is alkylated immediately under the superacid condition by a C$_6^+$ carbenium ion, forming a C$_9^+$ carbocation which is easily cracked to form a C$_4^+$ or C$_5^+$ ion and the corresponding C$_4$ or C$_5$ alkene. The alkenes are further alkylated by a C$_6^+$ carbenium ion in a cyclic process of alkylation and cracking reactions. The C$_4^+$ or C$_5^+$ ions also give the corresponding alkanes (isobutane and isopentane) by hydride transfer from the starting methylpentane. This scheme, which occurs under superacidic conditions, is at variance with the scheme that is generally adopted for the cracking of C$_6$ alkanes under less acidic conditions.

1. $\rightleftharpoons$ + H$^+$

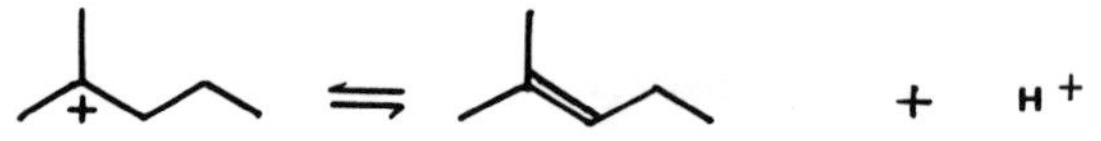

2. $\rightleftharpoons$ C$_{12}^+$ $\rightleftharpoons$ C$_4$ + C$_8$ etc.

β-Scission

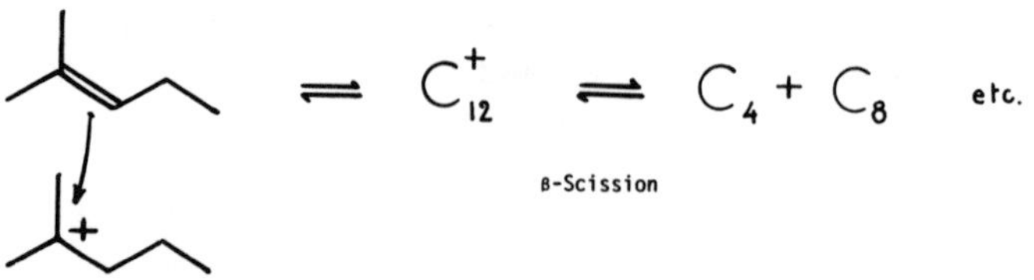

Under superacidic conditions, it is known that the deprotonation (first type) equilibria lie too far to the left (K = 10^{-16} for isobutane) (13) to make this pathway plausible. On the other hand, among the C_6 isomers 2MP is by far the easiest to cleave by β-scission. The 2-methyl-2-pentenium ion is the only species that does not give a primary cation by this process. For this reason, this ion is the key intermediate in the isomerization cracking reaction of C_6 alkanes.

3. REFERENCES

(1) C.D. Nenitzescu, A. Dragan, *Ber. Dtsch. Chem. Ges.*, 66, 1892 (1933).

(2) F. Asinger, *Paraffins, Chemistry and Technology*, Pergamon Press, New York, 1965.

(3) G.A. Olah, J. Lukas, *J. Am. Chem. Soc.*, 89, 2227 (1967).

(4) A.F. Bickel, G.J. Gaasbeek, H. Hogeveen, J.M. Oelderick, J.C. Platteuw, *Chem. Commun.*, 634 (1967).

(5) J.M. Lalancette, J. Lafontaine, *J. Chem. Soc., Chem. Commun.*, 815 (1973).

(6) J. Melin, A. Herold, *C.R. Acad. Sci.*, 280, 641 (1975) ; J. Melin, 'Thèse Université de Nancy I', 1976.

(7) K. Laali, M. Muller, J. Sommer, *J. Chem. Soc., Chem. Commun.*, 1088 (1980).

(8) N.L. Allinger, J.L. Coke, *J. Am. Chem. Soc.*, 81, 4080 (1959).

(9) N.L. Allinger, J.L. Coke, *J. Am. Chem. Soc.*, 82, 2553 (1960).

(10) J.J.L. Heinerman, J. Gaaf, *J. Mol. Catal.*, 11, 215 (1981).

(11) J. Sommer, R. Vangelisti, unpublished results.

(12) F. Le Normand, F. Fajula, F. Gault, J. Sommer, *Nouv. J. Chim.*, 6, 411 (1982) ; F. Le Normand, F. Fajula, J. Sommer, *Nouv. J. Chim.*, 6, 291 (1982).

(13) H. Hoffman, G.J. Gaasbeck, A.F. Bickel, *Rec. Trav. Chim. Pays-Bas*, 88, 703 (1969).

AN EXAMPLE OF CRYSTALLINE GROWTH IN A CONSTRAINED MEDIUM : THE INTER-
CALATION OF METALS AND METALLIC ALLOYS INTO GRAPHITE

P. Lagrange and A. Hérold
Laboratoire de Chimie du Solide Minéral
U.A. C.N.R.S. 158
Université de Nancy I
B.P. 239
54506 - Vandoeuvre-lès-Nancy Cédex   France

ABSTRACT. Graphite is a very strongly anisotropic material. Because of
this, it presents original chemical properties : the carbon planes can
be spread apart to receive foreign species in the intervals. Moreover,
chemical reactions can occur within these intervals between the already
intercalated species and a second reactive species. The constraints imposed by the bidimensionality of the medium in which the crystalline
growth or the chemical reaction occurs, and the energetic interaction
between the carbon planes and the foreign species, provoke particular
phenomena, which are illustrated here by choosing various metals and
metallic alloys as reactive species.
    The pure heavy alkali metals intercalate leading to monolayered
sheets, whose structure differs somewhat from that of the bulk metals.
    Under particular conditions, mercury can react with the already
intercalated alkali metal ; there also appears a new organization between the carbon planes : "sheets of the amalgam" are intercalated,
whose structure looks like that of free amalgams.
    When an alkali metal-mercury (or -bismuth) alloy reacts directly
with the graphite, these polylayered sheets also appear. However, the
formation mechanism is different according to the affinity of the
alloyed heavy metal with the alkali metals.

1.   FORMATION OF METALLIC MONOLAYERED SHEETS BY ACTION OF HEAVY ALKA-
      LI METALS ON THE GRAPHITE

The three heavy alkali metals crystallize in the body-centered cubic
system. In this structure, each metallic atom is surrounded by 8 adjacent atoms ; the metal-metal distance d_{M-M} is equal to $a\sqrt{3}/2$ (a = parameter of the unit cell). In table I, are collected the structural
data concerning these metals.
    Furthermore, these very strongly electropositive metals react ea-
sily with graphite leading to intercalation compounds. Each carbon
plane very solid because of its covalent bonds retains its structure ;
on the contrary, between these planes, there are only weak van der
Waals's bonds, so that the layers can be easily spread apart to enable

421

R. Setton (ed.), Chemical Reactions in Organic and Inorganic Constrained Systems, 421–428.
© 1986 by D. Reidel Publishing Company.

the intercalation of intermediary sheets, constituted of atomic single monolayers (1,2).

TABLE I - Parameters and interatomic distances for the three heavy alkali metals

	a	d_{M-M}
K	5.25 Å	4.55 Å
Rb	5.61 Å	4.86 Å
Cs	6.07 Å	5.26 Å

When the metal vapour arrives progressively on the graphite, one can see that the metallic atoms are not distributed in a homogeneous manner between the carbon planes. On the contrary, they are grouped in dense layers inside some intervals only, periodically stacked along the c-axis : thus one interval in s is regularly occupied by a metallic layer (figure 1). When the quantity of intercalated metal increases, a new arrangement of the layers appears (3), so that one interval in (s-1) is occupied, then one in (s-2), etc... until all the intervals are finally occupied ; the graphite is then saturated by the metal. One says that this reaction is an intercalation by successive stages, which are designated by the numbers s, (s-1), (s-2)...2, 1.

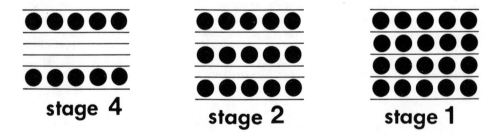

stage 4   stage 2   stage 1

Figure 1. Intercalation by successive stages in the graphite.

In the pristine graphite, the value of the distance between two successive carbon layers is 3.35 Å. In an intercalation compound, the distance, which separates the two carbon planes surrounding the inter-calated sheet, is called the "interplanar distance" (d_i). With the al-kali metals, its value does not vary with the stage.

Intercalation compounds of stage s greater than one possess in-tercalated sheets of relatively low density. Indeed, with respect to the adjacent carbon layer, there is one metallic atom for twelve car-bon atoms, that is an occupancy of 23.6 Å^2 for each metallic atom ; the formula of the compound is also written MC_{12s}. Moreover, the intercala-

ted sheet possesses a relatively complicated structure, since it is partially liquid.

On the other hand, when graphite is saturated by the metal (stage
1 compound), the intercalated sheets are denser : then there is one metallic atom for only 8 carbon atoms, or an occupancy of 15.7 $\mathring{A}^2$ for
each metallic atom. The formula of this compound is thus written MC_8.
Each metallic atom is situated in a prismatic hexagonal site made up of
12 carbon atoms (in fact, the carbon planes, initially stacked according to the alternated sequence ABAB..., are shifted to the AAA... stacking) (4). One prismatic site in four is thus occupied, so that the intercalated metallic layer forms a two-dimensional centered hexagonal
network, the parameter of which is equal to twice that of graphite (figure 2).

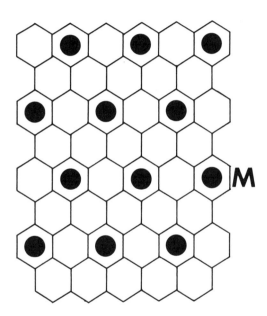

Figure 2. Projection on a carbon plane of the intercalated metallic
          layer in a stage 1 compound.

This basal plane structure, which is the same for KC_8, RbC_8 and
CsC_8, merits two remarks :
- The coordinence of the metal is now equal to 6 (no longer 8, as
in the bulk metal) ; it is equal to the coordinence of a face-centered
cubic metal in the (111) plane.
- The value of the distance between two adjacent atoms in the layer
is 4.26 $\mathring{A}$; it is practically the same for the three alkali metals, and
is always shorter than the interatomic distance in the bulk metals.

In table 2, are collected the interplanar distances and the relative decrease in the metal-metal distances for the three MC_8 compounds.

TABLE II - Structural data concerning the MC_8 compounds

	d_i	relative decrease of d_{M-M}
KC_8	5.35 Å	6.4 %
RbC_8	5.62 Å	12.3 %
CsC_8	5.92 Å	19.0 %

2. FORMATION OF POLYLAYERED SHEETS BY ACTION OF MERCURY ON THE INTER-CALATED MONOLAYERED SHEETS

A (K or Rb)-saturated intercalation compound (stage 1) is now submitted to mercury vapour (5, 6). This latter is unable to intercalate by it-self into graphite. On the other hand, it is very reactive with the al-kali metals.

If the quantity of mercury introduced is very large, the alkali metal is extracted from the graphitic intervals and a mercury-rich a-malgam appears on the edges of the carbon layers, which revert again their usual spacing of 3.35 Å. This reaction is a classical de-interca-lation.

If, on the contrary, the quantity of mercury introduced is calcu-lated (one mercury atom for one alkali metal atom), the alkali metal is no longer extracted from the graphitic intervals. The mercury pro-vokes a grouping of the intercalated monolayers in thicker sheets, constituted by two layers identical to the previous, exactly superim-posed and stabilized by the presence of an intermediate mercury layer (7).

The superposition of the two alkali metal layers creates triangu-lar prismatic sites, and each of them is occupied by a mercury atom (figure 3). Thus there are in the middle plane as many mercury atoms as there are alkali metal atoms in the two adjacent layers ; thus one writes the threelayered sheet : (MHg_2M). The formula of the compound is written $MHgC_8$.

When the alkali metal layers come togheter, they liberate graphi-tic intervals : one half of these intervals is empty and the other half contains the threelayers. The distribution of these empty and oc-cupied intervals is not arbitrary ; they are stacked alternately along the c-axis (stage 2). The new interplanar distance corresponding to the threelayered sheet is indicated in table III. Also the fixation of a new metal provokes an increase of the stage, and simultaneously a dilation along the c-axis (cf table III).

TABLE III - Structural data concerning the $MHgC_8$ compounds

M	d_i	dilation along the c-axis
K	10.16 Å	26.3 %
Rb	10.76 Å	31.9 %

On the contrary, mercury reacts with much difficulty with intercalated cesium of CsC_8 (8). Even in excess, it can never entirely extract the alkali metal from the graphitic intervals. However there sometimes appears a badly defined ternary compound of high stage, but always in the presence of a great quantity of the initial compound CsC_8.

One can explain these results by the great affinity of cesium for graphite (much greater than those of the two other metals). That renders very difficult the formation of intercalated polylayers.

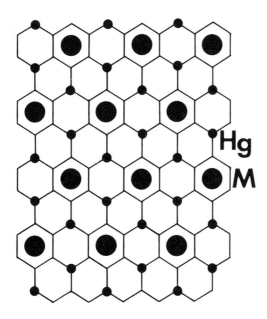

Figure 3. Projection on a carbon plane of the intercalated polylayered sheet in the $MHgC_8$ compounds.

It is interesting to notice here that the threelayered sheet, intercalated in $KHgC_8$, possesses a 2D structure similar to the 3D one of the KHg_2 amalgam (7). Indeed, the elementary group of atoms which constitutes this sheet is geometrically a mercury hexagon with which are associated two potassium atoms placed on both sides of its plane.

The 2D stacking of an infinity of these groups of atoms builds up the threelayered (KHg_2K) sheet (figure 4).

    The same group of atoms, very slightly distorted (inclination of the K-K axis with respect to that of the mercury hexagon ; dilation of the hexagon parameter ; decrease of the K-K distance) allows building up, by a 3D stacking, the KHg_2 amalgam, whose structure is monoclinic (figure 4). The interatomic distances concerning the intercalated three-layered sheet and the KHg_2 compound are compared in table IV.

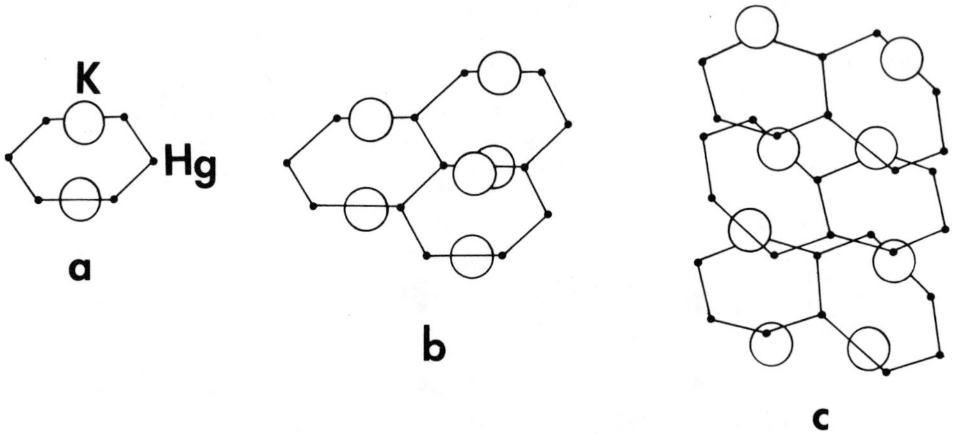

Figure 4. a) Elementary group of atoms (K_2Hg_6)
          b) 2D stacking of K_2Hg_6 in $KHgC_8$
          c) 3D stacking of K_2Hg_6 in KHg_2

    It appears that the growth of the amalgam in this constrained 2D space deforms relatively little the geometrical characteristics of this alloy : one can say that a sheet of the alloy is intercalated between the carbon planes.

TABLE IV - Structural data concerning the free amalgam KHg_2
           and the intercalated amalgam in $KHgC_8$

	Interatomic distances in	
	KHg_2	$KHgC_8$
Hg–Hg	3.00 Å	2.85 Å
K–K	4.13 Å	4.84 Å
K–Hg	3.52 to 3.74 Å	3.73 Å

3. FORMATION OF POLYLAYERED SHEETS BY DIRECT ACTION OF ALLOYS ON THE GRAPHITE

One starts now from pristine graphite and an amalgam of composition MHg (M = K, Rb), liquid or vapour (9).Under these conditions, the final product of the reaction is a compound of stage 1, and not 2 : all the graphitic intervals are occupied by threelayers of amalgam (MHg_2M), identical to those described above. The intercalation into graphite of these sheets is realized in two successive steps.

The first step is a quasi-selective intercalation of the alkali metal, which forms monolayers : this reaction is an intercalation by stages, which leads to the saturation of the graphitic intervals (compound of stage 1 : MC_8). We have showed that these monolayers contain small quantities of dissolved mercury.

The second step is a cooperative simultaneous intercalation of the alkali metal and the mercury. This latter begins to react with the monolayers, gathers them by superimposing and builds the threelayers (MHg_2M). Simultaneously, new alkali metal atoms come and occupy the liberated intervals, and these monolayers then become threelayered sheets (MHg_2M). Finally one obtains the saturation of all graphitic intervals by threelayers. This second step is "isostage" : from a saturated in monolayers compound, there appears at the end of the reaction a saturated threelayer compound.

As previously, if a liquid alkali metal-bismuth alloy of composition M_3Bi_2 (M = K, Rb, Cs) reacts with pristine graphite, an intercalation of the alloy also takes place (10). Here also the intercalated sheets are threelayered : two layers of alkali metal surrounding an intermediate layer of bismuth. However, this latter contains less bismuth than the quantity of mercury in the central layer of the previous ternaries. A first stage compound (saturated) is described by a formula $MBi_{0.6}C_4$; the threelayered sheet is written (MBi_xM) with $x \approx 1.2$. The mechanism of formation of the threelayers is different from what we observed with the amalgams. A significant example of this type of mechanism is observed when the cesium-bismuth alloy reacts with graphite. The intercalation reaction also occurs in two successive steps.

There first appears a quasi-selective intercalation of cesium, but instead of obtaining, at the end of this step, a stage one compound, therefore saturated in alkali metal, one observs only a stage 4 compound (of formula about CsC_{48}).

After this, begins the second step, which is a simultaneous cooperative intercalation of cesium and bismuth. Together both elements penetrate between the carbon planes and build intercalated threelayered sheets, by mean of a successive stage mechanism (and not "isostage") : there appear ternaries of stage 3, then 2, before obtaining the final product of the reaction, which is the ternary compound of stage 1.

Bismuth therefore reacts earlier than mercury. This phenomenon seems to indicate that the interaction between the alkali metal and bismuth is greater than that between the alkali metal and mercury.

In fact, for a binary compound, when the stage decreases, the interaction between graphite and the alkali metal also decreases. Since the bismuth can compete with the graphite, as soon as the stage is 4, there exists a great affinity between cesium and bismuth. On the other hand, mercury can compete with the graphite only when the stage is one.

The same reason probably explains the existence of graphite-cesium-bismuth ternaries. In fact, the interaction between graphite and cesium is greater than the interaction between graphite and rubidium or potassium : therefore, mercury does not compete with graphite to build a ternary ; on the contrary, bismuth is able to compete with graphite and build ternary compounds.

ACKNOWLEDGMENTS - We are grateful to Dr A. W. MOORE of Union Carbide for providing the HOPG samples used in these studies.

REFERENCES

(1) K. Fredenhagen and G. Cadenbach
    Z. Anorg. Allgem. Chem., 158, (1926), 249

(2) A. Hérold
    Bull. Soc. Chim. Fr., (1955), 999

(3) N. Daumas and A. Hérold
    C. R. Acad. Sc. Paris, série C, 268, (1969), 373

(4) W. Rüdorff and E. Schulze
    Z. Anorg. Allgem. Chem., 277, (1954), 156

(5) M. El Makrini
    Thèse de Doctorat ès-Sciences, Nancy, (1980)

(6) M. El Makrini, P. Lagrange, D. Guérard and A. Hérold
    Carbon, 18, (1980), 211

(7) P. Lagrange, M. El Makrini and A. Hérold
    Rev. Chim. Min., 20, (1983), 229

(8) P. Lagrange, M. El Makrini and A. Bendriss
    Synth. Met., 7, (1983), 33

(9) P. Lagrange, M. El Makrini, D. Guérard and A. Hérold
    Synth. Met., 2, (1980), 191

(10) A. Bendriss
     Thèse de 3ème Cycle, Nancy, (1984)

ORDER-DISORDER PHENOMENA IN GRAPHITE INTERCALATION COMPOUNDS AND RE-
DUCTION REACTIONS IN ALKALI METAL-INTERCALATED COMPOUNDS OF GRAPHITE

H.P.Boehm, Ko Y.-Sh., B.Ruisinger and R.Schlögl
Institute for Inorganic Chemistry of the University
Meiserstraße 1
D-8000 München 2
Federal Republic of Germany

ABSTRACT. Graphite was oxidized to graphite hydrogen sulfate with ammonium peroxodisulfate. With increasing oxidation time significant differences were found for the interlayer spacing, the ratio of intercalated
anions and free acid molecules, and for the density of the stage 1 compound. There was a minimum of the density of the intercalated layers
after ca. 15 minutes. - Perfluoro-n-butanesulfonic acid was intercalated electrochemically by galvanostatic and cyclovoltammetric methods.
In this case two phases with different thickness of the intercalated
layers were observed for the stage 2 and stage 1 compounds. At relatively low charge densities for a given stage the spacing is lower by ca.
320 pm than in the fully oxidized compound. There was evidence for a
disordered sequence of thick and thin interlayers in the transition.
- Potassium graphite, KC_8, was reacted with solutions in THF of $ZnCl_2$
$FeCl_3$, $MnCl_2 \cdot 4\ H_2O$, $CoCl_2 \cdot 6\ H_2O$ and $CuCl_2 \cdot 2\ H_2O$. Literature reports that transition-metal-intercalated graphite is formed could not
be confirmed with the possible exception of $ZnCl_2$ where chemical composition and line scans in energy-dispersive analysis gave some indication of intercalation.

1.   INTRODUCTION

A large variety of substances have been intercalated between the carbon
layers of the graphite structure (1). The distance between the carbon
layers is increased by the thickness of the intercalated layer. A pecu-
liarity of the graphite intercalation compounds (GICs) is the staging
phenomenon, i.e. the intercalated layers may be separated by 1,2,3 ...n
carbon layers, n being the stage number. In contrast to other intercalated layer structures, e.g. with metal dichalcogenides, quite high
stage numbers, exceeding ten, have been observed. It is customary to
separate the graphite intercalation compounds into so-called acceptor
compounds where electrons are transferred from the carbon layers to an
acceptor, and donor compounds with electrons transferred from intercalated donor atoms, e.g. alkali metal, to the carbon layers. In the graphite salts, e.g. with sulfuric acid, anions such as HSO_4^- are interca-

429

R. Setton (ed.), Chemical Reactions in Organic and Inorganic Constrained Systems, 429–440.
© 1986 by D. Reidel Publishing Company.

lated, and the remaining volume between the carbon layers is filled with neutral acid molecules. A more or less dense packing in the inter-calated layers seems possible and has indeed been observed. In princi-ple, such GICs should be anion exchangers.

It is notable that the observed interlayer spacings show in many cases a considerable variability (2). This is especially the case with graphite hydrogensulfate, $C_{24}^{+}HSO_4^{-} \cdot 2.4\ H_2SO_4$. This variation of the interlayer spacing, I_c, has been ascribed to differences in the charges residing on the layers (3,4), but we find evidence that differences in the organization of the intercalated layers could as well be responsi-ble. Pursuing these studies further, we have intercalated perfluoro-n-alkanesulfonic acids, especially perfluoro-n-butanesulfonic acid, and found large variations in the I_c values indicating that even two dis-tinct phases exist for a given stage with different packings of the in-tercalated anions and acid molecules. There is evidence for a more dis-ordered packing in the transition from one to the other phase.

It has been claimed that transition metal chloride graphites can be reduced in situ with metal-organic compounds to graphite compounds with intercalated transition metal atoms (5). The evidence that real GICs were formed is not convincing, however. Even more surprising was a report that first stage potassium graphite, KC_8, reacts with transi-tion metal chlorides dissolved in THF yielding transition-metal-inter-calated graphite (6,7). We cannot confirm this report, in agreement with Schäfer-Stahl (8). Only in the reaction with zinc chloride there was some evidence for the formation of a zinc intercalation compound from chemical analysis and from line scans over flakes of the reaction product in energy-dispersive analysis of X-ray emission.

2.   EXPERIMENTAL

2.1. Materials

Natural graphite purified to better than 99.5 % C was used, mostly in the form of RF1 or S 40 flakes from Kropfmühl, Bavaria (0.2 - 1mm dia-meter) or of Madagascar flakes with 1 - 2 mm size. A few electrochemi-cal experiments were also done with HOPG.

$C_4F_9SO_3$ and $C_6F_{13}SO_3H$ were liberated from their potassium salts with 100% H_2SO_4 and distilled in vacuo.

2.2. Chemical and electrochemical oxidation in acids

For the chemical oxidation a solution of ammonium peroxodisulfate in 98 % H_2SO_4 (7.36 g in 50 ml) was employed. X-ray powder diffraction, diffuse reflectance spectroscopy and buoyancy determinations have been described elsewhere (9).

Electrochemical oxidation in various acids was performed in the galvanostatic as well as in the linear sweep (cyclo)-voltammetric mode. Sample sizes were in the range of 1 to 10 mg, and the current density in galvanostatic experiments was kept below 500 µA per cm^2 of prism fa-ce. The sweep rate in the cyclovoltammetric experiments corresponded to

5 - 10 µV/s. Only at low current densities well-resolved curves were obtained. A platinum counter electrode and reference electrodes of graphite foil (Sigraflex, heat-treated at 2500°C) were employed. The reactions were interrupted at various, preselected times, and X-ray powder patterns were taken.

2.3. Reaction of KC_8 with transition metal chlorides

Potassium graphite, KC_8, was prepared by reacting the graphite (mostly RFl flakes) with potassium vapor and distilling off the excess at 10^{-4} Pa. For the reaction with the metal chlorides (anhydrous $ZnCl_2$ and $FeCl_3$, $MnCl_2 \cdot 4 H_2O$, $CoCl_2 \cdot 6 H_2O$ and $CuCl_2 \cdot 2 H_2O$) we followed closely the description by Braga et.al (7). 40 mmol of KC_8 were introduced into the reaction vessel under pure argon and covered with 25 ml of anhydrous THF. The color of the KC_8 remained unchanged for about one hour. 24 mmol of metal chloride dissolved in 125 ml of dry THF were added dropwise to the magnetically stirred suspension within one hour. In the case that not all of the metal halide dissolved in the solvent, the saturated solution was used, and the remaining halide was added in solid form after the solution. In one experiment with $ZnCl_2$ a solution containing 10 % HMPT was used. The reaction products were isolated by filtration after several hours and washed with water and methanol as described (7). A part of the material was washed with 2 N HCl. All operations and the vacuum drying were performed at room temperature. For analysis, the materials were oxidized by refluxing with H_2SO_4/HNO_3; the metals were determined by AAS after dilution and chloride by potentiometric titration with $AgNO_3$.

3.   RESULTS AND DICUSSION

3.1. Chemical oxidation in sulfuric acid

In an attempt to study the kinetics of the formation of graphite hydrogensulfate by chemical oxidation with the peroxodisulfate ion we have observed earlier that the stage 2 compound is formed quickly, within two minutes (9). Further oxidation to the first stage can be followed conveniently, it is completed after ca. 30 minutes. An interesting observation in these experiments was that the interlayer spacing of the stage 1 compound showed a remarkable variability as shown in Table I. We ascertained that no $S_2O_8^{2-}$ ions were intercalated in our experiments. These ions could be detected with titanyl sulfate only if the oxidation reaction was performed at low temperatures, i.e. below +7°C, and only then gas evolution was observed on heating rapidly a suspension of the graphite salt in fresh sulfuric acid.

The values observed by other authors are included in Table I for comparison. It has been established in electrochemical experiments that, for a given stage of the graphite salt, the charge density on the layers can cover a certain span (3,4). The I_c values contracted with continuing oxidation, and this was explained by the increased electrostatic force between the positive carbon layers and the anion layers (3). As plau-

Table I. Observed interlayer spacings of graphite hydrogensulfate, st.1 in the course of chemical oxidation.

Reaction time (resp.manner of oxidation)	degree of oxidation*)	I_c/pm	reference
10 min.	31.6	797.3	(9)
15 min.	28.6	803.6	
20 min.	26.4	801.1	
24 min.	25.5	788.6 and 801.1	
30 min.	24.7	801.1	
35 min.	24	801.1	
literature values:			
chemical	24	798	(14)
electrochemical		799	(15)
electrochemical	28	804	(10)
	21	792	
electrochemical	28	798	(3)
	21	788	

*) expressed as average number of C atoms per positive layer charge.

sible as this is, one must take also into account different ordering in the intercalated layers. In our chemical oxidation experiments, the interlayer distance increased with increasing oxidation time, i.e. with presumably increased layer charge. Bouayad et al. (10) showed that at a low layer charge, corresponding to C_{28}^+, the anion layers show some two-dimensional order which is disturbed, however. After oxidation to C_{21}^+ the layers were much better ordered (non-commensurate with the carbon layers), and there was even a three-dimensional order of the HSO_4^-/H_2SO_4 layers.

The relatively large I_C value found after 15 minutes oxidation in our experiments may be caused by some non-integrality of the X-ray reflections. After 10 minutes, we observed also reflections near the positions expected for the second stage compound, but shifted in an irregular manner - as one would expect for stacking disorder of stage 2 and stage 1 sequences. The second stage compound after 2 minutes oxidation was very well ordered, in contrast. Non-integral (001) reflections caused by a statistical, disordered intercalation of the $FeCl_3$ layers have been observed for ferric chloride graphite (11). The situation is analogous to that of so-called mixed-layer structures in sheet silicates (12).

After 24 minutes oxidation we observed two well-ordered stage 1 phases with different I_C spacings. The close electric contact of the particles in the X-ray capillary would preclude the simultaneous existence of two phases with a different degree of oxidation and, hence, different electrochemical potential. The different spacings can only be explained by differences in the packing in the intercalated layers.

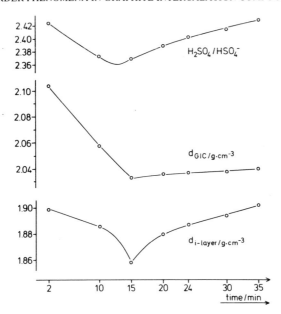

Fig.1.   Variation with reaction time of the ratio H_2SO_4/HSO_4^- in the intercalated layers, of the density of the intercalation compound and of the density of the intercalated layers. Oxidation with peroxodisulfate, S 40 flakes, 0.1 - 0.2 mm fraction. The corresponding average degrees of oxidation are listed in Table I.

The average degree of oxidation was determined from the position of the minima in optical reflectance as described in ref. (9). In parallel, the density of the samples was determined from buoyancy measurements under concentrated H_2SO_4. From these data the ratio of free acid molecules to anions and the density of the intercalated layers can be calculated (13). The results presented in Fig.1 show a distinct minimum of all three values after 15 minutes of oxidation.

These results, as well as those of Bouayad et al. (10) show clearly that different degrees of order are possible in the intercalated HSO_4^-/H_2SO_4 layers. Chemical intercalation is rapid, and the layers are incompletely filled in the beginning. The final, close packing is reached only after 20 - 30 minutes. In contrast, electrochemical intercalation is performed very slowly, near equilibrium, because only then the potential curves show well-resolved details. This explains the differences observed with the two methods.

3.2. Intercalation of perfluoroalkanesulfonic acids

We have described earlier that trifluoromethanesulfonic acid can be intercalated electrochemically into graphite as easily as sulfuric acid (16). The stage 1 compound has an I_C spacing of 804 pm. We have con-

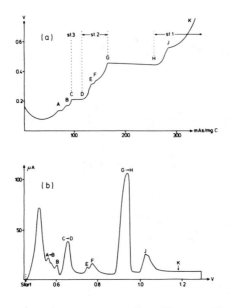

Fig.2. Galvanostatic (a) and linear-sweep voltammetric (b) oxidation of graphite (madagascar flakes) in $C_4F_9SO_3H$.

tinued this study using perfluoro-n-butanesulfonic acid. Galvanostatic oxidation was combined with linear-sweep voltammetry or cyclovoltammetry. The curves obtained with this acid showed much more detail than comparable curves with sulfuric or trifluoromethanesulfonic acid. Fig.2(a) and (b) shows the two curves, corresponding points are marked by the same letter. Also in this case a relative wide range of degree of oxidation was observed for each pure stage. A stage 3 compound was found exclusively at point C. As with sulfuric acid, it was not possible to observe pure higher stages. From C to D a second stage compound started to form. At G only a second stage phase was observed, as with H_2SO_4, and formation of a first stage started beyond this point. With H_2SO_4 and with CF_3SO_3H a small step occurs in the region E/F, accompanied by a current

Table II. Charge taken up at the points marked in Fig.2, expressed as n carbon atoms per positive charge.

Point	n in C_n^+	Point	n in C_n^+
A	117	F	56
B	94	G	48
C	85	H	30
D	70	I	28
E	60	K	24

Table III. Intercalation of $C_4F_9SO_3H$. Observed interlayer spacings at characteristic points of the potential curve, Fig.2(a).

Point	I_c/pm	stage
C	2764 ± 17	3
D	~ 2100	2 b
E,F	2114 ± 8	2 b
G	2434 ± 9	2 a
H	~ 1760	1 b
I	1773 ± 7	1 b
K	2099 ± 4	1 a

peak in cyclovoltammetry (4). With $C_4F_9SO_3H$, however, two small steps E and F are observed with a corresponding double peak in the voltammogram. Oxidation of the stage 2 compound is finished at H, and there is a step again at I. Such a step in the stage 1 regime has never been observed with H_2SO_4. The corresponding layer charges, expressed as carbon atoms per charge, are presented in Table II.

To our surprise, we found different interlayer spacings for stage 2 compounds at the points F and G, as shown in Table III. Analogous differences occured for the first stage phases at I and K. The difference in interlayer spacing is the same for both stages, it is 320 – 327 pm. Clearly, two distinct phases exist for each stage, one with a lower I_c spacing at lower layer charges, i.e. anion densities in the intercalated layers, and one with a high I_c spacing at higher charge densities. Immediately after point I, both phases were seen together.

Plotting of the reciprocal d values for each (001) series showed slight deviations of the reflection positions from the calculated values for integral ℓ indices. We are not sure, therefore, whether the small increases in interlayer spacing with continued oxidation for a given phase are real. At the break points in the potential diagram, Fig.2(a), when each phase is fully formed, the (001) reflections were practically integral. Distinct non-integrality immediately after passing the points F or I, when the "a" phases started to form, indicated a stacking disorder of "thin" and "thick" interlayers.

It follows from the observed I_c spacings that the anions and acid molecules must be ordered in bilayers, evidently with the charged or polar ends oriented towards the carbon layers. Fig.3 shows plausible models fitting the observed interlayer distances. In the phase with a high charge density the S-C bonds are oriented perpendicular to the layer planes, and the chains are inclined in an all-trans conformation, forming an angle (S to F) of 63° with the layer planes (Fig.3b). At lower charge densities, the chains seem to be more inclined, the S-to-F line forming an angle of approximately 45° with the layer planes (Fig.3a).

A similar orientation of n-alkyl chains has been observed with TaS_2 intercalated with n-alkylamines (17). The orientation of the chains is dependent on the charge density with clay minerals intercalated with n-alkylammonium ions. At low charge densities, they are oriented parallel to the silicate sheets, and at higher charge densities the inter-

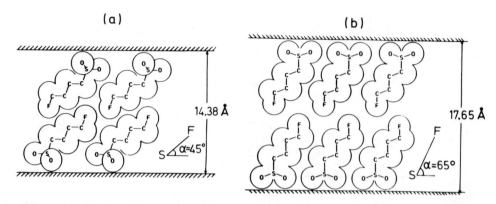

Fig.3. Models for the orientation of intercalated $C_4F_9SO_3H$ molecules and $C_4F_9SO_3^-$ ions, (a) in the 1773 pm phase, (b) in the 2100 pm phase.

calate forms monolayers of parallel chains which are inclined to the layer planes, the inclination angle depending on the charge density (18). Finally, if non-ionic compounds of equal chain length, e.g. n-alkanols or n-alkylamines, are added, a bilayer packing is established with the chains oriented perpendicular to the layer planes. The additional volume is filled by alcohol or amine molecules. The chains have a helical instead of the all-trans conformation in poly-(tetrafluoroethylene). With the relatively short C_4F_9 chains the all-trans conformation seems to be more stable, however.

No "a" and "b" phases were observed in the intercalation of $C_6F_{13}SO_3H$. There was only one phase for the stage 1 and stage 2 compounds with I_c values of 2600 - 2630 pm and 2915 pm, respectively, indicating that again a bilayer structure was formed.

Attempts to intercalate alkanesulfonic acids resulted in the formation of blue stage 1 compounds which decomposed with exfoliation on standing under the acid. The compound with CH_3SO_3H was relatively the most stable one of this series. Its I_c spacing of 883 pm (stage 1) suggests that there is a monolayer of $CH_3SO_3^-/CH_3SO_3H$ with the C-S bonds perpendicular to the layer planes.

3.3. Reaction of KC_8 with metal halides

The golden color of KC_8 turned blue and soon black after addition of the first drop of the metal chloride solution to the suspension in THF. Warming of the reaction mixture was most pronounced with the violet solution of $CoCl_2 \cdot 6\ H_2O$, but also noticeable with the dark green $FeCl_3$ solution. The color of the solution changed to brown after 30 minutes in the case of $FeCl_3$ but the solution was green again after 48 hours. Very little heat evolution was observed with $ZnCl_2$. Gas evolution followed the water content of the metal chlorides, almost none was observed with $ZnCl_2$. Analysis of the gas phase by mass spectrometry proved the presence of hydrogen when hydrated metal chlorides had been used, and of numerous organic compounds, e.g. C_2H_4 and C_3H_6.

Table IV. Chemical and XRD analyses of several reaction products, before and after washing with 2N HCl (RFl graphite flakes).

Reagent	HCl treatmt.	% K	% Cl	% trans. metal	X-ray diffraction identified*)	00l series not ident.**)
$ZnCl_2$	−	7.6	11.7	11.4	<u>KCl</u>, G, I	1042; 1774
	+	1.6	2.0	8.1	<u>G</u>, KC_{48}, KCl	963
$ZnCl_2$ (+HMPT)	−	13.8	13.4	11.8	<u>KCl</u>, G, II	974; 1038
$FeCl_3$	−	3.2	0.9	0.2	<u>G</u>, KC_{48}, II	−
$CoCl_2 \cdot 6\ H_2O$	−	3.4	15.0	4.6	<u>KCl</u>, G, II, III	1080
	+	3.3	2.3	0.2	<u>KC_{48}</u>, KCl, G	944
$CuCl_2 \cdot 2\ H_2O$	−	9.9	17.3	14.9	<u>KCl</u>, IV, V, KOH	948; 1112
	+	3.4	2.3	1.8	<u>KC_{60}</u>, G	956; 1421

*) G = graphite: I = $ZnCl_2 \cdot 4\ H_2O$: II = $K(THF)_2C_{48}$: III = higher stage potassium graphites (principal reflections near d = 335 pm): IV = CuCl, very likely present; V = $CuCl_2 \cdot 3\ Cu(OH)_2$.

**) (00l) series, I_c from at least 5 reflections.

The results of the chemical analyses and X-ray diffraction before and after washing with 2N HCl are presented in Table IV. The main constituents are underlined. Basic zinc chloride and basic copper chloride were identified using the JCPDS cards no. 7 – 155 and 18 – 439. The diffraction patterns were quite complicated and could be interpreted – with exception of the $FeCl_3$ preparation and the HCl-extracted manganese compound – to contain (00l) series of unidentified GICs. The I_c spacings (from at least five reflections) were 944 – 974 pm, 1038 – 1088 pm, 1421 – 1457 pm, and 1774 pm. Some of these values are near those experted for $K(THF)_{\sim 1}C_{48}$, st. 2 (1052 or 1073 pm) and $K(THF)_{\sim 1}C_{72}$, st. 3 (1387 pm) (19). The diffraction patterns of the substances not treated with 2 n HCl changed somewhat on exposure to air; the HCl-extracted products were stable in air, however.

Free graphite was always detected after extraction. Some of the KCl contained in all substances was in most cases resistant to washing out with 2N HCl. It is very difficult to be sure of the presence of GIC phases in such preparations with a large number of reflections. The ternary GICs with K and THF are often somewhat disordered and, very likely, there were also phases present with non-integral reflections, indicative of stacking disorder. Slight deviations from integrality may cause uncertainties in the I_c values. It is not permissible to establish the existence of a GIC phase on the basis of only two observed (00l) reflections as has been done in ref.(7). Our results are based on at least five (00l) reflections with all the uncertainties described.

We have studied earlier the reaction of potassium graphite with water (20) and with THF (21). With particle diameters in the range of 1 mm, considerable quantities of potassium were trapped in the reaction

products in the form of higher-stage potassium compounds, including ternaries with THF, of residue compounds and of encapsulated compounds such
as KOH. Hydrogen and other gases were trapped, too. It is not possible,
therefore, to calculate exact ratios of chlorine to transition metal
from the analytical data by correcting for KCl. However, there is only
with the zinc-containing samples a significant excess of transition metal with respect to the remaining chloride, if one assumes that all or
most of the potassium is present as KCl.

Scanning electron microscopy (SEM) showed the presence of particles
containing the transition metal and KCl to be situated mainly at the
edges of the graphite flakes. Particles were also deposited near cracks
formed by disruption of the flakes. Washing with hydrochloric acid led
frequently to increased exfoliation. The preparations were also examined
by energy-dispersive X-ray analysis line scans (EDA). With the exception
of zinc, the transition metals and chlorine were found mainly at the
periphery of the graphite flakes. The ratio of the Cl and Mn peaks in
the manganese preparation implied that at least half of the manganese
was present in the form of an obviously amorphous oxide-hydroxide. Mas-
sive precipitation of iron oxide-hydroxide at the edges of the flakes
was also observed when a brown $FeCl_3$ solution containing some water was
used as in the experiment of Braga et al.(7), as can bee seen in Fig.
4(a). Potassium, in contrast, is essentially found only in the interior
of the flake. Schäfer-Stahl (8) observed also γ-FeO(OH) in a similar

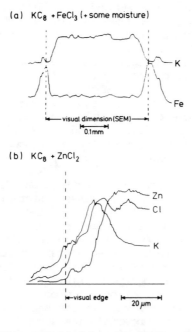

Fig.4. EDA line scans, (a) over flake of reaction product with a brown
$FeCl_3$ solution containing a little water, (b) over a flake of the reaction product with anhydrous $ZnCl_2$.

preparation. It cannot be excluded that the hydrous oxides formed only on access of air during the mounting for SEM and EDA. In contrast, anhydrous ferric chloride was reduced mainly to ferrous chloride which remains in solution, explaining the low iron content of the sample described in Table IV. In the case of copper chloride the ratio of the areas of the K, Cl and Cu peaks in EDA of the particles seen at the graphite edges was in agreement with a formulation as KCl + CuCl. XPS of the preparations gave no indication for the presence of Cu°. There was also indication of the deposition of K, Cu and Cl in localized spots within the graphite flakes. Apparently, KCl and CuCl, and possibly other compounds had been precipitated and encapsulated. The Cu intensity dropped to the detection limit in between such areas.

The situation was quite different with the zinc compounds. Here we find a high zinc concentration in the interior of the flakes whereas potassium is found near the edge (Fig.4b). The KCl seems to have precipitated near the periphery, but within the flake as it was resistent to extraction

When $ZnCl_2$ is added to THF a white crystalline precipitate of $[Zn(THF)_4]Cl_2$ forms which is sparingly soluble. On addition of metallic potassium a greyish mixture of this complex salt, metallic zinc and KCl was obtained. This complex salt was not present in the reaction product with KC_8 as follows from the lack of chlorine and of the corresponding X-ray reflections. However, we must conclude that zinc had been intercalated, either in atomar (or ionic) form or as the $[Zn(THF)_4]^{2+}$ complex ion. The zinc was resistant to acid extraction. The measured Zn contents seem, however, too high by far for all of the zinc to be intercalated as the THF complex. Furthermore, the observed I_c values do not fit well with estimated values for such compounds.

Our experiments did not give any evidence for the existence of transition metal-intercalated graphite compounds, with the possible exception of zinc compounds. However, the fact that some of the reaction products are precipitated and encapsulated in the edge zone of the graphite platelets could be interpreted to mean that not only potassium diffuses out of the interstitial planes but that also reagent is able to diffuse in and to react in situ. However, since the bonding energy of the elementary transition metals with unfilled d shells is much higher than that of potassium or zinc with a completely filled 3 d shell, the metals have a tendency to segragate as explained earlier (22). If this happens just at the edge of the platedets, the tiny crystals are highly reactive; they can react with the solvent mixture or with the washing fluid, i.e. they are reoxidized. In a few sites, e.g. at dislocations, they may precipitate within the layer structure leading to a disruption of the carbon layers. This is very difficult to detect in this system because potassium graphite is severely attacked already by the reaction with water and/or THF (20,21). The results with zinc chloride are interesting and warrant further studies.

ACKNOWLEDGEMENT. Financial support by the Fonds der Chemischen Industrie and by NATO (Research Grant 98/80) is gratefully acknowledged.

REFERENCES

1. A.Hérold, in: Intercalated Layered Materials (Ed. F.A.Levy), D.Reidel Publ.Co., Dordrecht, NL, Vol.6, 1979, p.323.

2. H.P.Boehm, Preprints Carbon'80, Internat.Carbon Conf., Baden-Baden 1980, p.73.

3. A.Métrot and J.E.Fischer, Synth.Met. 3, 201 (1981).

4. J.O.Besenhard, E.Wudy, H.Möhwald, J.J.Nickl, W.Biberacher and W.Foag, Synth.Met. 7, 185 (1983).

5. M.E.Volpin, Yu.N.Novikov, N.D.Lapkina, V.I.Kasatochkin, Yu.T. Struchkov, M.E.Kazakov, R.A.Stukan, V.A.Povitskij, Yu.S.Karimor and A.V.Zvarikina, J.Amer.Chem.Soc. 97, 3366 (1975).

6. D.Braga, A.Ripamonti, D.Savoia, C.Tromboni and A.Umani-Ronchi, J.Chem.Soc., Chem.Com. 1978, 927.

7. D.Braga, A.Ripamonti, D.Savoia, C.Tromboni and A.Umani-Ronchi, J.Chem.Soc., Dalton Trans. 1979, 2036.

8. H.Schäfer-Stahl, J.Chem.Soc., Dalton Trans., 1981, 328.

9. Ko, Y.-Sh. and H.P.Boehm, Z.Naturforsch. 39a, 768 (1984).

10. B.Bouayad, H.Fuzellier, M.Lelaurain, A.Métrot and F.Rousseaux, Synth.Met. 7, 325 (1983).

11. W.Metz and D.Hohlwein, Carbon 13, 87 (1975).

12. D.M.C. Mac Ewan, A.Ruiz-Amil and G.Brown, in: The X-ray Identification and Crystal Structures of Clay Minerals, Mineralogical Society, London, 1961, p.393.

13. Ko, Y.-Sh., doctoral thesis, Univ.München (1983).

14. W.Rüdorff and U.Hofmann, Z.Anorg.Allg.Chem. 238, 1 (1938).

15. M.J.Bottomley, G.S.Parry, A.R.Ubbelohde and D.A.Young, J.Chem.Soc. 1963, 5674.

16. D.Horn and H.P.Boehm, Mat.Sci.Eng. 31, 87 (1977).

17. R.Schöllhorn, E.Sick and A.Weiss, Z.Naturforsch. 28b, 168 (1973).

18. A.Weiss, Angew.Chem., Internat.Ed.Engl. 2, 134 (1963).

19. F.Béguin and R.Setton, Carbon 13, 293 (1975); F.Béguin, R.Setton, A.Hamwi and P.Touzain, Mat.Sci.Eng. 40, 167 (1979); F.Béguin, L.Gatineau and R.Setton, Ext.Abstr.Carbon'82, 6th Internat.London Carbon and Graphite Conf., 1982, p.97.

20. R.Schlögl and H.P.Boehm, Carbon 22, 351 (1984).

21. R.Schlögl and H.P.Boehm, Carbon 22, 341 (1984).

22. G.Bewer, N.Wichmann and H.P.Boehm, Mat.Sci.Eng., 31, 73 (1977).